Advances in Low-Carbon and Stainless Steels

Advances in Low-Carbon and Stainless Steels

Special Issue Editors

Nima Haghdadi
Mahesh Somani

MDPI • Basel • Beijing • Wuhan • Barcelona • Belgrade • Manchester • Tokyo • Cluj • Tianjin

Special Issue Editors
Nima Haghdadi
UNSW sydney
Australia

Mahesh Somani
University of Oulu
Finland

Editorial Office
MDPI
St. Alban-Anlage 66
4052 Basel, Switzerland

This is a reprint of articles from the Special Issue published online in the open access journal *Metals* (ISSN 2075-4701) (available at: https://www.mdpi.com/journal/metals/special_issues/stainless_steels).

For citation purposes, cite each article independently as indicated on the article page online and as indicated below:

LastName, A.A.; LastName, B.B.; LastName, C.C. Article Title. *Journal Name* **Year**, *Article Number*, Page Range.

ISBN 978-3-03928–869-4 (Pbk)
ISBN 978-3-03928-870-0 (PDF)

Cover image courtesy of Nima Haghdadi.

Contents

About the Special Issue Editors

Nima Haghdadi is a research fellow at UNSW Sydney, Australia. He was awarded his Ph.D. in 2017 from Deakin University (Australia) for a thesis on processing of duplex stainless steels. He is a former Victoria Fellow and Alfred Deakin Research Fellow, and the recipient of Thermo Fisher Scientific Cowley-Moodie and Acta Materialia student awards. He was also the runner up in the James Clerk Maxwell Writers Prize in the *Philosophical Magazine* in 2017. Dr Haghdadi's main interest is engineering of microstructures of metallic materials including steels during conventional processing and additive manufacturing for advanced mechanical and corrosion properties.

Mahesh Somani is an Adjunct Professor at the University of Oulu, Finland. During his academic and research career spanning over 35 years, he has authored over 150 technical papers. In addition, he has authored 20 technical reports and has 2 patents to his credit. Dr. Somani established his career at the Defence Metallurgical Research Laboratory (DMRL), Ministry of Defence, India, during 1982–1998 in the capacity of Senior Scientist working mainly in the field of aerospace materials for gas turbine engines. At the University of Oulu (since 1999), his academic and research interests have mainly focused on physical simulation and modelling of hot deformation processes of steels and other materials, accelerated cooling/direct quenching, and phase transformation in steels, technology of alloy steels, static and dynamic restoration mechanisms, dynamic materials modelling and development of processing maps, and strength of metallic materials.

Preface to "Advances in Low-Carbon and Stainless Steels"

Almost 80 wt % of all the metallic parts for engineering purposes are composed of steel. Steel has been studied for about 160 years. Despite such a long history of research, many aspects of steel remain unknown. This Special Issue of Metals is dedicated to the recent advances in low-carbon and stainless steels. Although these types of steels are not new, they are still receiving considerable attention from both research and industry sectors due to their wide range of applications and complex microstructure and behavior under different conditions. The microstructures of low-carbon and stainless steels resulting from solidification, phase transformation, and hot working are complex, which, in turn, affect their performance under different working conditions. A detailed understanding of the microstructure, properties, and performance for these steels has been the aim of steel scientists for a long time.

For this Special Issue, we invited academics and researchers to submit papers on different aspects of these steels including their solidification, thermomechanical processing, phase transformation, texture, etc., and their corrosion, wear, fatigue, and creep properties. We received many submissions and after a rigorous review process, 14 papers were accepted for publication, two of which were featured papers. The papers published in this Issue of the journal are from world class scientists from around the world including from Asia, Europe, South America, and North America, and display outstanding depth and novelty. This volume will be of use as reference for steel scientists and engineers in both academia and industry that would like to develop an understanding of the latest metallurgical advances in low-carbon and stainless steels.

Nima Haghdadi, Mahesh Somani
Special Issue Editors

Article

Hot Forming of Ultra-Fine-Grained Multiphase Steel Products Using Press Hardening Combined with Quenching and Partitioning Process

Esa Pirkka Vuorinen [1,*], Almila Gülfem Özügürler [1], John Christopher Ion [1], Katarina Eriksson [2], Mahesh Chandra Somani [3], Leo Pentti Karjalainen [3], Sébastien Allain [4] and Francisca Garcia Caballero [5]

1 Division of Materials Science, Luleå University of Technology, SE-97187 Luleå, Sweden; almilagulfem@hotmail.com (A.G.Ö.); John.Ion@mail.se (J.C.I.)
2 Gestamp HardTech AB, SE-97125 Luleå, Sweden; KaEriksson@se.gestamp.com
3 Centre for Advanced Steels Research, University of Oulu, FIN-90014 Oulu, Finland; Mahesh.Somani@oulu.fi (M.C.S.); Pentti.Karjalainen@oulu.fi (L.P.K.)
4 Institut Jean Lamour, Université de Lorraine, BP-50840 Nancy, France; sebastien.allain@univ-lorraine.fr
5 National Center for Metallurgical Research (CENIM-CSIC), E-28040 Madrid, Spain; fgc@cenim.csic.es
* Correspondence: Esa.Vuorinen@ltu.se

Received: 18 February 2019; Accepted: 18 March 2019; Published: 20 March 2019

Abstract: Hot forming combined with austempering and quenching and partitioning (QP) processes have been used to shape two cold rolled high silicon steel sheets into hat profiles. Thermal simulation on a Gleeble instrument was employed to optimize processing variables to achieve an optimum combination of strength and ductility in the final parts. Microstructures were characterized using optical and scanning electron microscopy and X-ray diffraction. Tensile strengths (R_m) of 1190 and 1350 MPa and elongations to fracture (A_{50mm}) of 8.5 and 7.4%, were achieved for the two high-silicon steels having 0.15 and 0.26 wt % C, respectively. Preliminary results show that press hardening together with a QP heat treatment is an effective method of producing components with high strength and reasonable tensile ductility from low carbon containing steels that have the potential for carbide free bainite formation. The QP treatment resulted in faster austenite decomposition during partitioning in the steels in comparison with an austempering treatment.

Keywords: hot forming; multiphase steel; quenching and partitioning; austempering; Gleeble simulation; press hardening

1. Introduction

Press hardening of boron alloyed steels has been used since the 1980s [1] to produce beams, pillars, and safety-related components for cars [2]. A six-fold increase in the adoption of the technique for component production was anticipated between 2006 and 2015 [3] and the production reached 360 million components in 2015 [4]. Strength levels achievable in boron steels in as quenched conditions are considered excellent ($R_m \approx 1500$ MPa) but the ductility is often limited ($A_{50\,mm} \approx 6\%$ or lower) as a result of the essentially martensitic microstructure of the steels [5]. Tailor-welded blanks and differentiation of heat treatment are the methods that can be used to tailor and optimize the properties in different parts of a component [6]. In addition, both ductility and toughness may be enhanced in these steels with the formation of carbide free bainitic (CFB) microstructures through austempering process and/or subjecting these steels to a novel concept of quenching and partitioning (QP) thermal treatment as described below. Formation of CFB microstructures can be facilitated in specially tailored steel compositions containing high levels of Si and/or Al (about 1.5–3 wt %), through austempering because both Si and or Al are strong graphitisers and hence, hinder or delay carbide formation in the steel structure. The microstructures of CFB steels comprise mainly of fine laths of bainitic ferrite and carbon enriched austenite divided finely between bainitic sheave [7,8] and martensite in some cases [9]. Likewise, the QP treatment first described by Speer et al. [10] also promotes formation of essentially martensitic microstructures with small fractions of finely divided, carbon-enriched interlath austenite [11], besides a small fraction of bainitic ferrite and in some cases also carbides [12]. Tensile properties typical of selected steels processed through QP technique are shown in Table 1.

Table 1. Typical tensile properties of quenching and partitioned (QP) steels vis à vis boron (22MnB5) and austempered bainitic (CFB) steels.

Steel (wt % C)	$R_{p0.2}$ (MPa)	R_m (MPa)	A (%)	Reference
22MnB5 (0.22)	1010	1480	6	Naderi 2007 [13]
CFB (0.2)	950	1020	19	Zhang 2008 [14]
CFB (0.2)	1180	1360	7	Putatunda 2011 [15]
QP (0.2)	1200	1400	12	De Moor 2011 [11]
CFB (0.3)	1028	1800	11	Caballero 2006 [7]
QP (0.3)	1100	1500	15	De Moor 2011 [11]
CFB (0.4)	1250	1400	12	Putatunda 2009 [9]
QP (0.4)	1400	1750	14	Li 2010 [12]

The twin benefits of the existing direct press hardening process applied to boron steels are (i) the combination of rapid forming through optimized processing and (ii) quenching of the component in the pressing tool. During austempering, austenite is isothermally transformed into lower bainite at a temperature slightly above the martensite start temperature (M_s) for a duration adequate enough for complete austenite decomposition. However, slow kinetics of the austenite to bainite transformation at temperatures close to M_s can have limitations in respect of the austempering process in combination with press hardening for commercial production of automotive components. In the QP process the steel is quenched to a temperature between the start (M_s) and the finish (M_f) of martensite reaction and subsequently either held at the quenching temperature or heated to just above or below M_s temperature to facilitate partitioning of carbon from transformed supersaturated martensite into austenite or from the bainite that may form during subsequent partitioning step. The transformation rate has been shown to increase when the austempering temperature is lowered just below the M_s temperature [16]. A quench stop below M_s allows a small amount of martensite to form prior to bainite transformation, thereby increasing the number of possible nucleation sites for bainite and thus its rate of formation [16]. It has also been shown that the transformation from austenite to bainite can be accelerated if a small fraction of martensite can be formed from the austenite [17] even though the rate of bainite formation following martensite formation remains unchanged and same amount of bainite would form following austenite decomposition [18]. Utilization of the QP heat treatment thus provides the possibility to shorten the production cycle time. Press hardening with a QP treatment of boron steel has been shown to improve the ductility of the steel but with a marginal loss in yield strength [19,20], as compared with the properties obtained through conventional press hardening of 22MnB5 steel. This process has been repeated for low-carbon Si-Mn steels and the maximum volume fraction of retained austenite reached 17.2% with corresponding total elongation of 14.5% when hot stamping is done at 750 °C [21]. Seo et al. [22], designed two types of modified press hardening steels (PHS) by adding Si, and Si + Cr to 22MnB5 steel, followed by optimized QP processing to achieve improved properties. In the best QP conditions the ductility improved to 17% total elongation with 1032 and 1098 MPa yield and tensile strengths respectively, for the Si + Cr added (PHS) grade.

The aim of this work was to produce components with properties equal to or better than conventional press hardened boron steels, within a reasonable processing time for improved productivity. Various thermal treatments following the forming stage were investigated to achieve a fine multiphase microstructure. The quench stop temperature in the die was identified as a variable of interest along with the furnace temperature and the holding time of the heat treatment. Two variants of quench stop temperature were investigated, above and below the M_s. Both isothermal heat treatment above and below M_s and QP were investigated using thermal simulations for two cold rolled Fe-(0.15 and 0.26)C-1.5Si-2Mn-0.6Cr alloys. This paper reports an account of the mechanical properties obtained after press hardening experiments to produce hat-shaped profiles using QP heat treatments for high-silicon steels, in comparison with those of commercial 22MnB5 profiles. The effect of using QP treatments on austenite decomposition kinetics in comparison with austempering treatment is also studied.

2. Materials

Two laboratory heats of 15 kg, coded here as CR1 and CR3, were produced in a vacuum induction furnace under an inert atmosphere. Alloying elements were added in sequence to pure (>99.9%) electrolytic iron. Carbon deoxidation was performed and an analysis of C, S, N, and O was made on line during the final adjustment of the composition. Samples of 40 mm thick plates were hot rolled to a final thickness of 3 mm in several passes finishing at 900 °C. The 3 mm thick strips were then cold rolled to blanks with a thickness of 1.3 mm. Table 2 shows the chemical compositions of the experimental steels determined by optical emission spectroscopy (ARL 4460, Thermo Fisher Scientific, Lausanne, Switzerland). Critical transformation temperatures M_s and A_{c3}, determined by high resolution dilatometry are also included in Table 2. It also shows the times required for completing bainitic transformation, determined by dilatometric analyses at temperatures at and above M_s. If isothermal treatment takes place at a temperature at or below M_s, athermal martensite forms before the isothermal transformation starts. For more information about design, processing, and properties of isothermally treated CR1 and CR3 steels, see Caballero et al. [23]. In addition, the composition range of the 22MnB5 reference steel used in final hat-profile pressing is given in Table 2.

Table 2. Chemical compositions (wt %) of experimental CFB steels and the reference 22MnB5 (B5) steel. Experimentally measured M_s and A_{C3} temperatures as well as bainite formation time (t_{Bf}) at select isothermal holding temperatures (T_B) are also included.

Steel	C	Mn	Si	Cr	P	S	M_s (°C)	A_{C3} (°C)	T_B (°C)	t_{Bf} (min)
CR1	0.15	2.01	1.45	0.62	0.016	-	400	895	400	7.7
CR3	0.26	2.02	1.47	0.62	0.017	-	322	853	350	23.1
B5	0.20–0.25	1.10–1.30	0.20–0.35	0.15–0.25	Max 0.025	Max 0.005	-	-	-	-

3. Methods

3.1. Gleeble Thermal Simulation

Experiments simulating press hardening conditions in respect of temperature—time cycles were carried out using a Gleeble 1500 simulator (Dynamic Systems Inc., Postenkill, NY, USA). Flat specimens with dimensions 1.3 × 10 × 70 mm^3 were subjected to thermal cycles that produced a uniform heat-affected zone of about 20 mm in width in the center of the samples. Thermal cycles were designed to simulate two industrial processing routes in which the following sequence of steps is used: (i) austenitization; (ii) forming (at a specified temperature); (iii) quenching to a specified temperature to simulate either an austempering treatment above the M_s temperature or a QP process below the M_s temperature (resulting in a certain amount of martensite formation); (iv) cooling the austempered samples or heating the QP samples to a specified temperature both above M_s respectively, followed by isothermal holding to facilitate transformation of, some or all of, the balance untransformed austenite into a very fine bainitic structure and; v) cooling to room temperature (after complete or partial transformation to bainite). The Gleeble simulations performed in this work however, did not include blank deformation, i.e., step (ii) in the sequence above.

All specimens were first reheated at 5 °C/s to 930 °C and held for 60 s before cooling at 20 °C/s to 770 °C. Two different sequences of heat treatment were then performed, directly without delay:

i Austempering: Quenching to temperatures above M_s followed by holding at Ms, M_s + 30 °C and M_s − 30 °C for 5, 10, 15, and 25 min.

ii QP process: Quenching to temperatures corresponding to M_s − 10 °C and M_s − 20 °C followed by heating to M_s or M_s + 30 °C and holding for 0.5, 1, 5, and 15 min.

Following the heat treatments, samples were cooled to room temperature at 5 °C /s. Referring to the QP process the athermal martensite fractions formed at temperatures of M_s − 10 °C and M_s − 20 °C were estimated to be ~10 and 20 vol %, respectively using the Koistinen-Marburger equation [24].

3.2. Press Hardening Trials of Hat Shaped Profiles

Steel blanks having dimensions 1.3 × 70 × 150 mm^3 were cut from cold rolled sheets, heated in a furnace to 930 °C and held for 4 min, prior to transferring them to the tooling within ~9 s. The dies were preheated to enable the blank temperature to be controlled during forming and simultaneous quenching of the hat profile. The dies were closed in 2.5 s and the blanks cooled in the dies for 8 s to a temperature corresponding to M_s − 10 °C. The blanks were then transferred to a second furnace within ~11 s and held at M_s + 30 °C for 5 min before air cooling to room temperature. The temperature was measured using a thermocouple welded to the blank. Examples of typical temperature–time curves recorded on the blanks during press hardening are shown in Figure 1. Six hat profiles were produced from steel CR1 and five from steel CR3. The quenching temperatures were controlled within 4 °C, for all CR1 samples and 2 °C for 4 out of 5 CR3 samples with regard to the set value (M_s − 10 °C). The measured temperature for the fifth CR3 sample was 9 °C below the set value. The heat created when austenite was transformed to martensite and bainite respectively increased the temperature in the samples, but the furnace temperature was controlled so that the set value (M_s + 30 °C) was reached within 5 min.

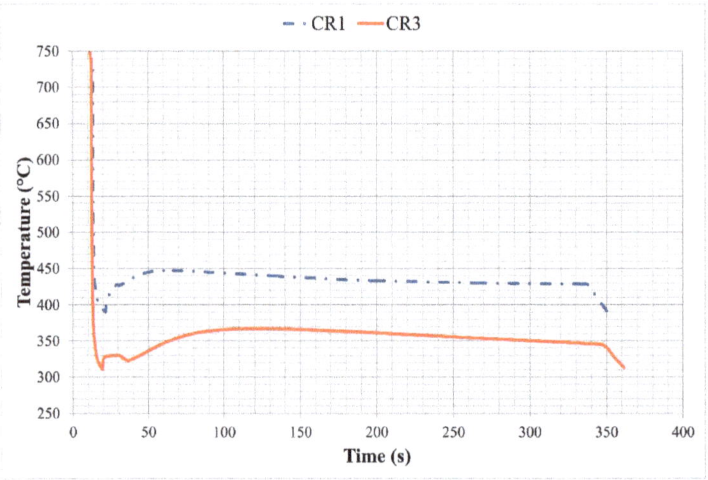

Figure 1. Typical temperature–time curves during press hardening of hat profiles of steels CR1 and CR3.

3.3. Characterization

Microstructural examinations were performed using light optical microscopy (LOM) (Olympus Vanox-T AH-2, Tokyo, Japan) and scanning electron microscopy (SEM) (JEOL JSM 6064LV, Tokyo, Japan). X-ray diffraction (XRD) (Siemens D5000 PANalytical Empyrean diffractometer, Munich, Germany) measurements were made with monochromatic CuKα radiation (40 kV and 45 mA). Rietweld analysis was used to establish the volume fraction of austenite present in the microstructures of the hat profiles.

Vickers hardness was measured on the Gleeble specimens using a load of 5 N. A load of 20 N was used for the hat profiles in which the hardness was determined at 5 positions—on the top, side walls 1 and 2, and flanges 1 and 2—as shown in Figure 2.

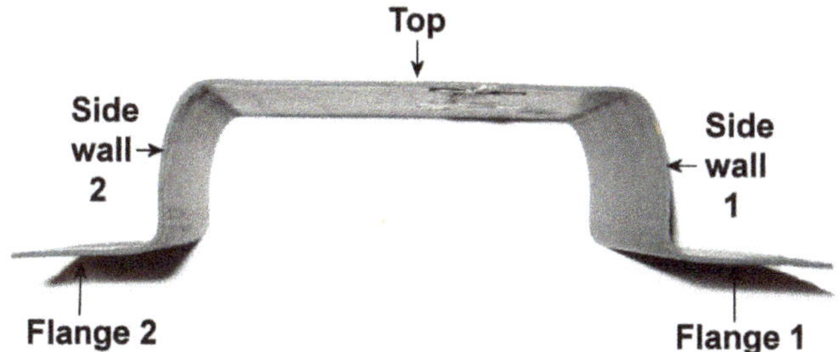

Figure 2. Schematic view of the hat shaped profile, after press hardening operation.

Tensile tests were carried out according to the standard EN ISO 6892-1:2009. One test specimen was laser cut from the top surface of each press hardened hat shaped profile. Reference measurements were performed on isothermally heat-treated samples for both steels but with a shorter gauge length of 25 mm instead of 50 mm because of the limited amount of material available. Steel CR1 was soaked at 400 °C for 15 min and steel CR3 at 350 °C for 30 min following austenitization at 890 °C for 100 s. In principle this temperature does not imply complete austenitization of steel CR1, as the A_{c3} temperature was determined to be 895 °C: Electron microscopy of a sample quenched from 890 °C confirmed the presence of a minute quantity of intercritical ferrite.

4. Results

Microhardness results together with standard deviation values of Gleeble simulated samples are plotted in Figures 3 and 4. Figure 3 presents data for samples that experienced a quench stop above M_s (i.e., austempering treatment) and Figure 4 shows data for samples with a quench stop below M_s (i.e., the QP treatment). The times for bainite formation at M_s for steel CR1 and at M_s + 30 °C for steel CR3 are given in Table 2. These values were determined by dilatometric measurements. The hardness values for fully bainitic structures were in the ranges 370–410 HV for steel CR1 and 450–470 HV for steel CR3 as measured from the dilatometric test samples. The hardness values of the Gleeble-treated samples in Figure 3 lay in the typical ranges of fully bainitic structures for the stated carbon levels in the steels. The effect of holding time on hardness remains difficult to analyze for 0.15C steel CR1 after isothermal transformation, but in general the lower the holding temperature is, the higher the hardness achieved. All hardness values of steel CR1 lay close to the expected values for a bainitic structure, Figure 3. For the 0.26C steel CR3, longer holding times (10–25 min) gave results expected for a fully bainitic structure, whereas the shortest (5 min) holding time resulted in hardness values that exceeded the expected level for just bainite and, most likely is a result of untempered martensite formation during final cooling, Figure 3.

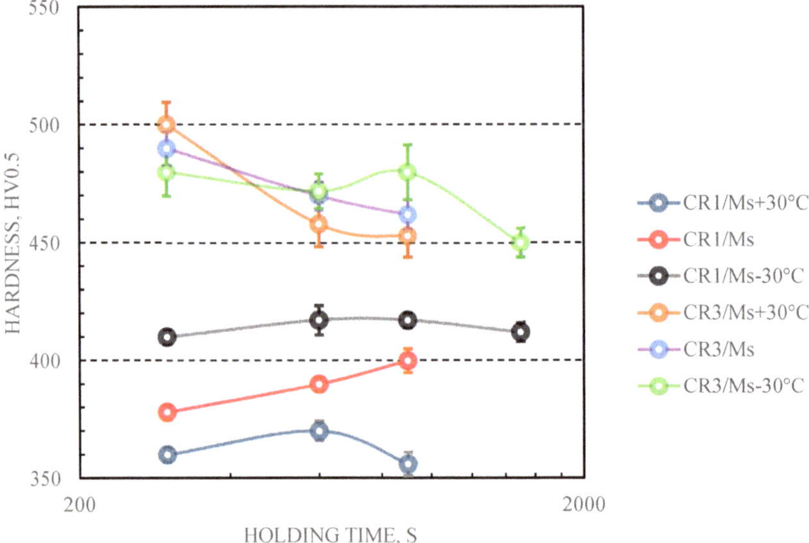

Figure 3. Hardness ($HV_{0.5}$) of Gleeble simulated CR1 steel and CR3 steel after quenching with a cooling stop above M_s followed by isothermal heat treatment for various times at temperatures relative to M_s as indicated.

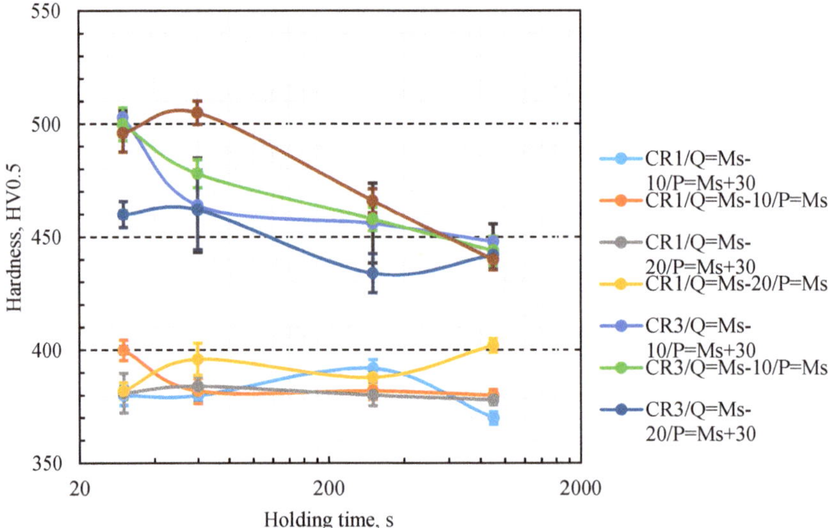

Figure 4. Hardness ($HV_{0.5}$) of Gleeble simulated CR1 steel and CR3 steel after quenching with a cooling stop below M_s followed by isothermal heat treatment for various times at temperatures relative to M_s as indicated.

Quenching with a cooling stop below M_s followed by isothermal heat treatment resulted in the expected hardness typical of lower bainite for all 0.15C CR1 specimens, Figure 4. For 0.26C CR3 steel the shortest holding time (0.5 min) appears to be insufficient for complete bainite transformation (Figure 4), whereas a longer holding time led to the expected hardness typical of lower bainite. Isothermal heat treatment at M_s appears to require a slightly longer time than the transformation at M_s + 30 °C, obviously due to slower kinetics at lower temperature. No appreciable difference can be seen following quenching to M_s − 10 °C or M_s − 20 °C in respect of final hardness. For the quenching and post heat treatment process the initial formation of martensite appears to accelerate bainite transformation noticeably. There are also indications that bainite transformation is more rapid at a holding temperature of M_s + 30 °C than at M_s, though the results can become complex with the occurrence of other microstructural mechanisms, such as carbon partitioning and stabilization of a small fraction of austenite. The process variant selected for further investigation was die quenching to M_s − 10 °C and subsequent post heat treatment at M_s + 30 °C for 5 min before cooling to room temperature. The microstructures of steels CR1 and CR3 after Gleeble simulation by quenching to M_s − 10 °C followed by holding at M_s + 30 °C for 5 min are shown in Figure 5. Both microstructures essentially consist of bainite and a small fraction of tempered martensite.

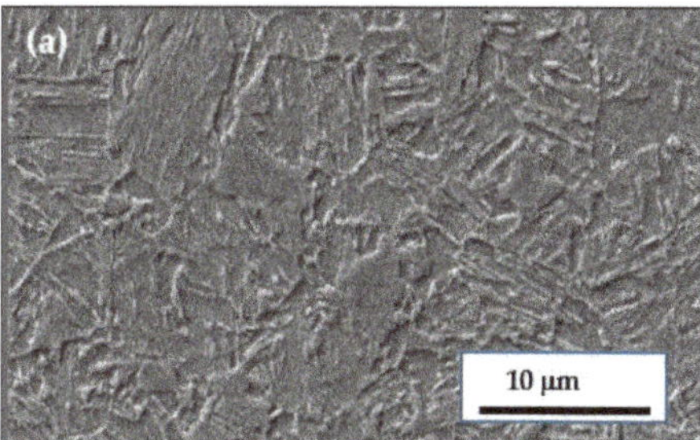

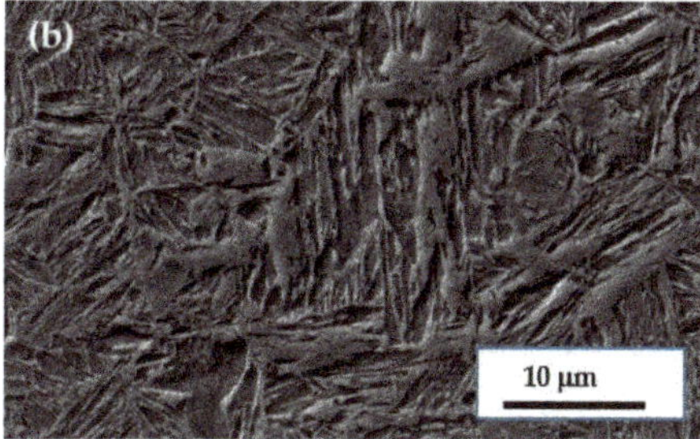

Figure 5. Scanning electron images of steel CR1 (**a**) and steel CR3 (**b**). Both samples quenched to M_s − 10 °C followed by holding at M_s + 30 °C for 5 min.

Temperature measurements from the pressed hat profiles showed that the latent heat of transformation is released during the formation of martensite. Some latent heat release was also observed during the bainite transformation, see Figure 1. However, the targeted quench stop temperatures were achieved within about 10 °C. For most of the trials, the partitioning temperature was set lower than the targeted temperature (M_s + 30 °C) to account for the latent heat generation and the corresponding increase in temperature. The highest temperatures caused by latent heat generation were M_s + 60 °C. The targeted temperatures were achieved after approximately 5 min of isothermal holding in the case for the CR1 steel profiles. However, the partitioning temperature for CR3 steel profiles were estimated to be about 5–15 °C lower than the target value.

The microstructure obtained at the top of the profiles in CR1 and CR3 steels are shown in Figure 6. Since it was not possible to distinguish between bainite, tempered martensite and retained austenite from the Nital etched samples, LePera etchant was used to reveal the microstructures. Following such etching, the white spots seen in the microstructure were identified as austenite or untempered martensite [25]. Tempered martensite appeared as slightly brown-colored constituent and bainite often appeared as the blue-colored constituent.

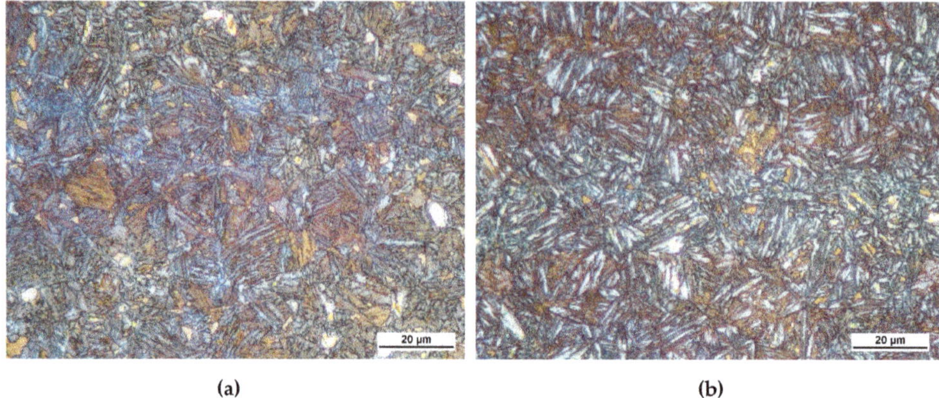

Figure 6. LePera etched optical micrographs from the top surfaces of hat profiles in (**a**) steel CR1 and (**b**) steel CR3.

Typical SEM images from hat profiles are displayed in Figure 7. It is seen that the microstructures comprise multiple phases such as a fine mixture of lath-like ferrite, retained austenite and some tempered martensite, and possibly also some untempered martensite formed during the final cooling.

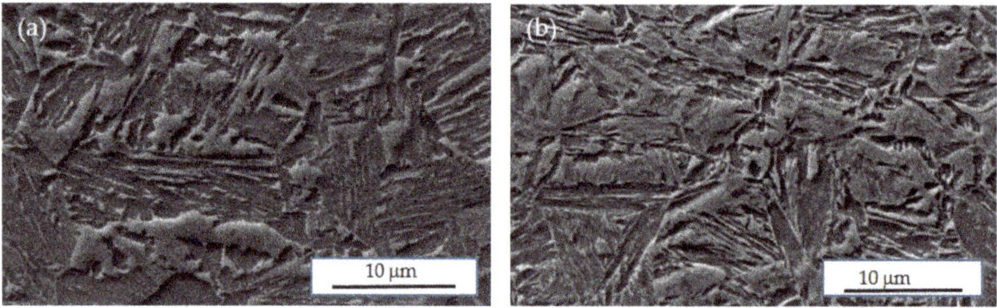

Figure 7. Scanning electron micrographs from the top surfaces of hat profiles in (**a**) steel CR1 and (**b**) steel CR3.

The volume fraction of retained austenite was determined by analysis the of X-ray diffraction data using two samples of each material. Accordingly, steel CR1 contained about 12% retained austenite and steel CR3 about 17%. Hardness values at different positions on the five hat-shaped profiles of each steel sort (according to Figure 2) are shown in Figure 8. Average values containing standard deviations are included for each position of the hat profiles in Figure 8.

The average values of three measurements at each position are presented. The expected hardness values for bainite were achieved at all locations in the hat profile specimens of steel CR1, except for flange 1. This was presumably caused by slower cooling in that part of the die, which resulted in some ferrite formation. For steel CR3 the hardness values lie between 455 and 475 HV_2, which are close to the expected hardness of bainite (~450–470 HV) for this steel, as measured on isothermally treated samples. A few specimens showed higher hardness, up to 488 HV_2. It can be seen that the part with the lowest quenching stop temperature, CR3-5, has the most uniform hardness, Figure 8.

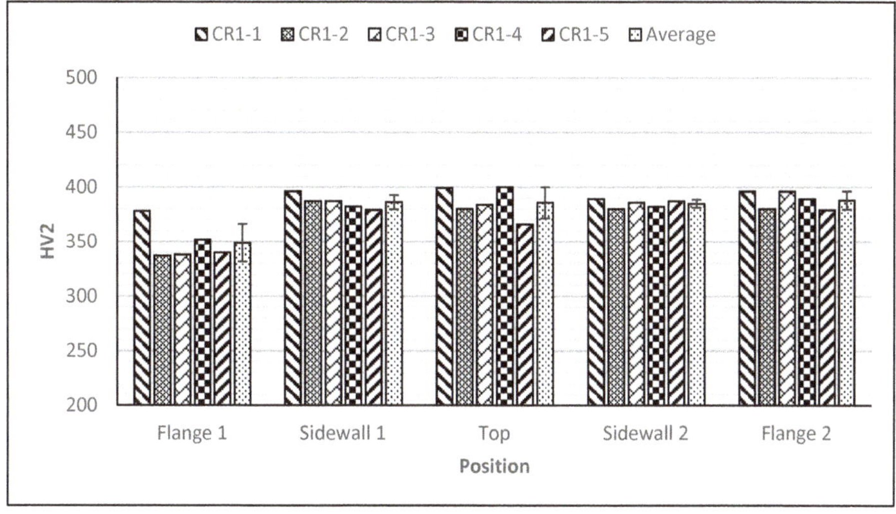

(a)

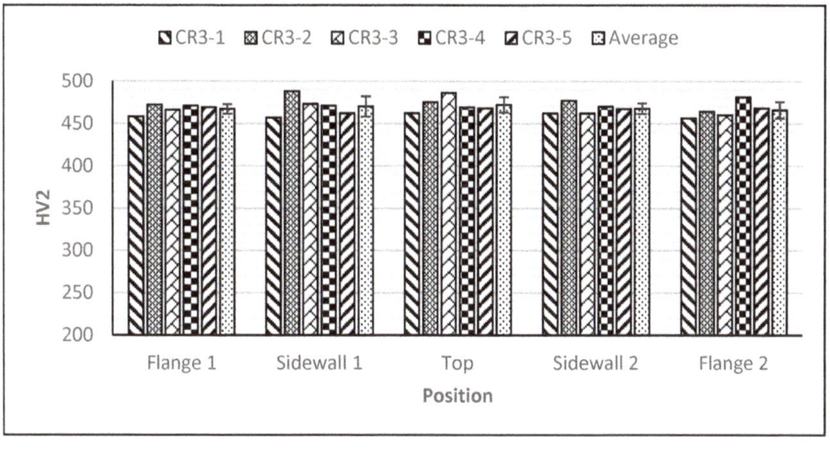

(b)

Figure 8. Hardness data (HV_2) measured at various positions in press hardened hat profiles, made of steel CR1 (**a**) and steel CR3 (**b**), as shown in Figure 2. Average values including standard deviation, are presented for each position of hat profile.

Tensile properties were determined for the top sections of all the manufactured hat profiles. Figure 9 shows a comparison between the tensile tests properties of CR1 and CR3 together with uncoated commercial boron 22MnB5 steel currently in use for press hardening. The presented values are the average of 10 measurements for steel CR1 and 5 measurements for steel CR3.

The targeted yield (1000–1300 MPa) and tensile (1400–1700 MPa) strength values were not achieved in the 0.15C CR1 steel. Though the targeted yield strength (1036–1093 MPa) was achieved in 0.26C CR3 steel, but the tensile strength was still below the target (1323–1404 MPa). On the other hand, the elongation to fracture for both steels (average values of 8.5% and 7.4% for CR1 and CR3 steels, respectively) was somewhat higher, and better than the targeted value of $A_{50mm} > 5\%$. The measured total elongation was 4.9% for the tested 22MnB5 reference steel, which agrees well with the values found in the literature, see Table 1. Furthermore, tensile tests were performed on sheets subjected to isothermal treatments, CR1 at 400 °C for 15 min and CR3 at 350 °C for 30 min to obtain reference values for bainitic structures of the steels. The elongation to fracture was higher for these samples in comparison to those of the hat profiles (CR1: 15.8% and CR3: 12.2%). The strength values of CR1 steel were approximately the same as for the hat profile, but for the CR3 steel the yield and tensile strength values were slightly higher for the isothermally treated samples. The shorter gauge length, 25 mm instead of 50 mm, for the austempered tensile test samples was identified as one reason for the difference in elongation to fracture.

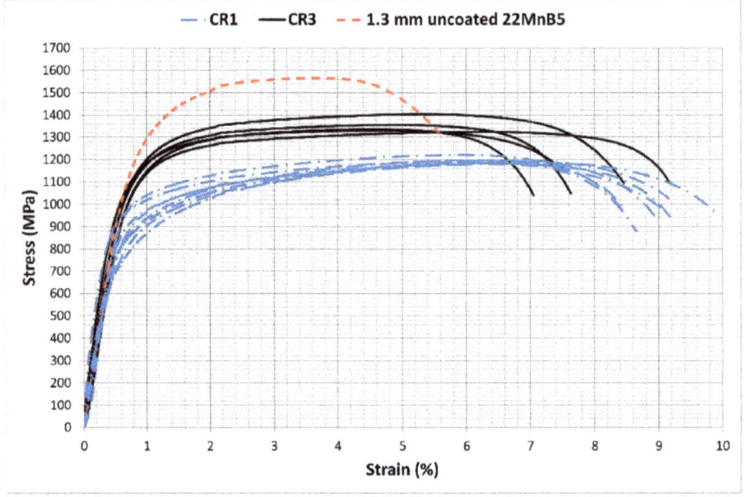

Figure 9. Tensile test curves of QP-processed hat profiles of CR1 and CR3 steels, alongside one tensile test curve for steel 22MnB5.

5. Discussion

The hardness values and microstructures of the Gleeble simulated samples have been found to be similar to those of the final press hardened hat profiles. This indicates that physical simulation is a reliable means of determining press hardening parameters to achieve a desired microstructure. One complication encountered in press hardening of hat profiles was the increase in temperature caused mainly due to the exothermic heat generated by the decomposition of austenite to martensite and/or bainite. The isothermal transformations at different temperatures presented in Figure 3 show that the time for completion of bainitic transformation is temperature dependent, and that longer holding time leads to higher fraction of bainite and it can also lead to more carbon enriched austenite and a decrease in the dislocation density by recovery mechanisms. Other microstructural mechanisms such as isothermal martensite formation below M_s, carbide precipitation and carbon partitioning from supersaturated martensite or bainite to austenite can also take place, as reported in literature [26]. The hardness values shown in Figure 4 for the Gleeble samples quenched to a temperature below M_s followed by a heating at a temperature above M_s, show that the initial martensite formation shortens the time necessary for the bainite transformation. Two mechanisms that also influence the hardness value are the tempering of the initial martensite formed and the martensite that can be formed after final cooling to room temperature.

It is reasonable to assume that microstructural changes occurring during isothermal holding are essentially diffusion controlled, thus following a relationship similar in form to that of the Larson–Miller parameter LMP [27]. Here the diffusion equation $D = D_0\exp(-Q/(RT))$ is used to calculate an effective 'time-diffusivity' t_D (the sum of the individual diffusivities over the thermal cycle following quenching to desired temperature). D is diffusivity (m^2/s), D_0 is a constant (m^2/s), Q is the activation energy of diffusion (J/mol), and R is the gas constant (8.31 J/molK). The time interval used in the calculation of t_D is from the point the samples reached their lowest temperature on quenching (Q_T) to the point after isothermal treatment when the specimens had cooled to 200 °C, see Equation (1).

$$t_D = \int_{Q_T}^{200} tDdT \quad (1)$$

A pre-exponential diffusivity constant D_0 of 2.3×10^{-5} m^2/s and an activation energy Q of 148 kJ/mol for carbon diffusion in austenite were used in these calculations [28]. The calculations may be used for an approximate comparison with the experimental data, but they do not take into account accurately the effect of the high silicon content on diffusivity in the steel. Figures 10 and 11 present variation of hardness with t_D for the Gleeble simulated QP type heat-treatments with a quench stop temperature of $M_s - 10$ °C and subsequent isothermal holding at $M_s + 30$ °C, carried out on CR3 and CR1 steels, respectively. The corresponding hardness values of the hat profiles are also included in the figures. As expected, a reduction in hardness with an increase in t_D is seen. It is clear that the hardness data for the press hardened hat profiles lie close to those obtained on the CR3 and CR1 steels subjected to equivalent Gleeble simulations, see Figures 10 and 11. The calculations confirm the usefulness of the method for obtaining a rough estimate of the achievable hardness, though the results can easily be affected by the deformation applied in press hardening. The hardness after press hardening trials are 5–10 HV points higher in comparison with the Gleeble simulated samples. As stated above, one possible explanation could be the influence of the forming operation on the M_s temperature during sheet deformation [29]. The large standard deviation for CR1 sample probably is, caused by slow cooling of flange 1, resulting in ferrite formation, see Figure 8a.

A comparison of the hardness data presented in Figure 3 for bainitic microstructures following austempering, with those presented in Figure 4 for samples subjected to prior QP type treatment and Figure 8 for different locations on hat profiles indicates that it is possible to shorten the isothermal holding time without the risk of obtaining excessive hardness in the steel. If the partitioning period is too short, during which only a small fraction of the austenite has transformed to bainitic ferrite and/or stabilized as high carbon austenite, untempered high carbon martensite may form on final cooling resulting in high hardness. The results for steel CR1 indicate that even the minimum holding time gives the targeted hardness value. By comparing the 'time-diffusivity' calculations for different samples, it can be seen that a holding time of 5 min for steel CR3 at 350 °C is equivalent to about 0.5 min holding time for steel CR1 at 430 °C. Similarly, a holding time of 15 min for steel CR3 is equivalent to a holding time of 1 min for steel CR1 at the same test temperatures (350 and 430 °C) for the two steels.

The quench stop temperatures and subsequent temperature–time holding combinations investigated in this study resulted in mechanical properties that are promising for industrial application. The strength of 0.15C steel CR1 hat profile is lower than that for a conventional press hardened profile of 22MnB5 boron steel. The yield strength of 0.26C steel CR3, however, is comparable with that of the boron steel, though the tensile strength is lower. The elongation to fracture is, however, superior in the CR3 steel compared with that of the press hardened boron steel. Liu et al. [19] applied hot stamping with QP type treatment and obtained tensile strength and ductility of the order of 1500–1600 MPa and 6.6–14.8%, respectively, though the yield strength was limited to 655–850 MPa. Hence in the present work, a higher yield strength was obtained. The improvement in ductility shown in the novel press hardening process provides the possibility to produce safety related components for cars with a possibility of reduction in weight. The hardness values after Gleeble simulated austempering cycles, Figure 3, show that the hardness values reach the same levels as the fully austempered samples measured by dilatometry after 10–15 min. The Gleeble cycles simulating different QP type cycles reach same hardness levels after 1–5 min. The possibility of time reduction for processing CR1 hat profile is from ca. 8 min to between 1 and 5 min, and for sample CR3 from 23 to 1–5 min. The special QP type process simulated in Gleeble experiments and the subsequent production of the hat profiles created opportunities to shorten the processing time considerably in comparison to typical times of the conventional austempering process. The tests have also shown that the inevitable variations in the processing cycle with regard to reaching the desired quench stop temperature and subsequent holding to complete bainitic transformation in actual industrial component forming operation do not influence the properties of the final product significantly. On the other hand, a quick comparison of the strength and ductility properties reported in the literature for CFB and QP steels, as shown in Table 1, reveals that the same levels of high strength and good ductility cannot be attained in press hardening, using the QP type treatment for realizing CFB microstructures in steels.

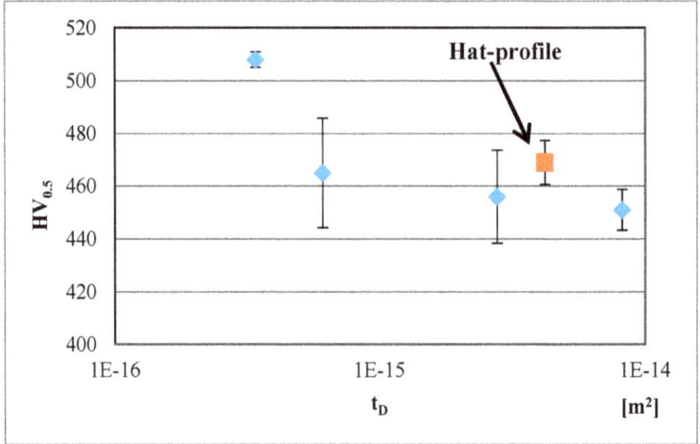

Figure 10. Hardness versus time-diffusivity t_D [m^2] for steel CR3 quenched to M_s − 10 °C followed by holding at M_s + 30 °C for 0.5, 1, 5, and 15 min by Gleeble simulation. The value for the press-hardened hat profile is included. Standard deviations, based on 5 values for each Gleeble treated sample and 25 values for the hat profile are presented.

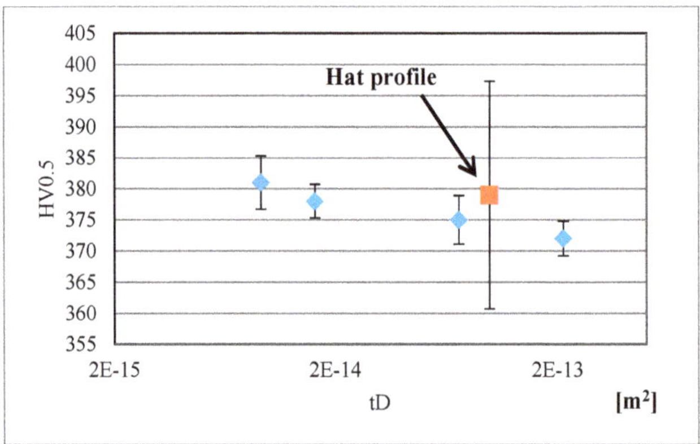

Figure 11. Hardness versus time-diffusivity t_D [m^2] for steel CR1 quenched to M_s − 10 °C followed by holding at M_s + 30 °C for 0.5, 1, 5, and 15 min. by Gleeble simulation. The value for the press-hardened hat profile is included. Standard deviations, based on 5 values for each Gleeble treated sample and 25 values for the hat profile are presented.

The novel press hardening technique for steels with increased Si content presented here together with the use of a QP type treatment to obtain a final CFB microstructure is shown to result in a combination of mechanical properties comparable with those of existing boron steels. It can be emphasized, however, that with further investigations of process variables, on the industrially produced sheets, and careful optimization of chemical composition to realize CFB microstructures for this kind of application are likely to provide property combinations far superior to those of boron bearing steels and closer to those obtained in CFB steels and QP-processed steel sheets themselves.

6. Conclusions

The results of simulation and press hardening experiments show that it is possible to produce complex steel sheet components with high strength and ductility by press hardening in combination with a controlled quenching and partitioning treatment. By quenching to a temperature below the M_s temperature of the steel, heating to a temperature, e.g., 30 °C above M_s and holding there, the phase transformation time is shortened, in comparison with a traditional austempering treatment. Consequently, the total processing time is shortened, benefiting productivity. Even though the steel with a carbon content of 0.15 wt % gave yield and tensile strength values lower than those of conventional press hardened boron steel, the steel with the carbon content of 0.26 wt % resulted in a yield strength comparable with that of the boron steel, although with a lower tensile strength. In addition, the elongation to fracture after press hardening in combination with quenching and partitioning is significantly higher than that of conventional press hardened 22MnB5 boron steel.

The microstructure achieved after pressing and QP treatment contains a very fine multiphase structure comprising lath-like ferrite, retained austenite and tempered martensite, which contribute to the good tensile properties achieved for the materials. In comparison with conventional martensitic microstructure achieved by press hardening of boron steels, the structure achieved by QP treatment in combination with pressing enabled the formation of a very refined structure containing a large amount of ferrite laths and interlath retained austenite, which rendered relatively higher ductility besides high strength in the produced components.

Author Contributions: The author contributions have been following; Conceptualization, E.P.V.; Methodology, E.P.V. and K.E.; Investigation, E.P.V., A.G.Ö., K.E., F.G.C., and S.A.; Data curation, F.G.C.; Writing—original draft preparation, E.P.V., M.C.S., and J.C.I.; Writing—review and editing, E.P.V., M.C.S., and L.P.K.; Visualization, A.G.Ö., K.E., and E.P.V.; Gleeble test planning and supervision, L.P.K.; Gleeble tests and data analysis, interpretation of hardness data in respect of quenching and partitioning process, M.C.S.; Alloy design based on thermodynamic and kinetics calculations, and microstructure characterization, F.G.C.

Funding: The European Research Fund for Coal and Steel, contract RFSR-CT-2008-00021, has funded this work.

Acknowledgments: The support of the European Research Fund for Coal and Steel for funding the contract RFSR-CT-2008-00021 is gratefully acknowledged by the authors. Michelle Nicolaus and Farnoosh Forouzan are acknowledged for their assistance in the work.

Conflicts of Interest: The authors declare no conflict of interest. The funders had no role in the design of the study; in the collection, analyses, or interpretation of data; in the writing of the manuscript, or in the decision to publish the results.

References

1. Berglund, G. The history of hardening of boron steel in Northern Sweden. In Proceedings of the 1st International Conference on Hot Sheet Metal Forming of High-Performance Steel, Kassel, Germany, 22–24 October 2008; pp. 175–177.
2. Jonsson, M. Press hardening, from innovation to global technology. In Proceedings of the 1st International Conference on Hot Sheet Metal Forming of High-Performance Steel, Kassel, Germany, 22–24 October 2008; pp. 253–256.
3. Karbasian, H.; Tekkaya, A.E. A review on hot stamping. *J. Mater. Process. Technol.* **2010**, *210*, 2103–2118. [CrossRef]
4. Bakewell, J. Hot Stamping Hits Heights. Automotive Manufacturing Solutions (AMS). 15 June 2016. Available online: https://automotivemanufacturingsolutions.com/process-materials/hot-stamping-hits-heights (accessed on 10 February 2019).
5. Fan, D.W.; Kim, H.S.; De Cooman, B.C. A review of the physical metallurgy to the hot press forming of advanced high strength steel. *Steel Res. Int.* **2009**, *80*, 241–248.

6. Steinhoff, K.; Saba, N.; Maikranz-Valentin, M.; Weidig, U. Optimized processes and products in hot sheet metal forming. In Proceedings of the 2nd International Conference on Hot Sheet Metal Forming of High-Performance Steel, Luleå, Sweden, 15–17 June 2009; pp. 29–42.
7. Caballero, F.G.; Santofimia, M.J.; Capdeville, C.; García-Mateo, C.; Garcia De Andrés, C. Design of advanced bainitic steels by optimization of TTT diagrams to T_0 curves. *ISIJ Int.* **2006**, *46*, 1479–1488. [CrossRef]
8. Caballero, F.G.; Santofimia, M.J.; García-Mateo, C.; Chao, J.; Garcia De Andrés, C. Theoretical design and advanced microstructure in super high strength steels. *Mater. Des.* **2009**, *30*, 2077–2083. [CrossRef]
9. Putatunda, S.K.; Singar, A.V.; Tackett, R.; Lawes, G. Development of a high strength high toughness ausferritic steel. *Mater. Sci. Eng. A* **2009**, *513–514*, 329–339. [CrossRef]
10. Speer, J.G.; Streicher, A.M.; Matlock, D.K.; Rizzo, F.; Krauss, G. Quenching and partitioning: A fundamentally new process to create high strength TRIP sheet microstructures. In *Austenite Formation and Decomposition*; Damm, E.B., Merwin, M.J., Eds.; TMS (The Minerals, Metals & Materials Society): Pittsburgh, PA, USA, 2003; pp. 505–521.
11. De Moor, E.; Speer, J.G.; Matlock, D.K.; Kwak, J.-H.; Lee, S.-B. Effect of carbon and manganese on the quenching and partitioning response of CMnSi steels. *ISIJ Int.* **2011**, *51*, 137–144. [CrossRef]
12. Li, H.Y.; Lu, X.; Jin, X.; Dong, H.; Shi, J. Bainitic transformation during the two-step quenching and partitioning process in a medium carbon steel containing silicon. *Mater. Sci. Eng. A* **2010**, *527*, 6255–6259. [CrossRef]
13. Naderi, M. Hot Stamping of Ultra-High Strength Steels. Ph.D. Thesis, RWTH Aachen, Aachen, Germany, 2007.
14. Zhang, M.R.; Gu, H.C. Fracture toughness of nanostructured railway wheels. *Eng. Fract. Mech.* **2008**, *75*, 5113–5121. [CrossRef]
15. Putatunda, S.K.; Martis, C.; Boileau, J. Influence of austempering temperature on the mechanical properties of a low carbon low alloy steel. *Mater. Sci. Eng. A* **2011**, *528*, 5053–5059. [CrossRef]
16. Yakubtsov, I.A.; Purdy, G.R. Analyses of transformation kinetics of carbide-free bainite above and below the athermal martensite-start temperature. *Metall. Mater. Trans. A* **2012**, *43*, 437–446. [CrossRef]
17. Kawata, H.; Hayashi, K.; Sugiura, N.; Yoshinaga, N.; Takahashi, M. Effect of martensite in initial structure on bainite transformation. In Proceedings of the 2009 6th International Conference on Processing & Manufacturing of Advanced Materials, Berlin, Germany, 25–29 August 2009; pp. 3307–3312.
18. Smanio, V.; Sourmail, T. Effect of partial martensite transformation on bainite reaction kinetics in different 1%C steels. *Solid State Phenom.* **2011**, *172–174*, 821–826. [CrossRef]
19. Liu, H.; Lu, X.; Jin, X.; Dong, H.; Shi, J. Enhanced mechanical properties of a hot stamped advanced high-strength steel treated by quenching and partitioning process. *Scr. Mater.* **2011**, *64*, 749–752. [CrossRef]
20. Liu, H.; Jin, X.; Dong, H.; Shi, J. Martensitic microstructural transformations from the hot stamping, quenching and partitioning process. *Mater. Charact.* **2011**, *62*, 223–227. [CrossRef]
21. Chang, Y.; Li, G.; Wang, C.; Li, X.; Dong, H. Effect of Quenching and Partitioning with Hot Stamping on Martensite Transformation and Mechanical Properties of AHSS. *J. Mater. Eng. Perform.* **2015**, *24*, 3194–3200. [CrossRef]
22. Seo, E.J.; Cho, L.; De Cooman, B.C. Application of Quenching and Partitioning (Q&P) Processing to Press Hardening Steel. *Metall. Mater. Trans. A* **2014**, *45*, 4022–4037.
23. Caballero, F.G.; Allain, S.; Cornide, J.; Puerta Velàsquez, J.D.; Garcia-Mateo, C.; Miller, M.K. Design of cold rolled and continuous annealed carbide-free bainitic steels for automotive application. *Mater. Des.* **2013**, *49*, 667–680. [CrossRef]
24. Koistinen, D.P.; Marburger, R.E. A general equation prescribing the extent of the austenite-martensite transformation in pure iron-carbon alloys and plain carbon steels. *Acta Metall.* **1959**, *36*, 59–60. [CrossRef]
25. Girault, E.; Jacques, P.; Harlet, P.; Mols, K.; Van Humbeeck, J.; Aernoudt, E.; Delannay, F. *Metallographic Methods for Revealing the Multiphase Microstructure of TRIP-Assisted Steels*; Elsevier Science Inc.: Amsterdam, The Netherlands, 1998; pp. 111–118.
26. Somani, M.C.; Porter, D.A.; Karjalainen, P.L.; Misra, D.K. On the decomposition of austenite in a high-silicon steel during quenching and partitioning. In Proceedings of the Materials Science &Technology, Pittsburgh, PA, USA, 7–11 October 2012; pp. 1013–1020.
27. Hertzberg, R.W. *Deformation and Fracture Mechanics of Engineering Materials*, 4th ed.; John Wiley & Sons: Hoboken, NJ, USA, 1996; ISBN 9780471012146.

28. Callister, W.D.; Rethwisch, D.G. *Materials Science and Engineering*, 8th ed.; John Wiley & Sons: Hoboken, NJ, USA, 2011.
29. Abbasi, M.; Saeed-Akbari, A.; Naderi, M. The effect of strain rate and deformation temperature on the characteristics of isothermally hot compressed boron-alloyed steel. *Mater. Sci. Eng. A* **2012**, *538*, 356–363. [CrossRef]

Article

Effects of Chemical Composition and Austenite Deformation on the Onset of Ferrite Formation for Arbitrary Cooling Paths

Aarne Pohjonen *, Mahesh Somani and David Porter

Materials and Production Technology Department, University of Oulu, 90014 Oulu, Finland; Mahesh.Somani@Oulu.fi (M.S.); David.Porter@Oulu.fi (D.P.)

* Correspondence: Aarne.Pohjonen@Oulu.fi; Tel.: +358-40-771-9225

Received: 5 June 2018; Accepted: 10 July 2018; Published: 12 July 2018

Abstract: We present a computational method for calculating the phase transformation start for arbitrary cooling paths and for different steel compositions after thermomechanical treatments. We apply the method to quantitatively estimate how much austenite deformation and how many different alloying elements affect the transformation start at different temperatures. The calculations are done for recrystallized steel as well as strain hardened steel, and the results are compared. The method is parameterized using continuous cooling transformation (CCT) data as an input, and it can be easily adapted for different thermomechanical treatments when corresponding CCT data is available. The analysis can also be used to obtain estimates for the range of values for parameters in more detailed microstructure models.

Keywords: austenite; steel; thermomechanical processing; phase transformation; nucleation; ferrite; CCT; TTT; incubation; transformation start

1. Introduction

Modern thermomechanically processed steels can be rolled and cooled to a variety of microstructures with properties tailored to many kinds of applications. To minimise scatter in microstructures and properties it is beneficial to have phase transformation models that are as accurate as possible. To be able to predict and understand the onset of phase transformations occuring during the complex cooling path of hot rolled steel for different compositions, a quantitative model is needed. To give a realistic estimate, the model should take into account the severe mechanical deformation occurring in hot rolling prior to cooling. In addition, it should be possible to extend the model and adapt it to specific steels and mechanical treatments prior to cooling. It is also desirable that the input data required for the model are easy to obtain, for example, by constant cooling rate experiments.

Several phase transformation models that can be used for simulations along arbitrary cooling paths have been introduced earlier, for example, references [1–11]. However, there is currently no method that includes the effect of austenite deformation on the transformation start for different steel compositions and which could be used to calculate the transformation start for arbitrary cooling path. Since the nucleation phenomenona during the onset of transformation is different from the successive growth stage, several models separate the growth kinetics of the phase transformation from the onset stage by considering an incubation time or delay for the transformation start [3–5,7,8,10,11]. However, these models do not consider the effect of deformation prior to cooling [6–9], or they are specific to a certain steel composition, or they only take in to account the effect of a few alloying elements [2–5,10,11].

Also, detailed models, such as [12–14], which include the description of the steel microstructure, can be used to calculate phase transformation phenomena. These detailed models include a number of parameters which can be either obtained experimentally [13,14] or calculated using theory [12]. Such detailed models could benefit from data that can be obtained directly from linear cooling rate experiments in order to find the range of possible values for the parameters.

The method for calculating the start time of transformation from temperature–time transformation (TTT) diagrams has been presented in several articles, e.g., [1,6–8,15], which could be used to calculate the transformation start by applying Scheil's additivity rule for an arbitrary cooling path. However, it has been shown that using such isothermal transformation data to calculate the transformation start for cooling conditions by applying the additivity rule can give incorrect estimates [16,17]. For cooling, the application of the so-called ideal (or true) incubation time [18,19] gives a better estimate for the transformation start for an arbitrary cooling path [20], as it gives exactly the correct continuous cooling transformation (CCT) diagram for constant cooling rates when Scheil's additivity principle is applied.

We present a computational method to calculate an estimate for the ferrite phase transformation start of an arbitrary cooling path and for varying steel compositions containing the following alloying elements: C, Si, Mn, Cr, Cu, Ni, Mo, V, Ti, B, and Nb. The model can be easily adapted for new steels and thermomechanical treatments by using constant cooling rate CCT data as the input for the model. The transformation start time includes the time required for the formation of critical-sized nuclei (incubation time) as well as the time required for the growth of the nuclei to the volume fraction required for the detection of the phase transformation start. The method comprises three parts as follows: (1) Calculation of the CCT start time for different steel compositions and phases using the results presented in [21,22] for steels subjected to strain relevant to hot rolling procedures; (2) construction of the so-called ideal TTT diagram [18–20,23,24] from a constant cooling rate CCT diagram; and (3) application of Scheil's additivity principle to calculate the phase transformation start for an arbitrary cooling path after the ideal TTT diagram has been calculated [18]. We show how it is used to quantitatively compare the manner in which different alloying elements affect the austenite to ferrite phase transformation start time and the corresponding activation energy for a given steel composition. The same calculation is done for steel which has been additionally strained below the no-recrystallization temperature, and the results are compared. We also calculate an estimate for the phase transformation start for different kinds of cooling paths.

Since the model presented in this work is based on experimental CCT diagrams [21], it can be easily adapted and extended for different steels when there is a need to obtain more detailed estimates for given steel compositions or mechanical treatments prior to cooling. If constant cooling rate CCT data is analyzed, it can be used as the basis of this method, and the corresponding results, as presented in this article, can be obtained. We separately analyze the austenite to bainite transformation onset and present the results in [25,26].

2. Theory

2.1. Calculation of CCT Start Time for Different Steel Compositions

An estimate for the phase transformation start point in CCT diagrams for different steel compositions can be calculated with Equations (1) and (2) which were presented in reference [21]. As described in the original works, the equations were obtained by linear regression analyses on the effect of alloying elements on the CCT diagrams, considering the following elements: C, Si, Mn, Cr, Cu, Ni, Mo, V, Ti, B, Nb. The experimental results were obtained with 34 different steel compositions by cooling specimens after thermomechanical treatments relevant to industrial hot rolling and cooling processes.

$$T_{\mathrm{s,cct}}(\mathrm{K}) = \underbrace{273.15 + B_0 + \sum_i B_i c_i}_{\mathrm{T_N}} + B_{\mathrm{t}} \operatorname{arcsinh}(t_{85} - t_{85,\mathrm{k}}) \tag{1}$$

$$\log_{10}(t_{85,\mathrm{k}}) = A_0 + \sum_i A_i c_i \tag{2}$$

where $T_{s,cct}(K)$ is the start temperature (in degrees Kelvin) for the phase transformation occurring with a constant cooling rate; c_i are the concentrations of different elements in the steel composition in weight %; A_i and B_i are the constants obtained from the regression analysis; t_{85} is the time elapsed while cooling the specimen from 800 °C to 500 °C; and $t_{85,k}$ corresponds to the critical cooling rate, which is the slowest cooling rate that does not produce a measurable phase transformation. T_N is used to denote $273.15 + B_0 + \sum_i B_i c_i$. Prior to cooling, the steel sample was subjected to a mechanical deformation schedule consisting of either (a) two 20% strain deformations above the recrystallization limit temperature, or (b) the same as (a) but with an additional 30% deformation below the non-recrystallization temperature. The different deformation schedules (a and b) yielded different CCT start curves and regression constants. The regression constants for ferrite transformation for both deformation schedules were given in reference [21] and are reproduced in Table 1.

Table 1. The regression constants given in [21], which were used in Equations (1) and (2). The constants are given for two different deformation schedules, (a) and (b) (see text).

	B_0	B_C	B_{Si}	B_{Mn}	B_{Cr}	B_{Cu}	B_{Mo}	B_V	B_{Ti}	B_B	B_{Nb}	B_t
(a)	834.8	−251	60.5	−103.2	−69.7	0	−105.5	0	202	9.0	205	11.86
(b)	884.2	−331	65.2	−98.7	−75.9	−97.4	−76.7	290	158	0	−322	9.28
	A_0	A_C	A_{Si}	A_{Mn}	A_{Cr}	A_{Cu}	A_{Mo}	A_V	A_{Ti}	A_B	A_{Nb}	A_{Ni}
(a)	−0.524	3.07	0	0.718	0.885	0	3.825	0	0	0	0	0
(b)	−0.319	3.45	−1.47	0	1.440	0	4.221	0	0	0	0	0

The composition limits (in wt %) for the experimental data used in the regression analysis are shown in Table 2 [21,27].

Table 2. The composition limits (in wt %) of the model.

	C	Si	Mn	P	S	N	Al	Cr
min	0.01	0.01	0.16	0.007	0.002	0.003	0.002	0.01
max	0.65	0.71	1.69	0.022	0.011	0.012	0.062	1.20
	Cu	Ni	Mo	V	Ti	B	Nb	
min	0	0	0	0	0	0	0	
max	0.30	1.32	0.51	0.10	0.03	0.0022	0.05	

A comparison with the experimental results showed that the functional form of Equation (1) represents the transformation to ferrite near the critical cooling rate (or $t_{85,k}$) well. For some steels whose CCT diagrams indicate a transformation start near the equlibrium $Ae3$ temperature, the functional form gives a slightly higher value for the transformation temperature near the equlibrium temperature than that observed experimentally. When the transformation temperature described by Equation (1) of the steel is well below the equilibrium temperature, the functional form represents the transformation temperatures for the whole experimental range, $t_{85} \in t_{85,k} ... 1000\,\text{s}$, well.

At a constant cooling rate, $\dot{\theta} = (300\,\text{K})/t_{85}$, the temperature of the specimen can be expressed as $T = Ae3 - \dot{\theta}t$, where $Ae3$ is the equilibrium ferrite formation temperature, and t is the time spent below the $Ae3$ temperature. By applying Equations (1) and (2), the time required for phase transformation to start during cooling, $t_{s,cct}$ is given by Equation (3):

$$t_{s,cct} = \frac{Ae3 - T_{s,cct}}{300} t_{85} = \frac{Ae3 - T_{s,cct}}{300} \left[\sinh \left(\frac{T_{s,cct} - T_N}{B_t} \right) + t_{85,k} \right]. \quad (3)$$

Equation (3) is valid for the temperatures where Equations (1) and (2) represent the experimental results well (see discussion above) and can be used to calculate $t_{s,cct}$ for different temperatures, $T_{s,cct}$. The coordinates $(t_{s,cct}(T_{s,cct}), T_{s,cct})$ then represent the CCT diagram in time-temperature coordinates. Following [28], we used Equation (4) to describe the $Ae3$ temperature (in degrees Kelvin):

$$Ae3(K) = (1184 - 29Mn + 70Si - 10Cr) - (418 - 32Mn + 86Si + 1Cr)C + 232C^2 \quad (4)$$

where the element symbols refer to their concentrations (in wt %). The equation includes the effects of C, Mn, Si and Cr. The effect of small amounts of microalloying with other elements is neglected for the equilibrium temperature calculation [28].

The CCT start curve was calculated using Equations (1)–(3) and is shown in Figure 1 with the critical cooling rate, which is the slowest cooling rate that does not produce the phase transformation. The plot was calculated for a steel composition of 0.09 C, 0.28 Si, 1.53 Mn, 0.012 P, 0.005 S, 0.03 Al, 0.05 Cr, 0.05 Cu, 0.035 Nb, 0.04 Ni, 0.02 Ti and 0.05 V (wt %) (named "TH16" in reference [22]) which was subjected to mechanical deformation but allowed time to recrystallize prior to cooling. To describe the "nose" of the CCT diagram without discontinuities in the temperature derivative, a piecewise polynomial was used for smooth transition from the CCT curve to the critical cooling rate curve (see Figure 1).

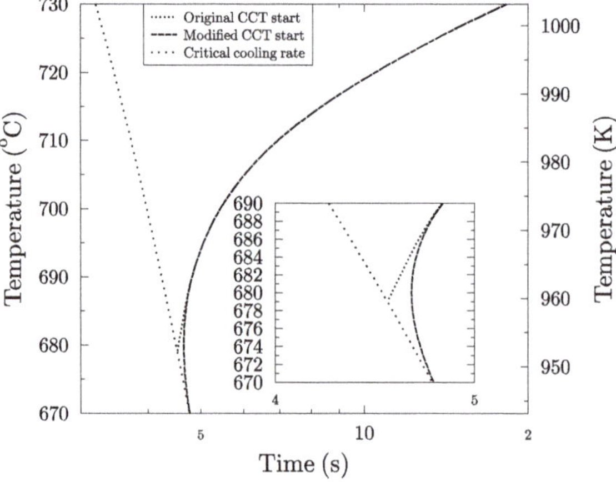

Figure 1. The continuous cooling transformation (CCT) diagram for the given steel composition can be calculated with the parameters reported in [21]. To describe the CCT "nose" without discontinuities in the derivative, a spline was fitted between the calculated CCT curve and the critical cooling rate.

2.2. Construction of the Ideal TTT Diagram from the Constant Cooling Rate CCT Diagram

The ideal isothermal transformation time, τ, can be calculated using Equation (5) [19,29]

$$\frac{1}{\tau(\Delta T_{CCT})} = \frac{d\dot{\theta}_c(\Delta T_{CCT})}{d\Delta T_{CCT}} \quad (5)$$

where the function $\dot{\theta}_c(\Delta T)$ is the constant cooling rate required to produce a transformation start exactly at the point of undercooling, ΔT_{CCT} (or, equivalently, ΔT_{CCT} is the amount of undercooling at which the transformation starts for constant cooling rate $\dot{\theta}_c$). Applying the chain rule of differentiation we obtain

$$\frac{1}{\tau(T_{s,cct})} = -\frac{d\dot{\theta}_c(T_{s,cct})}{dT_{s,cct}} \tag{6}$$

where $T_{s,cct}$ is the transformation start temperature in the CCT diagram.

2.2.1. Construction of the Ideal TTT Diagram from $(t_{85}(T_{s,cct}), T_{s,cct})$ CCT Diagram

The numerical value of the ideal isothermal start time can be obtained by taking the difference approximation of Equation (6) and solving for $\tau(T_{s,cct}) \approx -\Delta T_{s,cct}/\Delta\dot{\theta}_c(T_{s,cct})$, where $\Delta T_{s,cct} = (T_{s,cct} + h/2) - (T_{s,cct} - h/2) = h$ and $\Delta\dot{\theta}_c(T_{s,cct}) = \dot{\theta}_c(T_{s,cct} + h/2) - \dot{\theta}_c(T_{s,cct} - h/2)$ for a sufficiently small h value. The numerical value for the isothermal start time can be calculated using Equations (1), (2) and (7).

$$\begin{aligned} \tau(T_{s,cct}) &\approx -\frac{\Delta T_{s,cct}}{\Delta\dot{\theta}_c(T_{s,cct})} \\ &= \frac{h}{300\left[\frac{1}{t_{85}(T_{s,cct} - h/2)} - \frac{1}{t_{85}(T_{s,cct} + h/2)}\right]} \end{aligned} \tag{7}$$

where $t_{85}(T_{s,cct} \pm h/2)$ represent the function values of $t_{85}(T)$ at $T = T_{s,cct} \pm h/2$. When the value $h = 0.1$ K was used, the recalculation of the CCT diagram from the TTT diagram using Equation (12) successfully replicated the original CCT diagram (as shown in Figures 3 and 4).

2.2.2. Construction of the Ideal TTT Diagram from $(t_{s,cct}(T_{s,cct}), T_{s,cct})$ type CCT Diagram

The numerical value of the isothermal start time can also be obtained from the $(t_{s,cct}(T_{s,cct}), T_{s,cct})$ type constant cooling rate CCT diagram, where $t_{s,cct}$ and $T_{s,cct}$ are the time and temperature coordinates of the CCT curve and T_{start} is the cooling start temperature temperature, by using the difference approximation described in Section 2.2.1. Since the constant cooling rate is $\dot{\theta}_c = (Tstart - T_{s,cct})/t_{s,cct}(T_{s,cct})$, Equation (8) describes the ideal isothermal start time as a function of temperature.

$$\begin{aligned} \tau(T_{s,cct}) &\approx -\frac{\Delta T_{s,cct}}{\Delta\dot{\theta}_c(T_{s,cct})} \\ &= \frac{h}{\frac{T_{start} - (T_{s,cct} - h/2)}{t_{s,cct}(T_{s,cct} - h/2)} - \frac{T_{start} - (T_{s,cct} + h/2)}{t_{s,cct}(T_{s,cct} + h/2)}} \end{aligned} \tag{8}$$

where $t_{s,cct}(T_{s,cct} \pm h/2)$ represent the function values of $t_{s,cct}(T)$ at $T = T_{s,cct} \pm h/2$. Equation (8) can be used to calculate the ideal isothermal transformation start, $\tau(T)$, when the constant cooling CCT curve $(t_{s,cct}(T_{s,cct}), T_{s,cct})$ is known.

2.2.3. High Temperature Extrapolation of the TTT Start Time

For ferrite formation, we used the following high temperature extrapolation of the calculated ideal TTT start time. The function describing the isothermal transformation start time can be expressed by Equation (9) [20]

$$\tau = K\exp(\frac{Q}{RT})\Delta T^{-m} \tag{9}$$

where K and m are constants; R is the ideal gas constant; T is the absolute temperature; $\Delta T = Ae3 - T$ is the amount of undercooling; and Q is the activation energy to facilitate carbon mobility in the austenite–ferrite transformation.

Taking the logarithm of both sides of the Equation (9) and differentiating yields,

$$\frac{d\ln\tau}{dT} = -\frac{Q}{RT^2} + \frac{m}{Ae3 - T} \tag{10}$$

Since near the equilibrium, $Ae3$, the temperature is $\frac{m}{Ae3-T} >> \frac{Q}{RT^2}$, we approximate $\frac{d\ln\tau}{dT} \approx \frac{m}{Ae3-T}$. This allows us to use Equation (11) for high temperature extrapolation of the transformation start time:

$$\ln\tau \approx \ln B - m\ln(Ae3 - T) \Leftrightarrow \tau \approx B\Delta T^{-m} \tag{11}$$

where the constants B and m are determined from the conditions under which the functions $\ln\tau$ and $d\ln\tau/dT$ are continuous. The ideal TTT curve calculated from the CCT diagram as well as the high temperature extrapolation are shown in Figure 2.

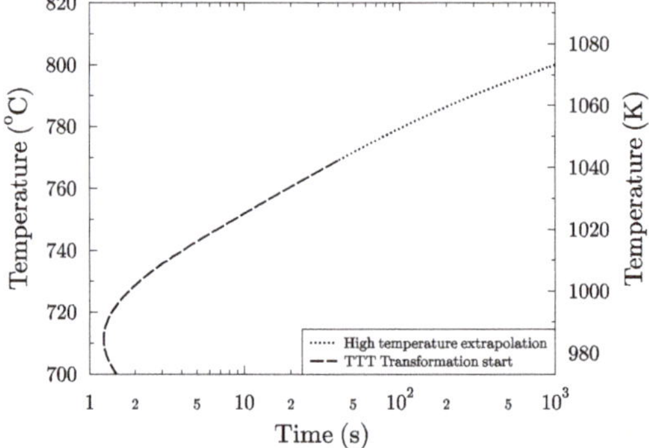

Figure 2. The high temperature extrapolation of the ideal temperature–time transformation (TTT curve), calculated using Equation (11), shown together with the ideal TTT start diagram calculated directly from the experimentally fitted CCT curve using Equation (8).

2.3. Calculation of the Phase Transformation Start for an Arbitrary Cooling Path

Once the ideal isothermal transformation time has been constructed, it can be used to calculate the transformation start time for an arbitrary cooling path using Scheil's additivity principle. The transformation is assumed to start when the sum of the fractional transformation times equals one, as shown in Equation (12) [18–20,30].

$$\sum_{i=1}^{n} \frac{t_i}{\tau_i(T)} = 1 \tag{12}$$

where t_i is the time spent at temperature T, and $\tau_i(T)$ is the time required to produce a measurable transformation at that temperature.

To check that the computational method gave the correct result, the CCT diagram, which was obtained using the regression Equations (1) and (2), was recalculated using the obtained ideal TTT diagram and Scheil's additivity principle, as shown in Equation (12). We also compared the results to the experimental data presented in reference [22] for steel "TH16". The comparison is shown in Figures 3 and 4. The agreement between the original CCT diagram obtained by the regression formulas and the recalculated CCT diagram is excellent. Also, the comparison with the experimental data [22] shows a good agreement, although the difference between the experimental result and the CCT curve calculated with the regression equations is slightly higher for the temperatures near the CCT "nose" for this steel, as shown in the logarithmic plot in the inset of Figure 4. In any case, at temperatures above 993 K (720 °C), the experimental values also agree well with the computed values.

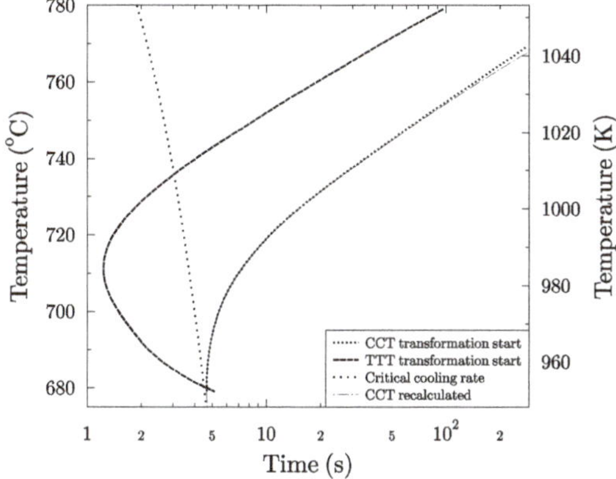

Figure 3. Ideal (or true) isothermal transformation diagram calculated with Equation (8), the original CCT diagram calculated with Equation (3), and the CCT diagram which was recalculated by applying the Scheil's additivity principle in Equation (12).

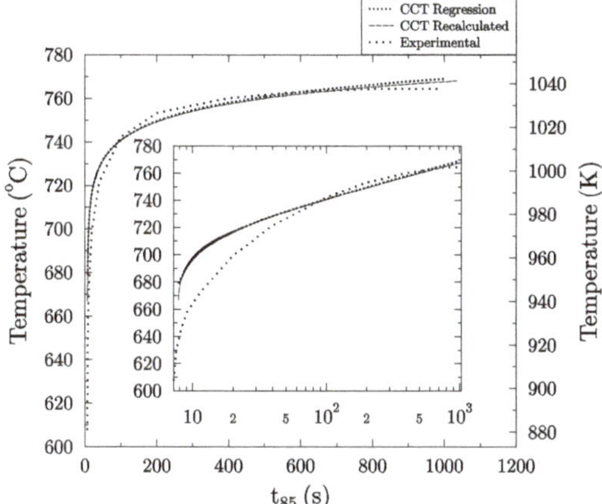

Figure 4. To validate the model, the CCT diagram was recalculated from the ideal (or true) TTT curve using Scheil's additivity rule. The result shows excellent agreement between the original CCT curve calculated from Equations (1) and (2) and the recalculated CCT curve. The results were also compared to the experimental data given in [22]. Note that the original CCT curve and the recalculated curve almost overlap.

2.4. Effective Activation Energy of the Transformation Start

Since ferrite nucleation is a thermally activated process, it can be described using the concept of activation energy [31–34]. The observed transformation start includes the nucleation of microscopic nuclei as well as the growth of the nuclei to a size at which the transformation can be observed by dilatometric measurement. The transformed fraction, χ, at a constant temperature can be described by an Avrami type Equation (13) [31–33]:

$$\chi = 1 - \exp[-(kt)^n] \approx (kt)^n \tag{13}$$

where the coefficient k is related to the nucleation and growth rates, and n is related to the shapes of the growing nuclei, and the approximation is valid for a small value of $(kt)^n$ (i.e., for the start of the transformation) [35]. We emphasize that during the intial stages of the transformation, before it has proceeded to the extent that it can be measured (e.g., 1% transformation), the values for k and n can be different than those produced later in the transformation [16]. This is because the energetically most favourable nucleation sites are consumed early in the process, and the transformed and diffusion regions do not overlap for the first ferrite regions that form on the austenite grain boundaries, whilst the mean carbon content of the austenite increases as the transformation proceeds.

Since both nucleation and growth are thermally activated processes, coefficient k can be described by the Arrhenius type Equation (14) [31–34]:

$$k = A\exp\left(\frac{-E_{\mathrm{A}}(T)}{RT}\right) \tag{14}$$

where $E_{\mathrm{A}}(T)$ is the effective activation energy, which depends on the temperature due to effects of undercooling (i.e., thermodynamic driving force) and atomic diffusion; R is the gas constant; $A = C\omega$ is a prefactor, which relates to the attempt frequency, ω, and the concentration of heterogenous nucleation sites, C. The effective activation energy is a linear combination of the activation energies of nucleation and growth [31–34].

By substituting Equation (14) in to Equation (13) and solving for t, we can obtain an expression for the time, τ, required to produce a measurable transformation, χ_{m}, at a given temperature, T, through Equation (15):

$$\tau = K\exp\left(\frac{E_{\mathrm{A}}(T)}{RT}\right) \tag{15}$$

where $K = \chi_{\mathrm{m}}{}^{\frac{1}{n}}/A$. Since the effective activation energy depends on the temperature, it is not straightforward to obtain its value. However, it is possible to show how different factors affect the transformation start from the differences in transformation start times, as described in Section 3.

3. Results

3.1. Model Validation

To demonstrate the benefits of using the model that has been specifically fitted to the experimental CCT data obtained after the thermomechanical processing described earlier, we compared the results obtained with Equations (1) and (2) to the corresponding simulations that were obtained with the commercial software JMatPro (TM) [1] as well as to the simulations based on the model described in [8]. This was possible, since the experimental data presented in the original reference [27] contained information on the austenite grain sizes of several steels, which could be used as input for the JMatPro (TM) software. The phase transformation model of the JMatPro (TM) software is a modified version of the Kirkaldy–Venugopalan phase transformation model [6], and it was fitted to isothermal transformation data [1]. The comparison is shown in Figure 5a,b. Experimentally-measured grain sizes were used as input in the JMatPro simulations. For the fully crystallized case, the DIN grain size number was 6, and for the strain hardened case, the DIN grain size number was 7.

To see how well the model can predict the transformation start for other steels that were not used in the regression fitting, we also compared the model result to CCT diagrams of other steels that were deformed in the austenite state [36]. Although the deformation conditions were not exactly the same as those used in the fitting model, the comparison shows that the model can give reasonable predictions (similar accuracy as in [6]) for the transformation start for some cases. The steel compositions and deformation schedules for these experimental data were as follows: (I) 0.08 C, 1.52 Mn, 0.37 Si, 0.007 S, 0.023 P, 0.21 Cr, 0.10 Ni, 0.10 V, 0.05 Nb, 0.34 Cu, 0.02 Al and 0.008 N (wt %), strained for 0.25 at 1000 °C followed by 0.45 strain at 850 °C [36]; and (II) 0.10 C, 0.87 Mn, 0.33 Si, 0.24 Mo, 0.002 B, 0.005 N, 0.48 Zr (wt %) strained for 0.25 at 830 °C [36,37]. The deformation schedule used in case (I) corresponds that the use of the model fitted to the CCT and most likely for this reason the results had better agreement than in case (II). It must also be mentioned that for one case, the comparison against the experimental CCT data described in reference [36] (page 600, upper diagram) matched poorly. Hence, when using the model to estimate the transformation start for steels that have been subjected to different thermomechanical treatments than the one described earlier, or that have chemical compositions that are much different from the data that was used in the fitting of the regression model [21,27], experimental verification should be performed. In any case, we have demonstrated that the use of the model fitted to the CCT data from exact thermomechanical processing conditions (Figure 5a,b) can give more accurate information on the process than using general transformation formulas that have been obtained by fitting to isothermal transformation data without deformation, as shown in Figure 5.

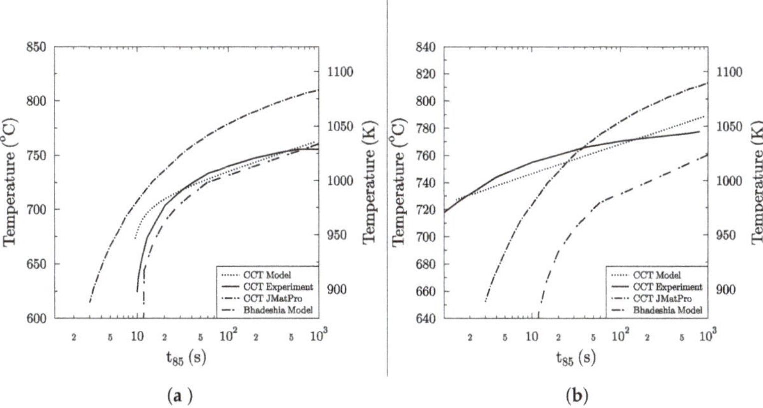

Figure 5. (**a**) CCT start temperature as a function of t_{85} calculated with Equations (1) and (2), JMatPro (TM) and the Bhadeshia model [8] and compared against the experimental data presented in reference [27] (steel TH15, 0.11 C, 0.36 Si, 1.57 Mn, 0.018 P, 0.006 S, 0.006 N, 0.038 Al, 0.04 Cr, 0.03 Cu, 0.03 Ni, 0.08 V, 0.045 Nb) for (**a**) fully recrystallized steel and (**b**) strain hardened steel. The experimentally-measured grain sizes were used as input in the JMatPro simulations.

3.2. Phase Transformation Start Calculation for Arbitrary Cooling Path

When the ideal TTT diagram has been determined for a given steel composition, it can be used to calculate the phase transformation start for an arbitrary cooling path. To give an example of this, we calculated transformation start diagrams for two different kinds of cooling paths: (1) for linear cooling paths, $T(t) = Ae3 - \dot{\theta}t$, where the constant cooling rate, $\dot{\theta}$, was varied; and (2) for modified cooling paths, $T(t) = Ae3(1 - (t/t_0)^2)$, where the time constant, t_0, was varied. The transformation start curves as well as two examples of the cooling paths are shown in Figure 6 for both types of cooling paths.

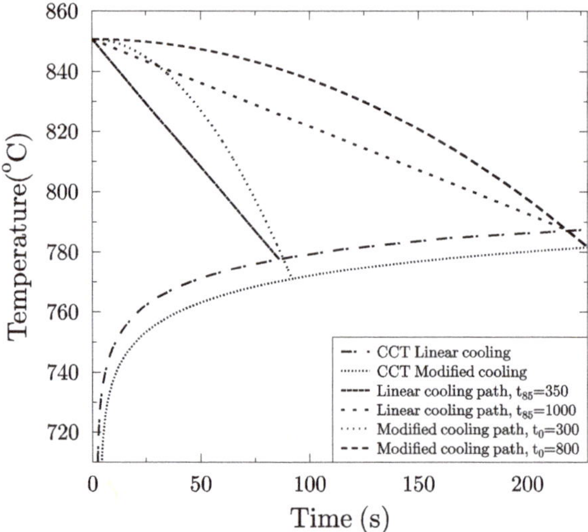

Figure 6. Phase transformation start calculated for two different types of cooling paths: (1) the linear cooling paths exemplified by linear plots and (2) the modified cooling paths given by the Equation $T(t) = Ae3(1 - (t/t_0)^2)$ shown by the curved plots.

3.3. How Different Alloying Elements Affect the Phase Transformation Start

The alloying elements affect the phase transformation start via two main mechanisms. The chemical composition directly affects the effective activation energy of the transformation. Also, because the chemical composition can affect the grain growth and recovery of dislocation structures after deformation, the chemical composition can also affect the transformation start by changing the concentration of heterogenous nucleation sites. In order to obtain a useful measure which takes into account both of the effects, we analyzed the effect of different elements on the transformation start in the following way. First, the ideal TTT start diagram was calculated for the original steel composition, $\tau_o(T)$. After this, the concentration of one alloying component was altered and the TTT start diagram was calculated for the altered composition, $\tau_*(T)$. By comparing the original TTT start diagram and the altered TTT start diagram, the extent to which the change in the alloying component affected the TTT start time could be observed. Figure 7 shows how changing the carbon concentration of the steel "TH16" [22] affected the transformation start time. The effect of alloying on the transformation start during constant cooling rate is given by Equations (1) and (2) [21], which formed the basis of our analysis.

Using the previously calculated incubation times τ_* and τ_o, we were able to calculate a useful measure to concisely describe both the effect of chemical composition on the transformation start as well as its effect on the number of nucleation sites. Since the effect of chemical composition on the thermal vibrations of the atoms is negligible, the fraction of the calculated transformation start times, Equation (16), gives a value that can be used to parameterize more detailed microstructure models. However, more information on the effect of chemical composition on the prior austenite microstructure and available nucleation sites is needed.

$$\frac{\tau_*}{\tau_o} = \frac{C_o}{C_*} exp\left(\frac{E_{A,*}(T) - E_{A,o}(T)}{RT}\right) \tag{16}$$

where C_* and C_o are the concentrations of the nucleation sites corresponding to τ_* and τ_o; and $E_{A,*}(T)$ and $E_{A,o}(T)$ are the effective transformation activation energies at different temperatures. The fractions of the transformation start times for elements that had non-negligible effects on the transformation start in the experimental analysis are shown in Figure 8 [21].

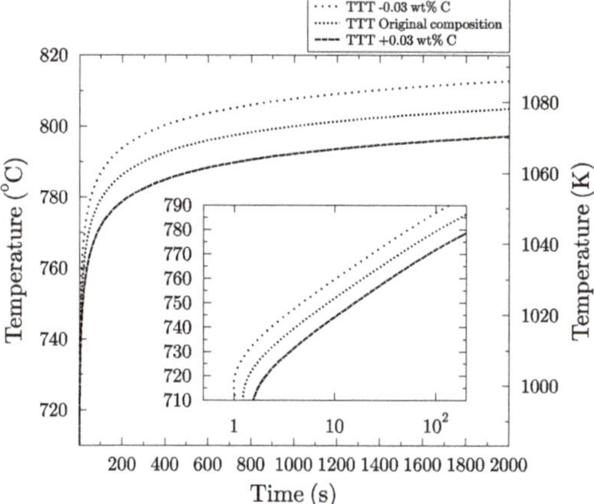

Figure 7. The start time of the original steel composition is compared to the compositions where the carbon wt % decreased/increased by 0.03%. The inset shows the change on a smaller scale near the ideal TTT "nose" in a logarithmic plot.

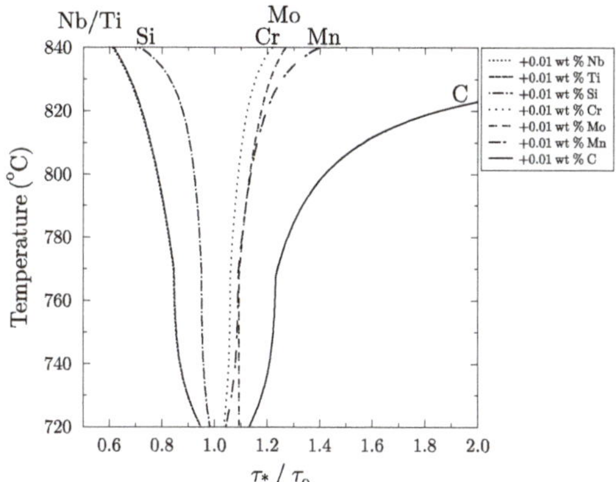

Figure 8. The effects of altering different alloying elements on the phase transformation start at different temperatures for fully recrystallized steel (deformation schedule a) is shown by calculating the fraction τ_{*}/τ_{o} (see text). The plots were calculated from the difference in the ideal TTT start times for steel "TH16" shown in reference [22] (0.09 C, 0.28 Si, 1.53 Mn, 0.012 P, 0.005 S, 0.03 Al, 0.05 Cr, 0.05 Cu, 0.035 Nb, 0.04 Ni, 0.02 Ti, and 0.05 V wt %).

3.4. *Effect of Strain Hardening on the Phase Transformation Start*

The microstructure of the steel and the phase transformation start are affected by the final rolling temperature and the time spent at elevated temperature prior to cooling. If the steel is rolled above the recrystallization limit temperature and kept at a high temperature prior to cooling, the mechanically-deformed grains have adequate time to recrystallize. In contrast, if the metal is additionally rolled at or below the temperature where no significant recrystallization occurs, the austeninte grains do not have time to recrystallize prior to cooling and are in a more unstable state. This causes the ferrite transformation to start earlier for strain hardened austenite. A comparison of numerical simulation studies to experimental data showed that deformation affects transformation mainly via reduction in the undercooling required for nucleation [38].

The original reference [21] that we used for calculating the transformation start temperatures and critical cooling rates via Equations (1) and (2), also provided the regression constants for strain hardened metal. It is therefore straightforward to calculate the same quantities for the strain hardened steel that were calculated for the fully recrystallized steel. The effects of different alloying elements on the transformation start of the strain hardened material are shown in Figure 9. By comparing the results, we can analyze the effect of strain hardening on the transformation onset.

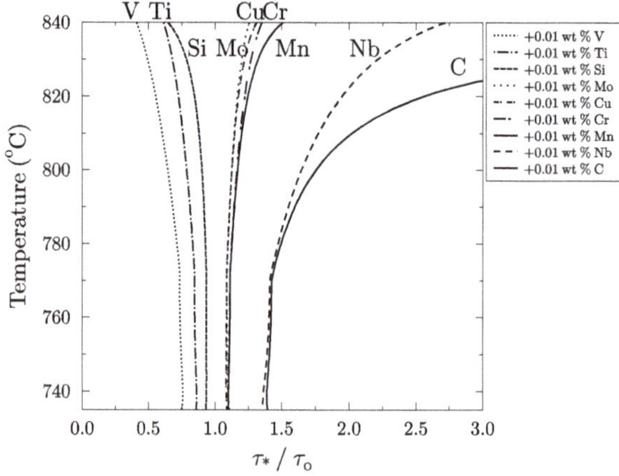

Figure 9. The effects of altering different alloying elements to the phase transformation start at different temperatures for strain hardened steel (deformation schedule b) is shown by calculating the fraction τ_* / τ_o (see text). The plots are calculated from the difference in the ideal TTT start times for steel "TH16" shown in reference [22] (0.09 C, 0.28 Si, 1.53 Mn, 0.012 P, 0.005 S, 0.03 Al, 0.05 Cr, 0.05 Cu, 0.035 Nb, 0.04 Ni, 0.02 Ti, and 0.05 V wt %).

We calculated the original ideal TTT transformation start time for the fully recrystallized steel, τ_r, and the strain hardened steel, τ_h, at different temperatures (Figure 10a). To quantitatively analyze the effect of strain hardening versus fully recrystallized steel, we also calculated the fraction τ_h / τ_r by applying Equation (16). The result is shown in Figure 10b. The steel composition used in the calculation was 0.096 C, 0.150 Si, 1.204 Mn, 0.018 P, 0.006 S, 0.008 N, 0.032 Al, and 0.036 Nb (wt %).

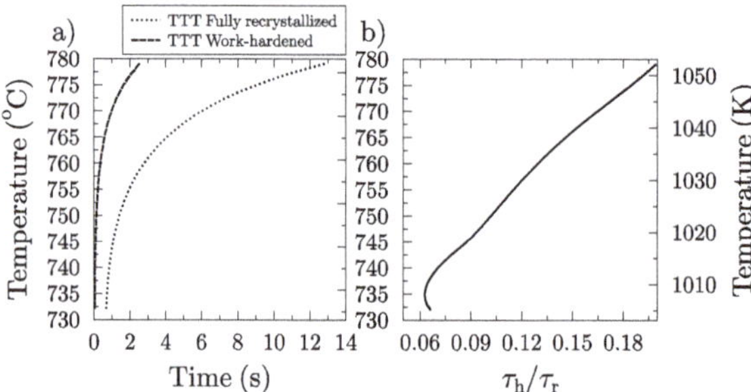

Figure 10. Comparison of the calculated ideal TTT curves of work hardened versus fully recrystallized austenite: (**a**) plots of the TTT curves; (**b**) the time required to start the transformation for strain hardened material divided by the corresponding time for fully recrystallized material.

4. Discussion

4.1. Model Validation

The results show that the use of a model that has been fitted to the CCT data from exact thermomechanical processing conditions (Figure 5a,b) gives more accurate results for the process than using general transformation formulas that have been obtained by fitting to isothermal transformation data without prior austenite deformation.

4.2. Phase Transformation Start Calculation for an Arbitrary Cooling Path

In Figure 6, it is clearly shown that for the modified cooling paths where the material spends more time at high temperatures, the transformation start occurs at lower temperatures than for the linear cooling paths. The reason for this is that at higher temperatures, where the undercooling is lower, the driving force for transformation is also lower. This result shows how the method can be quantitatively used to find a desired cooling path to fine-tune the final microstructure by calculating an estimate for the required cooling to achieve transformation start at lower or higher temperatures.

4.3. Effects of Chemical Composition on Transformation Start for Fully Recrystallized Steel

The effects of different alloying elements are qualitatively described in reference [39]. All the alloying elements, including the carbide-forming elements, were in solution at the start of the phase transformation, but following controlled deformation in the no-recrystallization regime, it is possible that some carbide-forming elements tended to promote strain-induced precipitation at some stage. Thermodynamically strain-induced ferrite formation is most likely to precede, or at least compete with, the precipitation process at some point. In any case, the precipitation state after the deformation schedules corresponds more closely to actual rolling conditions than the use of traditional CCT or TTT diagrams, as pointed out in [27]. The following descriptions in [39] agree well with our results and provide a physical explanation for some of the curves. Silicon and niobium promote ferrite formation, while carbon, manganese, chromium, and molybdenum retard the phase transformation. Carbon and manganese are austenite stabilizers. Manganese enhances carbon enrichment by increasing carbon solubility in austenite. Molybdenum exerts an important solute drag effect and delays the transformation of austenite to ferrite and to pearlite strongly. Contrary to its strong carbide forming tendency from a thermodynamic point of view, carbide precipitation is often retarded in presence of molybdenum [39].

4.4. Effects of Chemical Composition on Transformation Start for Strain Hardened Steel

For many alloying elements, the effects on transformation start is similar to those for the fully recrystallized case, but the absolute value of the effect is larger (i.e., the curves are spread wider apart from the zero axis). However, there are a few differences; for example, vanadium, which has a negligible effect for the fully recrystallized steel, promotes ferrite formation for the strain hardened steel. The effect of vanadium in promoting ferrite formation is mentioned in reference [39]. Also, copper, which has a negligible effect for the fully recrystallized steel, retards the transformation of strain hardened steel. Niobium, which promotes ferrite formation for the fully recrystallized material has the opposite effect for the strain hardened material. The effect of niobium in delaying the transformation for the strain hardened material can be explained by reduction of the prior austenite grain boundary energy [40]. After the additional deformation at a lower temperature, the niobium segregates more effectively to the grain boundaries than after high temperature deformation [35]. The reduction in the austenite grain boundary energy lowers the energy gain associated with ferrite nucleation on grain boundaries [40] and delays the nucleation. Niobium may also behave in a manner similar to molybdenum because of comparable atomic diameters and due to its ability to exert a strong solute drag pressure despite its carbide-forming tendency. However, this requires confirmation experiments with high resolution TEM studies to identify any possible precipitation.

According to the regression constants shown in Table 1 [21,27], boron should have almost negligible effect on transformation start. Since the study [27] was not specifically designed to show the effect of boron, this result may be misleading and has not been included in Figures 8 and 9. However, the regression constants fitted to describe the hardness of the steel after cooling in the same study [27] show that boron has a significant hardening effect on steel subjected to deformation schedule (b). The conclusion from [27], therefore, is that, for strain hardened steel, boron operates by reducing the increase in the ferrite volume fraction with time during cooling rather than through its effect on the start point of the transformation.

5. Conclusions

A computational model which can be used to analyze the onset of phase transformation for different steel compositions following different thermomechanical treatments prior to cooling was presented. The model can calculate the transformation start temperature for an arbitrary cooling path, and it can be used to analyze the effects of different alloying elements and mechanical straining on the transformation start at different temperatures. The calculation of transformation start for an arbitrary cooling path is of practical importance as it can be used in the design of cooling schedules. The estimate for the extent to which different alloying elements affect the phase transformation onset gives some indications of the underlying processes. These indications can be helpful in the construction of more detailed microstructure models, especially in finding ranges for different parameters in such models. Our computational method can be easily parameterized using constant cooling rate CCT data. When more detailed data is needed, for example, to consider the effect of different mechanical treatments prior to cooling, the method presented in this article can also be used.

Author Contributions: Methodology and Software A.P.; Writing—Original Draft Preparation, A.P.; Writing—Review & Editing, M.S., D.P.; Project Administration and Funding Acquisition, D.P.

Funding: This research was funded by TEKES through Finnish Metals and Engineering Competence Cluster (FIMECC) System Integrated Metals Processing (SIMP) Showcase 2.2 program as well as Digital Internet Materials & Engineering Co-Creation (DIMECC) Flex project.

Conflicts of Interest: The authors declare no conflict of interest. The founding sponsors had no role in the design of the study; in the collection, analyses, or interpretation of data; in the writing of the manuscript, and in the decision to publish the results.

Nomenclature

$T_{s,cct}$	Start temperature for the phase transformation occurring with a constant cooling rate.
c_i	The concentrations of different elements in the steel composition in weight %.
t_{85}	The time elapsed while cooling the specimen from 800 °C to 500 °C.
$t_{85,k}$	The t_{85} time corresponding to the slowest cooling rate that does not produce phase transformation.
A_i	Constant obtained from the regression analysis.
B_i	Constant obtained from the regression analysis.
T_N	$273.15 + B_0 + \sum_i B_i c_i$.
$\dot{\theta}$	Cooling rate.
$Ae3$	Equilibrium ferrite formation temperature.
t	time.
$t_{s,cct}$	Time required for phase transformation to start during cooling.
τ	Ideal isothermal transformation time.
CCT	Continuous cooling transformation.
TTT	Time–temperature transformation.
ΔT	Undercooling, i.e., the temperature below $Ae3$ temperature.
$\dot{\theta}_c(\Delta T)$	Constant cooling rate required to produce transformation start at undercooling ΔT.
$T_{s,cct}$	Transformation start temperature in the CCT diagram.
h	Value used in difference approximation.

K Constant used in describing ideal TTT start time.
m Constant used in describing ideal TTT start time.
R Ideal gas constant.
Q Activation energy.
B Constant used in high temperature extrapolation of ideal TTT start time.
t_i Isothermal timestep used in the application of Scheil's rule.

References

1. Saunders, N.; Guo, Z.; Li, X.; Miodownik, A.; Schillé, J.P. The Calculation of TTT and CCT Diagrams for General Steels. *JMatPro Softw. Literat*, **2004** . Available online: http://www.thermotech.co.uk/resources/TTT_CCT_Steels.pdf (accessed on 11 July 2018).
2. Kwon, O. A technology for the prediction and control of microstructural changes and mechanical properties in steel. *ISIJ Int.* **1992**, *32*, 350–358. [CrossRef]
3. Chen, X.; Xiao, N.; Li, D.; Li, G.; Sun, G. The finite element analysis of austenite decomposition during continuous cooling in 22MnB5 steel. *Model. Simul. Mater. Sci. Eng.* **2014**, *22*, 065005. [CrossRef]
4. Wang, X.; Li, F.; Yang, Q.; He, A. FEM analysis for residual stress prediction in hot rolled steel strip during the run-out table cooling. *Appl. Math. Model.* **2013**, *37*, 586–609. [CrossRef]
5. Woodard, P.; Chandrasekar, S.; Yang, H. Analysis of temperature and microstructure in the quenching of steel cylinders. *Metall. Mater. Trans. B* **1999**, *30*, 815–822. [CrossRef]
6. Kirkaldy, J.S.; Venugopalan, D. Prediction of microstructure and hardenability in low alloy steels. In Proceedings of the International Conference on Phase Transformations in Ferrous Alloys, Philadelphia, PA, USA, 4–6 October 1983; Marder, A.R., Goldstein, J.I., Eds.; Metallurgical Society of AIME: Warrendale, PA, USA, 1983; pp. 125–148.
7. Lee, J.; Bhadeshia, H. A Methodology for the Prediction of Time-Temperature Transformation Diagrams. *Mater. Sci. Eng. A-Struct.* **1993**, *171*, 223–230. [CrossRef]
8. Bhadeshia, H. Thermodynamic Analysis of Isothermal Transformation Diagrams. *Met. Sci.* **1982**, *16*, 159–165. [CrossRef]
9. Li, M.; Niebuhr, D.; Meekisho, L.; Atteridge, D. A computational model for the prediction of steel hardenability. *Metall. Mater. Trans. B* **1998**, 29, 661–672. [CrossRef]
10. Fernandes, F.; Denis, S.; Simon, A. Mathematical model coupling phase transformation and temperature evolution during quenching of steels. *Mater. Sci. Technol.* **1985**, *1*, 838–844. [CrossRef]
11. Pandi, R.; Militzer, M.; Hawbolt, E.; Meadowcroft, T. Modelling of austenite decomposition kinetics in steels during run-out table cooling. In *Proceedings of the International Symposium on Phase Transformations During the Thermal/Mechanical Processing of Steel*; Hawbolt, E.B., Yue, S., Eds.; Canadian Institute of Mining, Metallurgy and Petroleum: Vancouver, BC, Canada, 1995; pp. 459–471.
12. Umemoto, M.; Hiramatsu, A.; Moriya, A.; Watanabe, T.; Nanba, S.; Nakajima, N.; Anan, G.; Higo, Y. Computer modeling of phase transformation from work-hardened austenite. *ISIJ Int.* **1992**, *32*, 306–315. [CrossRef]
13. Enomoto, M.; Yang, J.B. Simulation of nucleation of proeutectoid ferrite at austenite grain boundaries during continuous cooling. *Metall. Mater. Trans. A* **2008**, *39*, 994–1002. [CrossRef]
14. Militzer, M.; Pandi, R.; Hawbolt, E. Ferrite nucleation and growth during continuous cooling. *Metall. Mater. Trans. A* **1996**, *27*, 1547–1556. [CrossRef]
15. Enomoto, M. Prediction of TTT-Diagram of Proeutectoid Ferrite Reaction in Iron Alloys from Diffusion Growth Theory. *ISIJ Int.* **1992**, *32*, 297–305. [CrossRef]
16. Hawbolot, E.; Chau, B.; Brimacombe, J. Kinetics of Austenite-Pearlite Transformation in Eutectoid Carbon Steel. *Metall. Trans. A* **1983**, *14*, 1803–1815. [CrossRef]
17. Zhu, Y.; Lowe, T. Application of, and precautions for the use of, the rule of additivity in phase transformation. *Metall. Mater. Trans. B* **2000**, *31*, 675–682. [CrossRef]
18. Wierszyllowski, I. The effect of the thermal path to reach isothermal temperature on transformation kinetics. *Metall. Mater. Trans. A* **1991**, *22*, 993–999. [CrossRef]
19. Pham, T.; Hawbolt, E.; Brimacombe, J. Predicting the onset of transformation under noncontinuous cooling conditions 1. Theory. *Metall. Mater. Trans. A* **1995**, *26*, 1987–1992. [CrossRef]

20. Pham, T.; Hawbolt, E.; Brimacombe, J. Predicting the onset of transformation under noncontinuous cooling conditions: Part 2. Application to the austenite pearlite transformation. *Metall. Mater. Trans. A* **1995**, 26, 1993–2000. [CrossRef]
21. Herman, J.-C.; Thomas, B.; Lotter, U. *Computer Assisted Modelling of Metallurgical Aspects of Hot Deformation and Transformation of Steels*; Publications Office for the European Union: Luxembourg, 1997; pp. 63–64
22. Herman, J.-C.; Donnay, B.; Schmitz, A.; Lotter, U.; Grossterlinden, R. *Computer Assisted Modelling of Metallurgical Aspects of Hot Deformation and Transformation of Steels (Phase 2)*; Publications Office for the European Union: Luxembourg, 1999; pp. 15, 72–73.
23. Wierszyllowski, I. Discussion of "predicting the onset of transformation under noncontinuous cooling conditions: Part I. Theory and predicting the onset of transformation under noncontinuous cooling conditions : Part II. Application to the austenite pearlite transformation". *Metall. Mater. Trans. A* **1997**, *28*, 251. [CrossRef]
24. Pham, T.; Hawbolt, E.; Brimacombe, J. Authors' reply. *Metall. Mater. Trans. A* **1997**, *28*, 1089. [CrossRef]
25. Pohjonen, A.; Somani, M.; Pyykkönen, J.; Paananen, J.; Porter, D. The onset of the austenite to bainite phase transformation for different cooling paths and steel compositions. *Key Eng. Mater.* **2016**, *716*, 368–375. [CrossRef]
26. Pohjonen, A.; Kaijalainen, A.; Somani, M.; Larkiola, J. Analysis of Bainite Onset During Cooling Following Prior Deformation at Different Temperatures. *Comput. Methods Mater. Sci.* **2017**, *17*, 30–35.
27. Lotter, U. *Aufstellung von Regressionsgleichungen zur Beschreibung des Umwandlungsverhaltens Beim Thermomechanische Walzen, Abslussbericht, Kommission der Europäischen Gemeninschaften, Generaldirektion Wissenschaft, Forshcung und Entwicklung*; EUR 13620 DE; Publications Office for the European Union: Luxembourg, 1991; p. 17.
28. Donnay, B.; Herman, J.; Leroy, V.; Lotter, U.; Grossterlinden, R.; Pircher, H. Microstructure Evolution of C-Mn Steels in the Hot Deformation Process: The Stripcam Model. In Proceedings of the 2nd International Conference on Modeling of Metal Rolling Processes, London, UK, 9–11 December 1996; Beynon, J.H., Ingham, P., Teichert, H., Waterson, K., Eds.; Institute of Materials: Cambridge, UK, 1996.
29. Kirkaldy, J.; Sharma, R. A New Phenomenology for IT and CCT Curves. *Scr. Metall. Mater.* **1982**, *16*, 1193–1198. [CrossRef]
30. Scheil, E. Anlaufzeit der austenitumwandlung. *Arch. Eisenhuttenwes.* **1935**, *8*, 565–567. [CrossRef]
31. Farjas, J.; Roura, P. Modification of the Kolmogorov–Johnson–Mehl–Avrami rate equation for non-isothermal experiments and its analytical solution. *Acta Mater.* **2006**, *54*, 5573–5579. [CrossRef]
32. Christian, J. Chapter 12-Formal Theory of Transformation Kinetics. In *The Theory of Transformations in Metals and Alloys*; Christian, J., Ed.; Pergamon: Oxford, UK , 2002; pp. 529–552.
33. Mittemeijer, E. *Fundamentals of Materials Science*; Springer Science & Business Media: Berlin/Heidelberg, Germany, 2010.
34. Liu, F.; Sommer, F.; Bos, C.; Mittemeijer, E.J. Analysis of solid state phase transformation kinetics: models and recipes. *Int. Mater. Rev.* **2007**, *52*, 193–212. [CrossRef]
35. Porter, D.; Easterling, K.E.; Sherif, M.Y. *Phase Transformations in Metals and Alloys*; CRC Press: Boca Raton, FL, USA, 2009.
36. Vander Voort, G.F. *Atlas of Time-Temperature Diagrams for Irons and Steels*; ASM International: Almere, The Netherlands, 1991.
37. Smith, Y.E.; Siebert, C.A. Continuous cooling transformation kinetics of thermomechanically worked low-carbon austenite. *Metall. Trans.* **1971**, *2*, 1711–1725.
38. Hanlon, D.; Sietsma, J.; van der Zwaag, S. The effect of plastic deformation of austenite on the kinetics of subsequent ferrite formation. *ISIJ Int.* **2001**, *41*, 1028–1036. [CrossRef]
39. Bleck, W. Using the TRIP effect—The dawn of a promising group of cold formable steels. In *Proceedings of the International Conference on TRIP-Aided High Strength Ferrous Alloys*; De Cooman, B., Ed.; Wissenschaftsverlag Mainz GmbH: Aachen, Germany, 2002; pp. 13–23.
40. Yan, P.; Bhadeshia, H.K.D.H. Austenite-ferrite transformation in enhanced niobium, low carbon steel. *Mater. Sci. Technol.* **2015**, *31*, 1066–1076. [CrossRef]

Article

Rational Alloy Design of Niobium-Bearing HSLA Steels

Rami A. Almatani [1] **and Anthony J. DeArdo** [1,2,*]

[1] The Basic Metals Processing Research Institute, Department of Mechanical Engineering and Materials Science, University of Pittsburgh, Pittsburgh, PA 15261, USA; raa148@pitt.edu

[2] Finland Distinguished Professor, Department of Mechanical Engineering, University of Oulu, P.O. Box 4200 (Linnanmaa), FIN-90014 Oulu, Finland

* Correspondence: deardo@pitt.edu; Tel.: +1-412-996-6770

Received: 20 February 2020; Accepted: 18 March 2020; Published: 23 March 2020

Abstract: In the 61 years that niobium has been used in commercial steels, it has proven to be beneficial via several properties, such as strength and toughness. Over this time, numerous studies have been performed and papers published showing that both the strength and toughness can be improved with higher Nb additions. Earlier studies have verified this trend for steels containing up to about 0.04 wt.% Nb. Basic studies have shown that the addition of Nb increases the recrystallization-stop temperature, $T_{5\%}$ or T_{nr}. These same studies have shown that with up to about 0.05 wt.% of Nb, the $T_{5\%}$ temperature increases in the range of finish rolling, which is the basis of controlled rolling. These studies also have shown that at very high Nb levels, exceeding approximately 0.06 wt.% Nb, the recrystallization-stop temperature or $T_{5\%}$ can increase into the temperature range of rough rolling, and this could result in insufficient grain refinement and recrystallization during rough rolling. However, the question remains as to how much Nb can be added before the detriments outweigh the benefits. While the benefits are easily observed and discussed, the detriments are not. These detriments at high Nb levels include cost, undissolved Nb particles, weldability issues, higher mill loads and roll wear and the lessening of grain refinement that might otherwise occur during plate rough rolling. This loss of grain refinement is important, since coarse grained microstructures often result in failure in the drop weight tear testing of the plate and pipe. The purpose of this paper is to discuss the practical limits of Nb microalloying in controlled rolled low carbon linepipe steels of gauges ranging from 12 to 25 mm in thickness.

Keywords: HSLA steel; alloy design; grain refinement of austenite; Zener pinning force; recrystallization; Niobium Nb

---

1. Introduction

It has been over 61 years since the first production heat of a niobium (Nb) bearing HSLA strip steel was commercially produced in 1958 [1,2]. Since that time, microalloying with Nb has been extended to virtually all product classes and forms. Perhaps nowhere has Nb been more beneficial than in linepipe steels for high pressure oil and gas transmission at low temperatures. The contribution of Nb to high strength by grain refinement; solid solution strengthening of the various types of ferrite formed; and further strengthening of the ferrite by precipitation and dislocation-hardening through increased hardenability have all been chronicled in numerous papers and research studies [3–8]. In a similar fashion, the improvement in lowering the ductile-brittle transition temperature (DBTT) through the use of niobium is critically important in pipelines intended for low temperature service [9]. Much of this improvement is caused by the grain refinement of both the austenite during hot rolling, and the grain refinement of the ferrite during transformation upon cooling. Several studies have shown that strength and toughness can be improved by increasing the level of Nb used, from zero to the conventional 0.03–0.04 wt.%, to very high levels near 0.1% Nb [10,11]. However, there are detriments beyond cost to the use of Nb, especially at higher levels, such as reduced weldability, large undissolved particles and higher mill loads.

When a 25 mm thick plate intended for pipe applications is hot rolled on a modern 5 m plate mill, it does so in three stages after slab reheating: rough rolling, finish rolling and accelerated cooling. For example, after reheating at 1200 °C, a continuously cast slab of 200 mm thickness might be rolled to 75 mm transfer bar in the roughing mill at temperature from 1150–1050 °C. The bar is then transported to the finishing stands, where it is further rolled to the final gauge of 25 mm at temperatures ranging from below 900 to 750 °C. Because of the large width and potentially high mill loads, the rough rolling and finishing rolling take place in numerous, light passes, often between 10 and 15 passes in each case, with pass reductions between 10–15% each in the roughing mill, and 5–10% each in the finishing mill. Since these are reversing mills, the interpass times can range from 10–30 s depending on conditions.

After final rolling, the plate is water spray cooled at about 30 °C/s to the water end temperature (WET) which is dictated by the continuous cooling transformation (CCT) diagram, the cooling path and the required microstructure and properties [9]. The WET for achieving ferrite–pearlite microstructures in steels of high hardenability, i.e., CE_{II} over 0.40, is in the range of 600–650 °C. When a bainitic ferrite microstructure is desired, a WET near 450–550 °C would be used. For even higher strengths, requiring martensite, a WET around 300 °C would be used. Normally, the steels are air cooled to room temperature from the WET.

The microstructure of austenite during hot rolling of Nb-bearing HSLA steels is believed to behave according to Figure 1 [12]. For a given strain, above $T_{95\%}$, austenite grains are repeatedly recrystallized and grain refined (complete recrystallization; RXN), while below $T_{5\%}$, the austenite grains are unrecrystallized and pancaked. These critical temperatures are themselves dependent on the content of Nb of steels and increase with higher Nb levels, as shown schematically in the red-dashed curves in Figure 1.

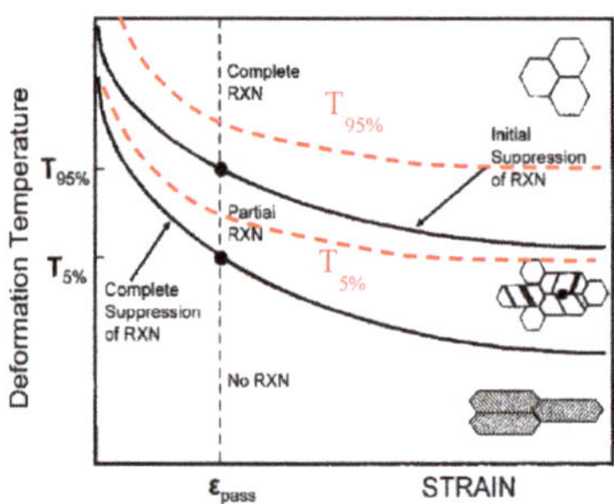

Figure 1. Schematic diagram of the resulting austenite microstructures from different deformation conditions, after reference [12]. $T_{5\%}$ and $T_{95\%}$ are temperatures for 5% and 95% recrystallization, respectively.

The hot processing of Nb steels is based on most of the Nb being in solution in the austenite during slab reheating and rough rolling, and some of the dissolved Nb being reprecipitated as strain-induced precipitate in deformed austenite during finish rolling. The nose of the C-curve for this precipitation in austenite is around 900 °C for 0.04 wt.% Nb, but is raised to 950 °C or higher at larger levels of Nb [3,13]. As mentioned above, studies of the recrystallization-stop-temperature, $T_{5\%}$ or T_{nr}, indicate that this temperature increases with Nb content and can reach or exceed 1050 °C in typical pipe steels containing 0.1% Nb [6,14]. Since the last roughing passes occur in this temperature range, it is quite possible that there might be strain-induced precipitation of NbC, even in the late roughing passes. Since the grain refinement expected during roughing requires multiple static recrystallization events in the interpass times, perhaps complete recrystallization may not occur by the exit of the roughing mill.

Therefore, there might be consequences for the grain refinement needed during rough rolling when the bulk Nb level is too high. For example, this lack of sufficient grain refinement is important, since coarse grained microstructures often result in failure in drop weight tear testing of plate and pipe.

Whether static recrystallization occurs or not depends on the comparison of the driving force for static recrystallization and the retarding Zener pinning force caused by the interaction of the moving austenite boundaries with particles formed earlier on the defect structure of the deformed austenite in the roll gap; these defects include grain and subgrain boundaries and deformation bands [15].

The potential for strain-induced precipitation of NbC or NbCN will depend on the bulk composition of the steel, the relevant solubility products for NbC or NbCN in austenite and the applicable rolling practice. The carbon content will be governed by the final strength needed in the final pipe, which can be X50–60 for ferrite–pearlite (F–P) microstructures, X70–100 for bainitic microstructures and X120 and above for martensitic microstructures [16,17]. While changes in %C have only a slight influence on the strength of F–P steels, they can have dramatic effects on the strength of bainite or martensite found in direct quenched steels. Therefore, the carbon content can be expected to influence both the strength of the plate or pipe through hardenability effects, and the toughness through its effect on the precipitation in austenite and the control of grain size.

The effect of hot rolling on toughness or lowering the DBTT, through grain refinement of the prior austenite grain size (PAGS), is the result of two sequential events. The first is the combination of the elimination of both the remaining as-cast structure and the large grain size that result from slab reheating. These occur during the repeated static recrystallization that takes place between the rough rolling passes that arise when the pass strains happen above the $T_{95\%}$ temperature. This leads grain refinement during multiple waves of recrystallization, where the PAGS might be reduced from 300 μm to 50 μm going into the finishing passes where pancaking occurs.

Since the finishing passes for controlled rolling are usually considered to occur below the $T_{5\%}$ or recrystallization stop temperature, normally about 900 °C in a steel containing 0.04% Nb, the final austenite will be heavily strained, elongated or pancaked [18]. This austenite is highly deformed with high strength and contains numerous crystallographic defects, such as deformation twins, deformation bands, subgrain boundaries and elongated grain boundaries. These near-planar crystalline defects contribute to what is call the *Sv* value, an index used to judge the effectiveness of a given rolling process in thermomechanical processing (TMP) [19]. In low hardenability or slowly cooled F–P steels, the nucleation of polygonal ferrite occurs on the *Sv*, where the high density of high angle ferrite grain boundaries resulting from ferrite grain refinement can act as crack arresters for potentially growing cleavage cracks. In higher hardenability or faster cooled steels, the defects themselves can act as cleavage crack arresters in bainitic or martensitic steels. In either case, these near planar defects can act as sites for crack arresters for the growth of cleavage cracks; hence, lowering the DBTT.

To investigate the rough rolling process, the possibility of strain-induced precipitate was explored in this research in the temperature range for rough rolling, i.e., 1150–1000 °C, for a series of steels with two carbon levels: a conventional one of 0.06 wt.% C and a lower carbon version of 0.03 wt.% C. Three levels of Nb were also studied—0, 0.04 and 0.08 wt.%.

The purpose of the current study was threefold: (i) to determine whether NbC could form during rough rolling and (ii) whether this precipitate might cause less than complete recrystallization during rough rolling, and (iii) therefore, lead to a larger slightly pancaked as-roughed austenite grain size resulting in non-optimum final austenite microstructure prior to transformation. To achieve these goals, two investigations were conducted. The first was a theoretical study of potential strain induced precipitation of NbC in the temperature range in which roughing passes would normally occur in a modern 5 m wide reversing plate mill. This required solubility calculations, the use of the subgrain boundary Zener pinning model and parameters taken from similar, earlier studies [3,18,20,21]. The second was the study of the austenite grain size and shape in quenched specimens after laboratory hot deformation experiments conducted in the temperature range in which rough rolling would normally occur. It was expected that this research would help answer the question as to whether high Nb steels can be expected to be successfully processed on modern plate mills in a range of strength levels.

2. Research Rationale

2.1. Alloy Design Used in Study

The experimental alloys were chosen to represent a typical linepipe steel, near X100 composition [16]. As mentioned earlier, the carbon and niobium levels were chosen to demonstrate the effects of NbC pinning forces in the temperature range of rough rolling. Figure 2 shows the temperature profile of a modern, 5 m wide reversing plate mill where the end of roughing rolling is approximately at 1050 °C [22].

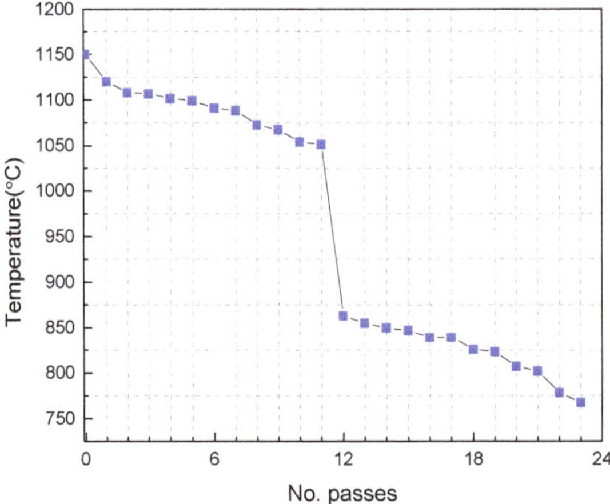

Figure 2. Temperature profile of a modern 5 m wide reversing plate mill, after reference [22].

NbC in solution was calculated using the equilibrium solubility product proposed by Palmiere et al. [20]:

$$\log[\mathrm{Nb}][\mathrm{C}] = 2.06 - \frac{6700}{\mathrm{T}} \tag{1}$$

Figure 3 shows the calculated dissolution temperatures of alloys with two Nb levels, namely, 0.04Nb and 0.08Nb, in wt.% while keeping carbon as a constant at 0.06 wt.%. It is clear that all the NbC available in 0.04 wt.% Nb steel will be dissolved at/or close to 1170 °C, while the higher Nb content will not be fully dissolved below a slab reheating temperature of 1265 °C, which indicates more stable precipitates at higher temperatures. It should be noted that this temperature far exceeds the normal reheating temperature of 1150–1200 °C used in practice. The dashed line represents the deviation between the two alloys in terms of soluble NbC; thus, the reheating temperature is not a variable, but is restricted to 1150–1200 °C in practice. Therefore, the question is how much Nb can be dissolved at perhaps 1200 °C. According to Figure 3, all of Nb in the 0.04 steel and 0.55% Nb in the 0.08 steel would be dissolved at 1200 °C. In normal practice with a commercial steel of similar composition containing 0.04Nb, all of the Nb is taken into solution during reheating, and very little if any is reprecipitated during rough rolling. Therefore, recrystallization goes to completion in roughing. However, in the 0.08Nb steel, only 0.055 wt.%. Nb is taken into solution and the remaining 0.025 wt.% Nb is left undissolved. The question is: will any of the 0.055 wt.% Nb dissolved in austenite reprecipitate during roughing, and if it does, will it suppress complete recrystallization during roughing? Higher Nb levels indicate that some precipitates could not be taken into solution based on solubility considerations in the range applicable to the rough rolling process.

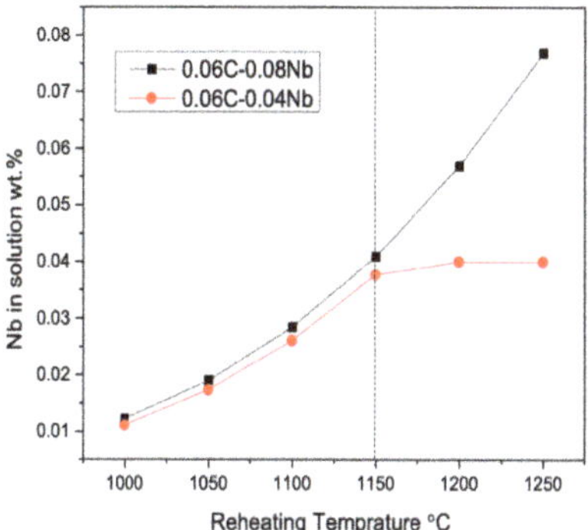

Figure 3. Calculated amount of dissolved Nb at equilibrium as a function of reheating temperature.

2.1.1. Phase I: Theoretical Study

The assumptions made here were that with the addition of Nb into the steel, the $T_{5\%}$ and $T_{95\%}$ temperatures will both be increased [18], and subsequentially the processing window above $T_{95\%}$ will be reduced. In addition, the $T_{5\%}$ temperature might be increased to where it enters the later portion of the rough rolling temperature range. Figure 4 shows a schematic of two Nb steels, namely, 0.1Nb and 0.04Nb, in wt.%. For a given strain (i.e., Σn_{Passes}) and temperature, and as the temperature decreases, the processing window at the end of roughing for a uniform austenite grain size distribution is reduced in the 0.1Nb steel, below which only partial recrystallization and grain refinement may occur; this is shown in the hatched area above 0.1Nb $T_{95\%}$. On the other hand, the processing window for the 0.04Nb steel to produce a uniform grain size distribution is higher even below 1050 °C, which is here designated as the end of rough rolling temperature.

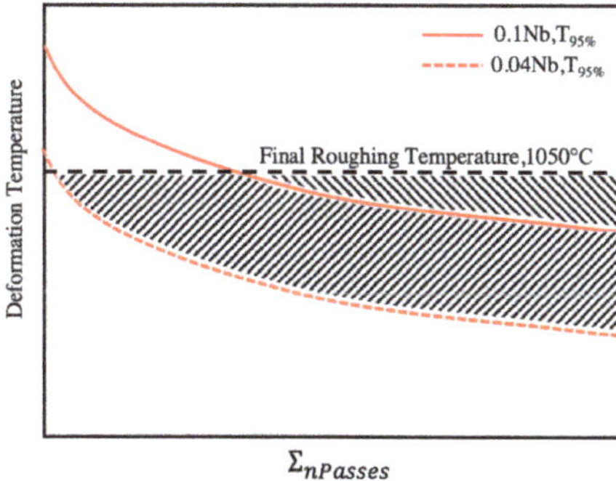

Figure 4. Schematic showing the processing window of high and low Nb steels; hatched areas show the window for which a uniform grain refined austenite microstructure can be obtained by rolling above the respective $T_{95\%}$ temperature.

In the context of hot rolling, the typical value of the recrystallization driving force of austenite subjected to a single low temperature rolling pass is found to be about 22 MPa [12]. This was for 900 °C; at 1100 °C it would be much lower, near 10 MPa. Thus, to retard recrystallization, a necessary pinning force is needed and several models were proposed to calculate the pinning force of precipitates in microalloyed steels as a result of the interaction between precipitates and austenite grain boundaries [18,23]. However, in this study, as mentioned earlier, the subgrain boundary model [3] was employed to estimate, theoretically, the pinning forces at different deformation temperatures and specifically during roughing passes. The reason for using this model is mainly that NbC particles are not randomly distributed but isolated on the subgrain boundaries of deformed austenite [24,25].

The estimations of pinning forces caused by strain induced precipitation of NbC were calculated using (i) the subgrain boundary model, together with (ii) the volume fractions of NbC estimated for the compositions and temperatures using the equilibrium solubility relation mentioned previously; (iii) particle sizes of NbC observed; and (iv) the dislocation cell sizes observed in the TEM in recovered austenite in earlier similar studies [17,20]. The subgrain boundary pinning force model used was [3]:

$$F = \frac{3\gamma f_v l}{2\pi r^2} \tag{2}$$

where γ is the high angle austenite grain boundary energy, f_v is the volume fraction of precipitates, l is the subgrain size and r is the particle size. The parameters used in this investigation were $\gamma = 0.8$ J/m^2, particle size = 2 nm and subgrain size = 0.5 μm; and f_v can be calculated from the solubility product.

It should be mentioned that the measured volume fractions of NbC for subgrain and grain boundaries were higher than predicted by equilibrium considerations [20]. It appeared that there was a segregation of Nb to the boundaries leading to higher than expected local volume fractions of NbC. This means that the local pinning forces are actually larger than those which would be calculated from equilibrium considerations

Table 1 shows the estimated pinning forces for two alloys, during roughing passes. The results indicate that higher Nb level gives the higher pinning force magnitudes at the end of roughing passes (i.e., 41 MPa at 1050 °C), due to higher volume fraction of NbC to precipitate at lower roughing temperature.

Table 1. NbC pinning force at the respective deformation temperature, MPa.

Deformation Temperature, °C	0.06C,0.08Nb	0.06C,0.04Nb
1150	27.5	3.4
1100	35.4	10.7
1050	41.0	17.0

2.1.2. Phase II: Validation Study Based on Experiment

To validate the theoretical predictions, deformation studies were conducted in the temperature range where rough rolling would be expected. The purpose of the experiments was to determine the extent of recrystallization of the austenite in the steels with different Nb levels and to compare these to the predicted values.

3. Materials and Methods

3.1. Experimental Materials

Laboratory heats of mass 45 Kg were vacuum melted and solidified in cast iron molds. The chemical compositions of the lab heats are listed in Table 2. These were hot rolled to 25 mm "slabs" in five passes with finishing temperature of 890 °C and air cooled to room temperature. The Nb-bearing heats were designated by L4, L8, H4 and H8, where the letter indicates the carbon level (i.e., L = 0.03 wt.% and H = 0.06 wt.%) and the number indicates Nb level (0 = Nb-free, 4 = 0.04 wt.% and 8 = 0.08 wt.%).

Table 2. Chemical composition of the experimental alloys in wt.%.

Alloy	C	Nb	Base
L0	0.032	-	1.88Mn-0.01P-0.002S-0.30Si-0.20Cu-0.20N-0.50Cr-0.10Mo-0.0145Ti-0.030Al-0.004N
L4	0.031	0.0429	
L8	0.029	0.0837	
H0	0.061	-	
H4	0.062	0.0413	
H8	0.059	0.07827	

From these slabs, 12 mm blanks were cut for producing cylinders for hot compression testing. Selected samples were machined to right cylindrical shapes of dimensions 12 mm (diameter) × 19 mm (height), as shown in Figure 5. A thermocouple hole 1.6 mm in diameter and approximately 5 mm deep was drilled into the cylindrical samples to control and monitor the temperature.

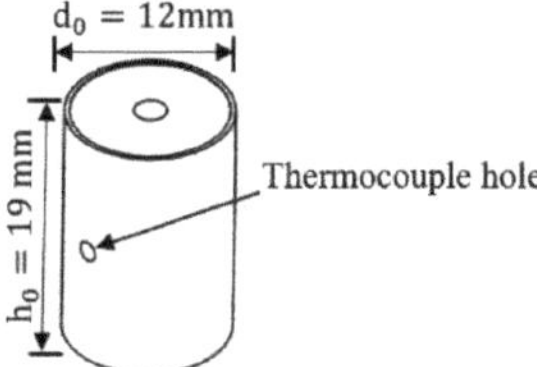

Figure 5. Schematic of the cylinders used in this study for hot compression testing.

3.2. *Studies of Grain Coarsening during Reheating*

In order to quantify austenite grain size as a function of reheating temperature, grain coarsening experiments were conducted using isothermal holding at various reheating temperatures using cube samples having dimensions of 12 mm on a side. Prior to each soaking, samples were encapsulated in evacuated quartz tubes and filled with pure argon to minimize any oxidation effects. Samples were placed in a box furnace and were soaked for 60 min at different temperatures starting from 900 °C to 1300 °C in increments of 50 °C. Following that, the capsules were broken and the samples were quenched in water.

3.3. *Hot Compression*

To verify the predictions of the calculated Zener pinning forces, hot compression tests were performed in the temperature range where the roughing passes would be expected to occur; i.e., 1150–1000 °C. The right cylinders were reheated to 1200 °C for two minutes and then cooled to various temperatures for recrystallization studies. The deformations were applied using an MTS-458 unit designed for deformation under constant true strain rate conditions equipped with a radiation furnace and a temperature controller. Glass lubricant was used to suppress barreling in the multiple hit, axisymmetric hot compression tests. If the pinning forces were low in the rough rolling range, then repeated recrystallization would be expected with accompanying grain refinement. On the other hand, if the pinning forces were high, then complete recrystallization would not occur, and the grain size would stay large, reflecting the reheated grain size, and the grain shape would not be equiaxed. Figure 6 shows an example of the thermomechanical path used in the first series of samples to closely simulate roughing passes in a typical hot rolled plate. Two reduction levels of 15% and 25% were used in this study and the strain rate ($\dot{\varepsilon}$) was set to be 10 s^{-1} for all hot compression experiments.

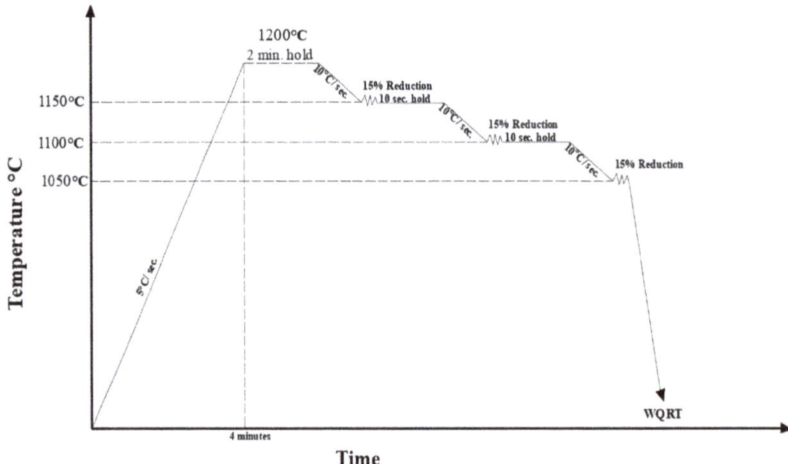

Figure 6. Thermomechanical schedules used in hot compression deformations with 15% reductions.

3.4. Characterization Methods

Samples from the as-received heats, grain coarsening studies and hot compression studies were analyzed using optical microscopy Nikon FX-35WA (Nikon Corp., Tokyo, Japan). Standard metallographic techniques were employed by mounting in Bakelite, grinding samples using waterproof abrasive silicon carbide papers gradually from grit 180 to 600 and polishing using alumina powders of 1 and 0.05 µm. The prior austenite grain boundaries were revealed using a special reagent consisting of 100 mL of saturated aqueous picric acid with the addition of 1 mL hydrochloric acid (HCL) and 10 g of sodium dodecylbenzene sulfonate as a wetting agent at 70 °C. The mean austenite grain sizes were measured using ImageJ (1.52a, National Institutes of Health, Bethesda, MD, USA), by manually outlining the austenite grains and calculating the equivalent grain diameter for each grain. From each sample, at least 400–500 grains were measured. Additionally, selected as-received samples were characterized using Vickers hardness tester LECO LM310AT (LECO Corp., St. Joseph, MI, USA). At least 5 Vickers hardness measurements with a load of 300 Kgf were taken from each direction in each of the selected samples. Further, in this study, EBSD Mapping was done in FEI Scios FIB/SEM Dual Beam (Thermo Fisher, Hillsboro, OR, USA) equipped with EDAX EBSD camera and TEAM software (V4.3, EDAX Inc., Mahwah, NJ, USA). Thin foils were examined under transmission electron microscope (TEM) using JEOL JEM2100F operating at an accelerated voltage of 200 kV.

4. Results and Discussion

4.1. As-Received Samples

Figure 7 shows optical micrographs of the air-cooled plates. It can be seen for the base alloys L0 and H0 in Figure 7a,b that the microstructures consisted mainly of polygonal ferrite (PF), and carbon-enriched constituents including acicular ferrite and/or martensite-austenite (MA) microconstituents. When the carbon is increased from 0.03 wt.% to 0.06 wt.%, the fraction of these carbon-enriched constituents was increased, as shown in Figure 7b. For the same cooling rate, the addition of Nb to the base alloys results in a complex microstructure, as can be observed in Figure 7c through Figure 7f. These complex microstructures consist of mainly acicular ferrite, MA and/or bainite. Another observation is that polygonal ferrite in the high Nb steels is suppressed significantly.

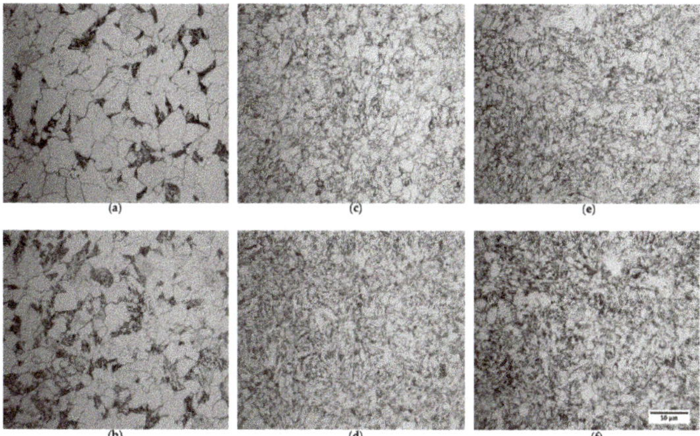

Figure 7. Optical metallographic microstructures of air-cooled samples: (**a**) L0, (**b**) H0, (**c**) L4, (**d**) H4, (**e**) L8 and (**f**) H8.

A TEM micrograph for H4 is shown in Figure 8. It is clearly evident that the steel contains a complex microstructure consisting of MA microconstituents and acicular ferrite. Additionally, a coarse, titanium-rich precipitate can be seen in the air cooled condition.

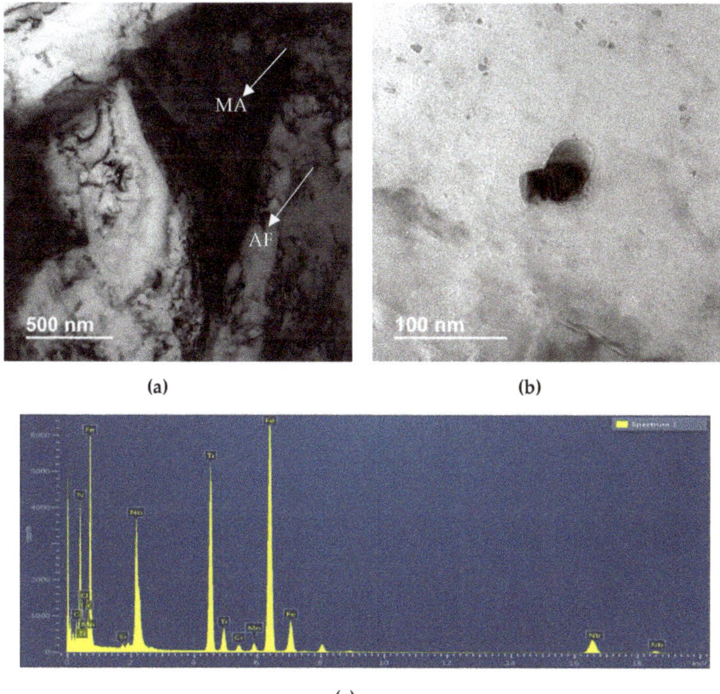

Figure 8. (**a**) Bright field image of H4 showing ferrite and carbon-enriched constituents. (**b**) Titanium-rich precipitate and the corresponding (**c**) EDS spectrum of the precipitate.

The hardness values of microstructures taken in the three normal directions, namely, normal direction (ND), rolling direction (RD) and transverse direction (TD), are shown in Figure 9. As expected, in general, as the carbon and/or niobium content increases, the hardness value increases, although for different reasons. For example, the higher Vicker's hardness number (VHN) values observed with increasing the carbon content appeared to be due to the fraction of carbon-enriched constituents. Although there is no clear evidence of which direction gives the highest hardness values in the investigated samples, the TD sample of H8 shows the highest value. It can be inferred from this that pancaking of the low carbon austenite in the roughing mill can occur easily in this direction, meaning that there is a higher tendency to nucleate hard carbon-enriched microconstituents in this direction.

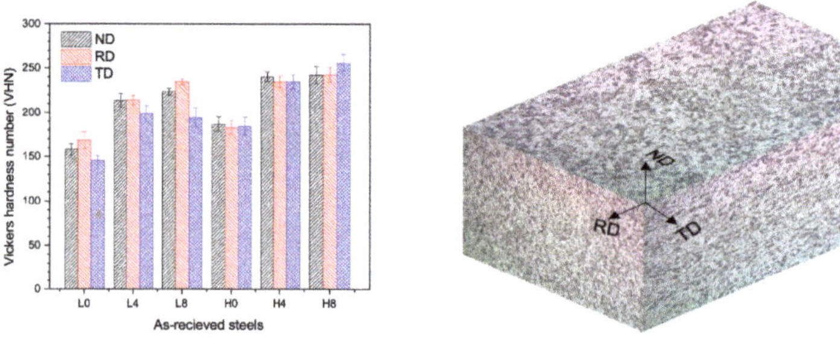

Figure 9. Vicker's hardness number (VHN) values of air-cooled samples in three orthonormal directions.

Another observation was made using EBSD mapping. Figure 10 depicts inverse pole figure data and misorientation angles of H0 and H8 taken from the transverse direction. For the same carbon level, there was a higher fraction of low angle grain boundaries (red boundaries) observed in H8 compared to H0, which is a typical characteristic of acicular ferrite.

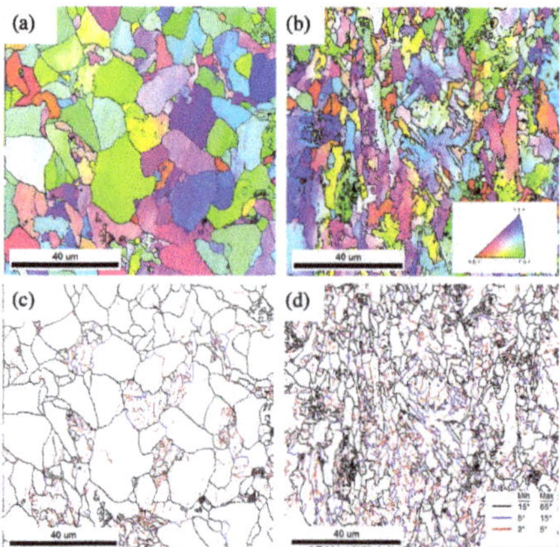

Figure 10. Inverse pole figure data and misorientation mapping of H0 (**a**,**c**) and H8 (**b**,**d**).

4.2. Coarsening of Prior Austenite Grain Size

Average austenite grain sizes as a function of reheating temperature for all six alloys are shown in Figure 11. Error bars represent 95% confidence intervals about the mean grain size. In general, austenite grain sizes increase with increasing the reheating temperatures. The Nb-bearing steels show a higher tendency for abnormal grain growth, a characteristic of microalloyed steels [25–27]. As an example, a series of microstructures of L8 is depicted in Figure 12, and it can be seen that abnormal grain growth occurs at/or close to 1100 °C.

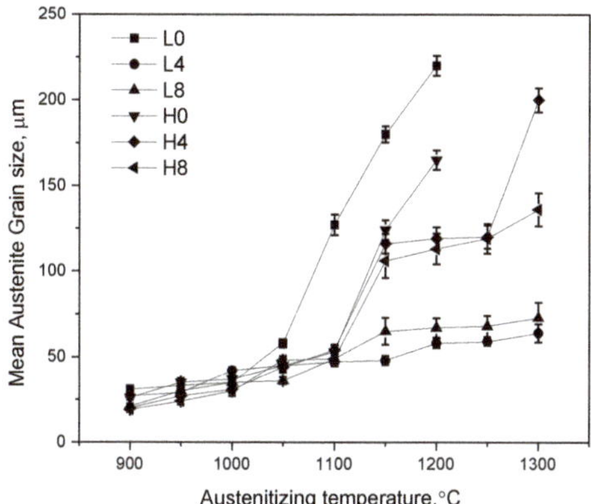

Figure 11. Mean austenite grain size as a function of reheating temperature.

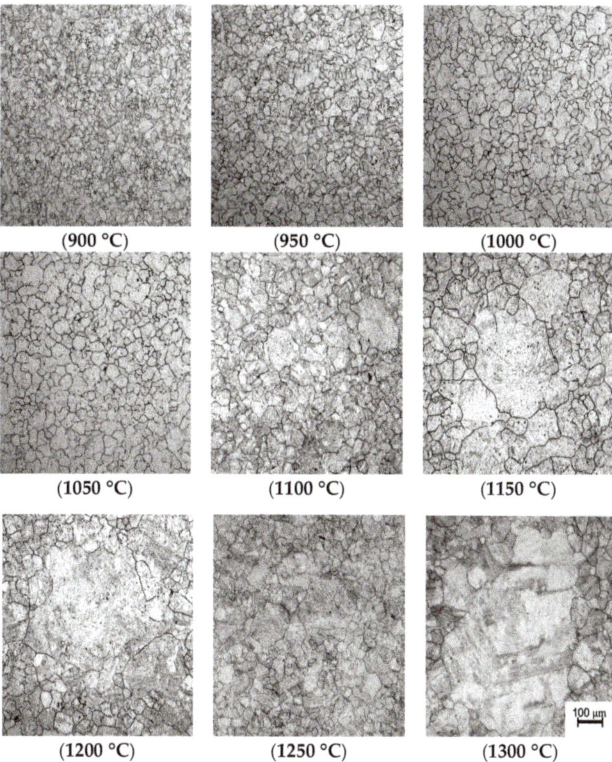

Figure 12. Light optical micrographs of prior austenite grains for steel L8 after soaking to the temperatures indicated.

4.3. Rough Rolling Simulations

The first series of samples was subjected to hot deformation of either two passes or three passes in the temperature range between 1150 and 1050 °C, where each pass was under 15%. Figure 13a–c shows samples subjected to two-pass deformation at 1150 °C and 1100 °C of H4, H8 and L8, respectively. Complete recrystallization is observed in all three samples under two passes and three passes. It is expected from solubility products that the fraction of NbC or NbCN in H8 during cooling is higher than those in H4 and L8. For the three pass microstructures, it can be seen in Figure 13d–f, that the average austenite grain sizes are reduced, and the highest reduction in grain size was observed in L8 samples during hot deformation. Figure 14 summarizes the average austenite grain sizes of steels H4, H8 and L8.

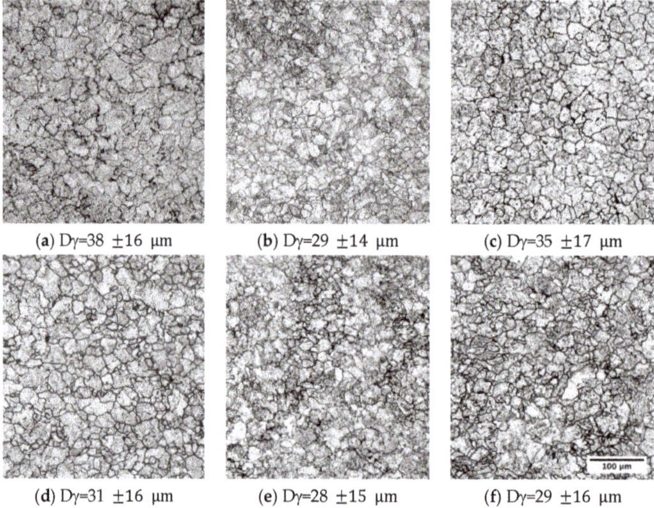

Figure 13. Two-pass reduction of 15% at 1150 and 1100 °C for (**a**) H4, (**b**) H8 and (**c**) L8; three pass reduction at 1150, 1100, and 1050 °C for (**d**) H4, (**e**) H8 and (**f**) L8.

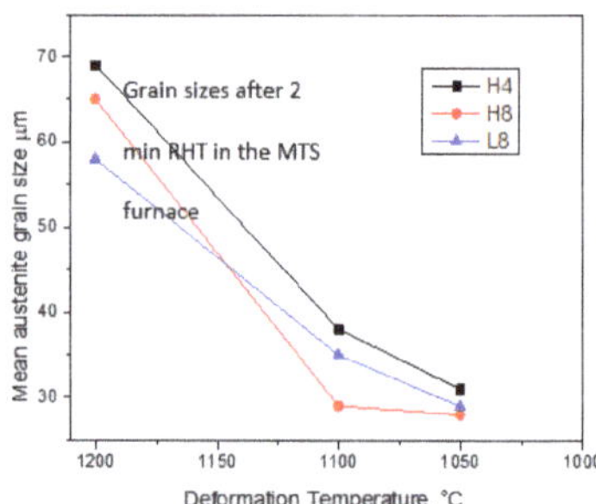

Figure 14. Mean austenite grain size after two passes with 15% reduction/pass at the respective deformation temperature; each sample was quenched, and prior austenite grain size was analyzed.

The second series of samples was subjected to three passes under 25% reduction in each pass. The micrographs of these samples are shown in Figure 15. It is evident from these micrographs that pancaking of austenite did not occur, even at 1050 °C. However, a non-uniform austenite grain distribution can be observed at temperatures at/or below 1050 °C in the high Nb steels. This leads to a mixture of recrystallized and unrecrystallized austenite grains (i.e., duplex austenite grains) in the roughing stage of plate steels. To demonstrate this idea, the L8 sample was reheated at 1200 °C for two minutes and subjected to three 25% passes, but the last pass was done at 1030 °C, which is near the end of the roughing passes. The interpass time was set to be 15 s after each pass in order to see whether complete static recrystallization could be observed. Figure 16 shows the prior austenite grain size of L8 subjected to the previously mentioned schedule, and the fraction of unrecrystallized grains was measured to be 15% based on the non-circularity of the grains. This may show that the recrystallization stop temperature is close to the range of 1030–1000 °C.

(**a**) Dγ=26 ±12 µm (**b**) Dγ=20 ±9 µm (**c**) Dγ=18 ±8 µm

Figure 15. Three-pass reduction of 25% each at 1150, 1100 and 1050 °C for (**a**) H4, (**b**) H8 and (**c**) L8.

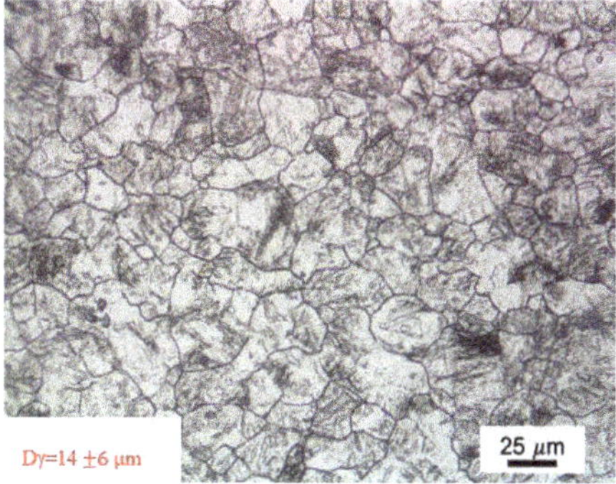

Figure 16. L8 subjected to three passes of 25% each at 1150, 1100 and 1030 °C; the interpass time was 15 s. The compression axis is vertical.

This also indicates that the NbC pinning force is large enough to retard recrystallization in the roughing stage. The resulting duplex microstructure at the end of the roughing passes and can deteriorate strength and toughness in the final product.

It is immediately apparent that higher pinning forces of NbC on austenite grain boundaries will increase the $T_{5\%}$ to higher temperatures, which may lead to a duplex microstructure at the end of rough rolling. Thus, it is necessary to choose Nb levels perhaps well below 0.1 wt.% to have uniform austenite grain sizes at the end of roughing rolling. This is true for both plate mills and thin slab casting mills, where the few, if any, roughing passes or the early finishing passes acting as roughing passes occur at temperatures above 950 °C, thereby eliminating or reducing the possibility of grain refinement before experiencing the low temperature finishing passes.

While further investigation is needed to confirm the effects of C and Nb levels on grain coarsening and recrystallization behaviors based on precipitate analyses, this study draws attention to the necessity to keep Nb levels optimum to produce a uniformly refined austenite grain size during the rough rolling of plate steels. Additionally, the study of the interaction between precipitation and recrystallization in the finishing passes of hot rolled plates needs to be considered in future studies.

5. Conclusions

From the present investigation, the following conclusions can be drawn:

For the same thermomechanical processing schedule, the as-received samples show different microstructures, depending on carbon and niobium contents. The Nb-free alloys mainly consist of polygonal ferrite and some carbon enriched constituents, while the Nb-bearing alloys exhibit acicular microstructures with significantly reduced amounts of polygonal ferrite.

Based on equilibrium thermodynamic relations, it appears that strain-induced NbC can precipitate in the roughing temperature range in high Nb steels.

Depending of the amount of strain, the suppression of some recrystallized austenite grains would be expected to cause duplex grain structures, which is believed to be responsible for inferior mechanical properties.

Author Contributions: Conceptualization, A.J.D.; methodology, A.J.D. and R.A.A.; validation, A.J.D. and R.A.A.; formal analysis, R.A.A. and A.J.D.; writing—original draft preparation, R.A.A.; writing—review and editing, A.J.D.; supervision, A.J.D.; project administration, A.J.D.; funding acquisition, A.J.D. All authors have read and agreed to the published version of the manuscript.

Funding: The Basic Metals Processing Institute, Department of Mechanical Engineering and Materials Science, University of Pittsburgh thanks its industrial sponsors for partially funding this study. In addition, the United States Steel Corporation (U.S. Steel) should be thanked for providing the experimental materials and processing. Finally, one of the authors (Almatani), was supported by a scholarship from King Abdulaziz City for Science and Technology (KACST), Saudi Arabia.

Acknowledgments: The authors would like to thank the Basic Metals Processing Institute, Department of Mechanical Engineering and Materials Science, University of Pittsburgh and its industrial sponsors for partially funding this study. In addition, the United States Steel Corporation (U.S. Steel) should be thanked for providing the experimental materials and processing. Finally, one of the authors (Almatani), was supported by a scholarship from King Abdulaziz City for Science and Technology (KACST), Saudi Arabia.

Conflicts of Interest: The authors declare no conflict of interest.

References

1. Beiser, C.A. The Effect of Small Columbium Additions to semi-killed Medium-Carbon Steels. ASM Preprint No. 138. In Proceedings of the Regional Technical Meeting, Buffalo, NY, USA, 17–19 August 1959.
2. Stuart, H. Niobium-Proceedings of the International Symposium. Metallurgical Society of AIME: Warrendale, PA, USA, 1984.
3. Hansen, S.; Vander Sande, J.; Cohen, M. Niobium carbonitride precipitation and austenite recrystallization in hot-rolled microalloyed steels. *Metall. Trans. A* **1980**, *11*, 387–402. [CrossRef]
4. Cuddy, L. Microstructures developed during thermomechanical treatment of HSLA steels. *Metall. Trans. A* **1981**, *12*, 1313–1320. [CrossRef]
5. Sobral, M.; Mei, P.; Kestenbach, H.-J. Effect of carbonitride particles formed in austenite on the strength of microalloyed steels. *Mater. Sci. Eng. A* **2004**, *367*, 317–321. [CrossRef]
6. Vervynckt, S.; Verbeken, K.; Lopez, B.; Jonas, J.J. Modern HSLA steels and role of non-recrystallisation temperature. *Int. Mater. Rev.* **2012**, *57*, 187–207. [CrossRef]
7. Priestner, R.; Hodgson, P. Ferrite grain coarsening during transformation of thermomechanically processed C–Mn–Nb austenite. *Mater. Sci. Technol.* **1992**, *8*, 849–854. [CrossRef]
8. Miao, C.; Shang, C.; Zurob, H.; Zhang, G.; Subramanian, S. Recrystallization, precipitation behaviors, and refinement of austenite grains in high Mn, high Nb steel. *Metall. Mater. Trans. A* **2012**, *43*, 665–676. [CrossRef]
9. Iron and Steel Institute of Japan. *Proceedings of the Conference Thermec 88, Tokyo, Janpan 1988*; Iron and Steel Institute of Japan: Tokyo, Janpan, 1988; Volume 1, pp. 330–336.
10. Tanaka, T. Controlled rolling of steel plate and strip. *Int. Met. Rev.* **1981**, *26*, 185–212. [CrossRef]
11. Shanmugam, S.; Misra, R.D.K.; Mannering, T.; Panda, D.; Jansto, S.G. Impact toughness and microstructure relationship in niobium-and vanadium-microalloyed steels processed with varied cooling rates to similar yield strength. *Mater. Sci. Eng. A* **2006**, *437*, 436–445. [CrossRef]
12. Palmiere, E.J. Suppression of Recrystallization during the Hot Deformation of Microalloyed Austenite. Ph.D. Thesis, University of Pittsburgh, Pittsburgh, PA, USA, 1991.

13. Watanabe, H.; Smith, Y.; Pehlke, R. Precipitation kinetics of niobium carbonitride in austenite of high-strength low-alloy steels. In *The Hot Deformation of Austenite*; TMS-AIME: New York, NY, USA, 1977; pp. 140–168.
14. Rajinikanth, V.; Kumar, T.; Mahato, B.; Chowdhury, S.G.; Sangal, S. Effect of Strain-Induced Precipitation on the Austenite Non-recrystallization (Tnr) Behavior of a High Niobium Microalloyed Steel. *Metall. Mater. Trans. A* **2019**, *50*, 5816–5838. [CrossRef]
15. Palmiere, E.J.; Garcia, C.I.; DeArdo, A.J. The influence of niobium supersaturation in austenite on the static recrystallization behavior of low carbon microalloyed steels. *Metall. Mater. Trans. A* **1996**, *27*, 951–960. [CrossRef]
16. Koo, J.Y.; Luton, M.J.; Bangaru, N.V.; Petkovic, R.A.; Fairchild, D.P.; Petersen, C.W. Metallurgical design of ultra high-strength steels for gas pipelines. *Int. J. Offshore Pol. Eng.* **2004**, *14*, 2–10.
17. Graf, M.K.; Hillenbrand, H.G.; Peters, P. *Accelerated Cooling of Steel*; TMS-AIME: Warrendale, PA, USA, 1986; pp. 165–179.
18. Cuddy, L. The Effect of Microalloy Concentration on the Recrystallisation of Austenite During Hot Deformation. In *Thermomechanical Processing of Microalloyed Austenite*; AIME: Pittsburgh, PA, USA, 1981; pp. 129–140.
19. Kozasu, I.; Ouchi, C.; Sampei, T.; Okita, T. Hot rolling as a high-temperature thermo-mechanical process. In *Proceedings of the Conference on Microalloying 75*; Union Carbide Corp.: New York, NY, USA, 1977; pp. 120–135.
20. Palmiere, E.J.; Garcia, C.I.; DeArdo, A.J. Compositional and microstructural changes which attend reheating and grain coarsening in steels containing niobium. *Metall. Mater. Trans. A* **1994**, *25*, 277–286. [CrossRef]
21. Murr, L.E. *Interfacial Phenomena in Metals and Alloys*; Advanced Book Program, Reading, Mass; Addison-Wesley Pub. Co.: New York, MA, USA, 1974.
22. Danieli, C. *Start up Model for a Five Meter Plate Mill*; Private Communication: SpA. Buttrio (UD), Italy, 2010.
23. Gladman, T. On the theory of the effect of precipitate particles on grain growth in metals. *Proc. R. Soc. Lond. A* **1966**, *294*, 298–309.
24. Palmiere, E.J.; Garcia, C.I.; DeArdo, A.J. The influence of niobium supersaturation in austenite on the static recrystallization behavior of low carbon microalloyed steels. *Metall. Mater. Trans. A* **1996**, *27*, 951–960. [CrossRef]
25. Kwon, O.; DeArdo, A.J. Interactions between recrystallization and precipitation in hot-deformed microalloyed steels. *Acta Metall. Mater.* **1991**, *39*, 529–538. [CrossRef]
26. Cuddy, L.; Raley, J. Austenite grain coarsening in microalloyed steels. *Metall. Trans. A* **1983**, *14*, 1989–1995. [CrossRef]
27. Fernández, J.; Illescas, S.; Guilemany, J.M. Effect of microalloying elements on the austenitic grain growth in a low carbon HSLA steel. *Mater. Lett.* **2007**, *61*, 2389–2392. [CrossRef]

Article

Investigation on the Formation of Cr-Rich Precipitates at the Interphase Boundary in Type 430 Stainless Steel Based on Austenite–Ferrite Transformation Kinetics

Tao Jia *, Run Ni, Hanle Wang, Jicheng Shen and Zhaodong Wang

The State Key Lab of Rolling and Automation, Northeastern University, Shenyang 110819, China; RunNi3017@163.com (R.N.); stuwanghanle@aliyun.com (H.W.); shenjicheng@163.com (J.S.); zhdwang@mail.neu.edu.cn (Z.W.)

* Correspondence: jiatao@ral.neu.edu.cn or tao.jia.81@gmail.com; Tel.: +86-24-8368-1190

Received: 30 August 2019; Accepted: 19 September 2019; Published: 26 September 2019

Abstract: The Cr-rich precipitates at the interphase boundary in stainless steels not only lead to the sensitization, which further induces the intergranular corrosion and intergranular stress corrosion cracking, but also significantly deteriorate the ductility and toughness. In this work, the formation of Cr-rich precipitates at the interphase boundary in type 430 stainless steel was investigated from the perspective of austenite–ferrite transformation kinetics. Cyclic heat treatment was firstly conducted to reveal the kinetic mode of transformation behavior, i.e., local equilibrium or para equilibrium. Subsequently, interrupted quenching during continuous cooling was carried out, which illustrated clearly the relevance of the formation of interphase Cr-rich precipitates to the Cr enrichment adjacent to the interphase boundary as revealed by line scanning of energy dispersive spectroscopy (EDS). Finally, this enrichment of Cr was interpreted by DICTRA simulation, which is based on the determined kinetic mode for austenite–ferrite transformation. This work has, for the first time, established the correlation between the formation of interphase Cr-rich precipitates and the austenite–ferrite transformation kinetics.

Keywords: transformation kinetics; local equilibrium; para equilibrium; Cr-rich precipitate; interphase boundary; type 430 stainless steel

---

1. Introduction

Intergranular corrosion (IGC) and intergranular stress corrosion cracking (IGSCC) are the main corrosion modes of stainless steels when exposed to an aggressive environment. They have been long recognized to be induced by the boundary sensitization, i.e., the existence of a chromium (Cr) depleted zone adjacent to boundaries [1–3]. Even though other underlying mechanisms [4–6] for the formation of a Cr-depleted zone are found, the precipitation of Cr-rich carbide and nitride at boundaries is certainly the major one [7,8]. Provided a sufficient chemical driving force for precipitation, this is conceivable since the interface energy for nucleation is comparatively large at boundaries and the subsequent growth would drain Cr atoms from neighboring areas alongside the boundaries [9]. Thus, from the kinetic perspective, the precipitation of Cr-rich precipitates would be easier in ferritic stainless steel (FSS) [10] or at the ferrite side of the interphase boundary in duplex stainless steel (DSS) [11] due to the lower solubility of C and N and the fast diffusivity of Cr in the ferrite phase.

In contrast to the well-investigated IGC and IGSCC, the loss in ductility and toughness caused by the Cr-rich precipitates at boundaries has drawn much less attention. Shankar et al. [12] attributed the deterioration of ductility in 316LN stainless steel to the Cr-rich precipitates at grain boundaries, their interaction with dislocations, and the associated stress buildup at the grain boundaries. Ghosh [13] found that the fracture mode changed from transgranular to intergranular with increasing formation of grain boundary precipitates, and the ductility and fracture toughness decrease significantly. Hilders et al. [14] also related the decrease in toughness of 304L stainless steel to the increasing volume fraction of voids nucleated at the grain boundary precipitates formed during sensitization. Kumar Subodh and Shahi [15] revealed the detachment at heavily precipitated grain boundaries in the heat-affected zone of AISI 304L welds after post-weld thermal aging.

Considering the adverse effect on the in-use properties of stainless steels, investigation on the mechanism for the formation of Cr-rich precipitates would be of great importance. The present work focuses on two aspects, i.e., the austenite–ferrite transformation kinetics and the formation of Cr-rich precipitates at prior austenite/ferrite interphase boundaries in type 430 stainless steel. The experimental studies and DICTRA simulation have, for the first time, enabled the establishment of the correlation between these two physical metallurgical behaviors.

2. Materials and Experiments

Two steels obtained from a steel company, i.e., type 430 (8 mm hot-rolled plate) and 410S (6 mm hot-rolled plate) stainless steel were used in this study. Type 410S stainless steel was selected as a comparison for the investigation of continuous cooling transformation kinetics. Their chemical compositions are listed in Table 1.

Table 1. Chemical composition of experimental steels (wt %).

Stainless Steel	C	N	Si	Mn	Cr	Ni
430	0.04	0.04	0.25	0.32	16.32	0.16
410S	0.026	0.025	0.28	0.27	12.6	0.15

Similar constituent phases appear on both phase diagrams in Figure 1 which include ferrite, austenite, chromium nitride, and carbide with body-centered cubic (BCC), face-centered cubic (FCC), hexagonal close packed (HCP), and $M_{23}C_6$ crystal structure, respectively. The enlarged lower left region of the phase diagram is shown in the inset. Chromium nitride and carbide would precipitate at ~850 °C and below. Formation of cementite is thermodynamically unfavorable. The most noticeable difference in the phase diagram was the austenite single phase region between 901 and 1037 °C in type 410S stainless steel while the maximum volume fraction of austenite in the dual phase region is 44.7% in type 430 stainless steel.

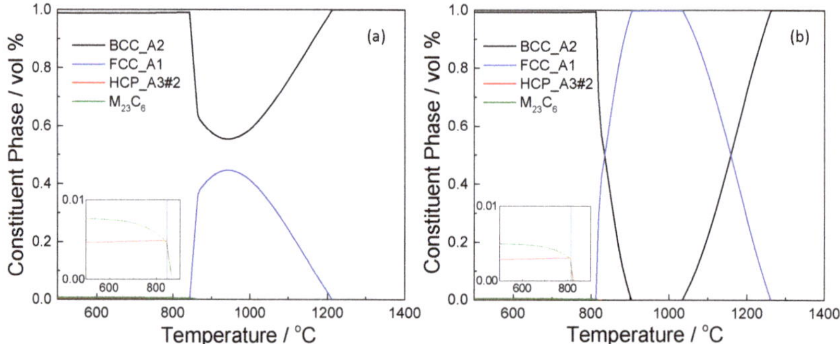

Figure 1. Phase diagram of type 430 (**a**) and 410S (**b**) stainless steels.

The heat treatment experiments in this work were conducted on a DIL 805A/D dilatometer (TA Instruments, New Castle, DE, USA). The sample size was Φ 4 mm × 10 mm. After machining, samples were all homogenized in a sealed quartz tube at 1200 °C for 120 min. Figure 2 shows the employed heat treatment procedure. The sample was firstly held at 1200 °C for 5 min. Then, in the cyclic heat treatment where austenite–ferrite transformation kinetics were studied, a 30 min isothermal holding at 950 °C was carried out to create a "ferrite + austenite" dual-phase microstructure. Subsequently, one cycle of heating and cooling between 950 and 1150 °C was applied to the sample before quenching to room temperature. The corresponding rate of temperature change (RTC) was 10, 100, and 200 °C/min. In the continuous cooling experiment, which was targeted for the investigation on the formation of interphase Cr-rich precipitates, the sample was cooled at 30 °C/min from 1200 °C. Interrupted quenching was respectively conducted at 850 and 200 °C to examine the resulted microstructure.

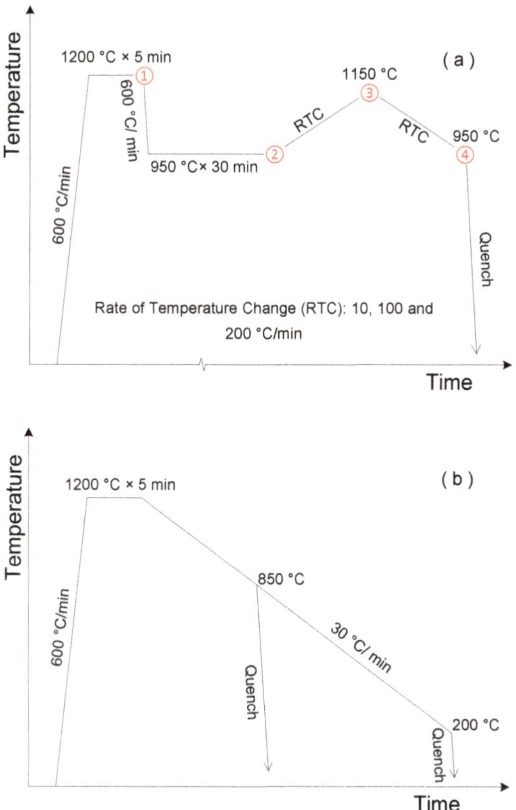

Figure 2. Schematic illustration of heat treatment procedure for (**a**) cyclic and (**b**) continuous cooling experiments.

Microstructure examination was made by optical microscopy (OM) and scanning electron microscopy (SEM, GeminiSEM 300, ZEISS, Oberkochen, Germany) with energy dispersive spectroscopy (EDS, Ultim Max, Oxford Instruments, Abingdon, UK). The heat-treated samples have gone through the standard metallographic preparation procedure, including grinding, polishing, and etching with 2% Nital solution.

3. Modeling of Austenite–Ferrite Transformation Kinetics

Over the past few decades, extensive research has been devoted to the study of austenite–ferrite transformation kinetics. Among the several proposed theories, diffusion-controlled theory is the most important one. For the Fe-C-M (M stands for the substitutional element) system, there are two proposed thermodynamic equilibrium conditions, i.e., para equilibrium (PE) and local equilibrium (LE).

3.1. Para Equilibrium

PE [16] describes the equilibrium state where only interstitial atoms are free to redistribute while substitutional atoms remain configurationally frozen during transformation, i.e.,

$$\frac{u_M^\alpha}{u_{Fe}^\alpha} = \frac{u_M^\gamma}{u_{Fe}^\gamma} = \frac{u_M^0}{u_{Fe}^0} \tag{1}$$

where u_{Fe} and u_M are molar fractions of Fe and M with respect to substitutional sites which are termed u-fraction. The superscript 0 stands for bulk concentration and α or γ denotes the ferrite or austenite phase. PE is a constrained equilibrium which is defined as

$$\begin{cases} \mu_C^\alpha = \mu_C^\gamma \\ \left(\mu_{Fe}^\gamma - \mu_{Fe}^\alpha\right) + \frac{u_M^0}{u_{Fe}^0}\left(\mu_M^\gamma - \mu_M^\alpha\right) = 0 \end{cases} . \tag{2}$$

3.2. Local Equilibrium

In LE [17], the chemical potential μ of carbon and substitutional element across the interface is constant, i.e.,

$$\mu_i^\alpha = \mu_i^\gamma \tag{3}$$

where μ is the chemical potential; the subscript i represents C or M. Due to the large difference in diffusivity between C and M, the LE is further classified into two types: negligible partition local equilibrium (NPLE) and partition local equilibrium (PLE). The Fe-C-Cr system, where Cr is a ferrite stabilizer, is used here for further elaboration.

In NPLE, as shown in Figure 3, the specific tie-line always connects the product phase with the Cr concentration of uCr0. Therefore, the product phase achieves the same Cr content as that in the bulk of the parent phase, and a positive or negative "spike" exists in front of the moving interface. When the interface is under NPLE, only local redistribution of Cr is required, and the transformation kinetics are controlled by carbon diffusion in the parent phase.

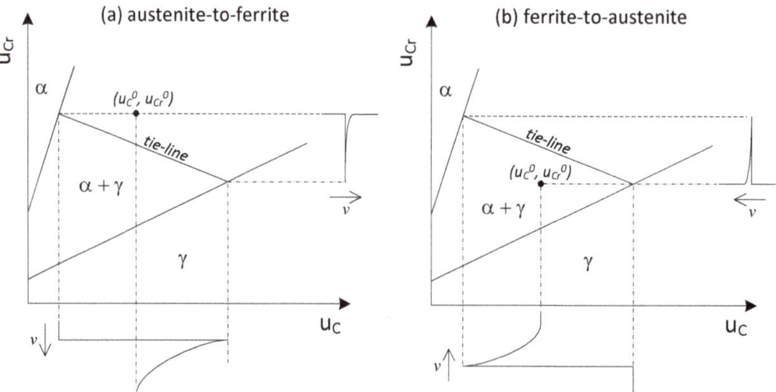

Figure 3. Schematic isotherm and concentration distributions depicting the (**a**) austenite-to-ferrite; (**b**) ferrite-to-austenite transformation under the negligible partition local equilibrium (NPLE) condition.

On the contrary, when partition of Cr takes place between the parent and product phase, as shown in Figure 4, long diffusion of Cr in the parent phase is necessary while a constant carbon activity is achieved from the interface to the bulk of the parent phase. In this case, the transformation is under

PLE mode and the sluggish diffusion of Cr in the parent phase becomes the decisive step in controlling the kinetics.

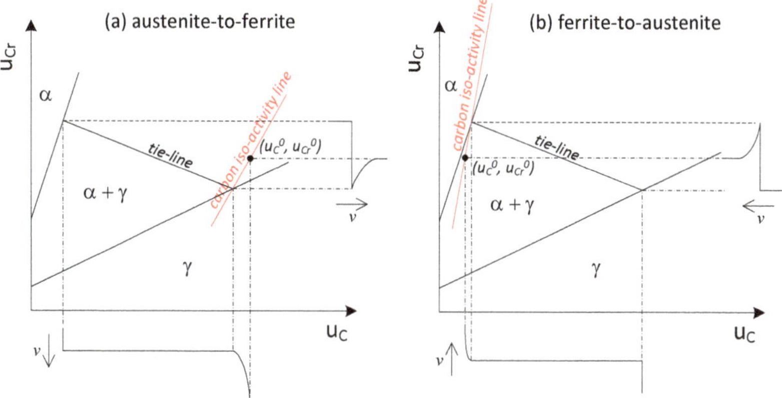

Figure 4. Schematic isotherm and concentration distributions depicting the (**a**) austenite-to-ferrite; (**b**) ferrite-to-austenite transformation under the partition local equilibrium (PLE) condition.

4. Results and Discussion

4.1. Determination of Austenite–Ferrite Transformation Kinetic Mode

Figure 5 shows the interruptedly quenched microstructure of type 430 stainless steel from the cyclic heat treatment with the RTC of 10 °C/min. The observed martensite, as indicated by the arrow, was transformed from the prior austenite by quenching. It is seen that, at each stage, i.e., ①, ②, ③, and ④ in Figure 2a, the prior austenite existed in the form of strips or islands in the ferritic matrix. The measured area fraction of austenite phase at ①, ②, ③, and ④ is 13.6%, 27.3%, 12.8%, and 23.7%, respectively. When the RTC increased to either 100 °C/min or 200 °C/min, the area fraction of austenite at the end of cyclic heat treatment decreased largely to about 16.5%, as shown in Figure 6, which suggests the characteristic RTC dependence of phase transformation kinetics.

To interpret the observed results, DICTRA simulation [18–20] under the assumption of one-dimensional planar geometry was carried out, where Tcfe9 thermodynamic and Mob4 mobility databases were used. In order to reduce the amount of calculation while ensuring the quality of simulation, type 430 stainless steel was simplified to the Fe-C-N-Cr system. The simulation was initiated from the beginning of isothermal holding at 950 °C and a domain size of 50 μm was used. The equilibrium constituent phases at 1200 °C were set as the starting point, i.e., ferrite and austenite with a chemical composition of Fe-0.039C-0.038N-16.34Cr (wt %) and Fe-0.118C-0.198N-14.837Cr (wt %), respectively. The initial ferrite/austenite interface was located globally at 49.35 μm. The simulation was carried out under both LE and PE conditions.

Figure 7 presents the evolution of a Cr profile during the heating and cooling stage under the LE condition. By the end of isothermal holding at 950 °C, the Cr profile exhibited a zigzag shape at the interface position, suggesting partitioning behavior of Cr from austenite to ferrite. During heating to 1150 °C, the zigzag shape of the Cr profile shrank when the interface migrated towards the austenite region. Even though the rate of change in the Cr gradient at the interface decreases with the increasing of the heating rate, a negative Cr spike in front of the moving interface was formed by the end of the heating stage, i.e., at 1150 °C, indicating a shift in transformation kinetics from a slow PLE mode to a fast NPLE mode. A similar ferrite/austenite interface location was achieved irrespective of the heating rate. In contrast, the enrichment of Cr at the ferrite side and the depletion of Cr at the austenite side of the interface gradually built up when the interface was moving backward during the cooling stage, suggesting the transformation kinetics switched from fast NPLE mode to slow PLE mode. At this stage, the cooling rate exerts a noticeable effect on the interface migration since the diffusion of Cr is very time-consuming compared with that of C. Finally, the one-dimensional austenite fraction, which is defined as the length of the austenite region divided by total domain size, reached 29.1%, 24.3%, and 23.6% at the RTC of 10, 100, and 200 °C/min, respectively.

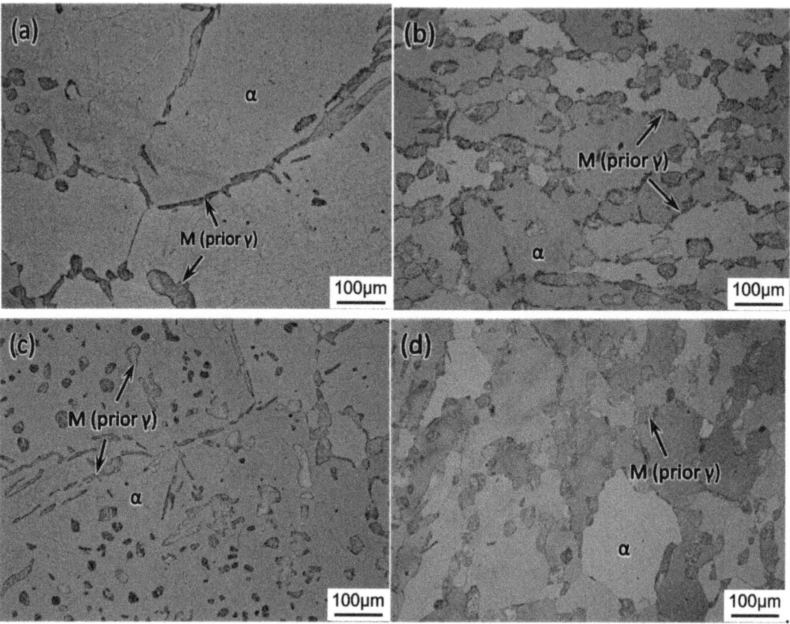

Figure 5. Optical micrograph showing the microstructure of an interruptedly quenched sample at (**a**) ①; (**b**) ②; (**c**) ③; (**d**) ④ during the cyclic heat treatment with the RTC of 10 °C/min.

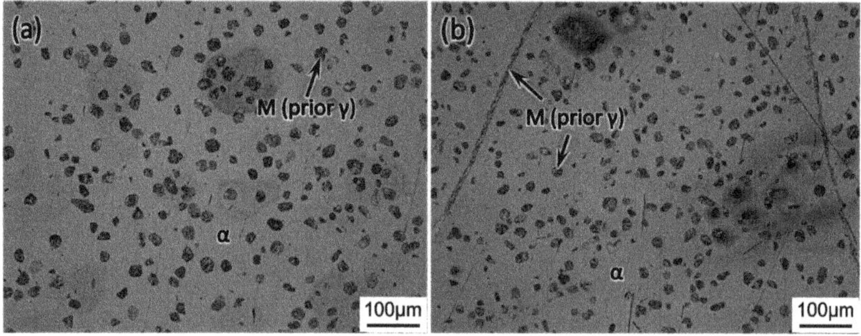

Figure 6. Optical micrograph showing the final microstructure of cyclic heat treatment with the RTC of (**a**) 100 °C/min; (**b**) 200 °C/min.

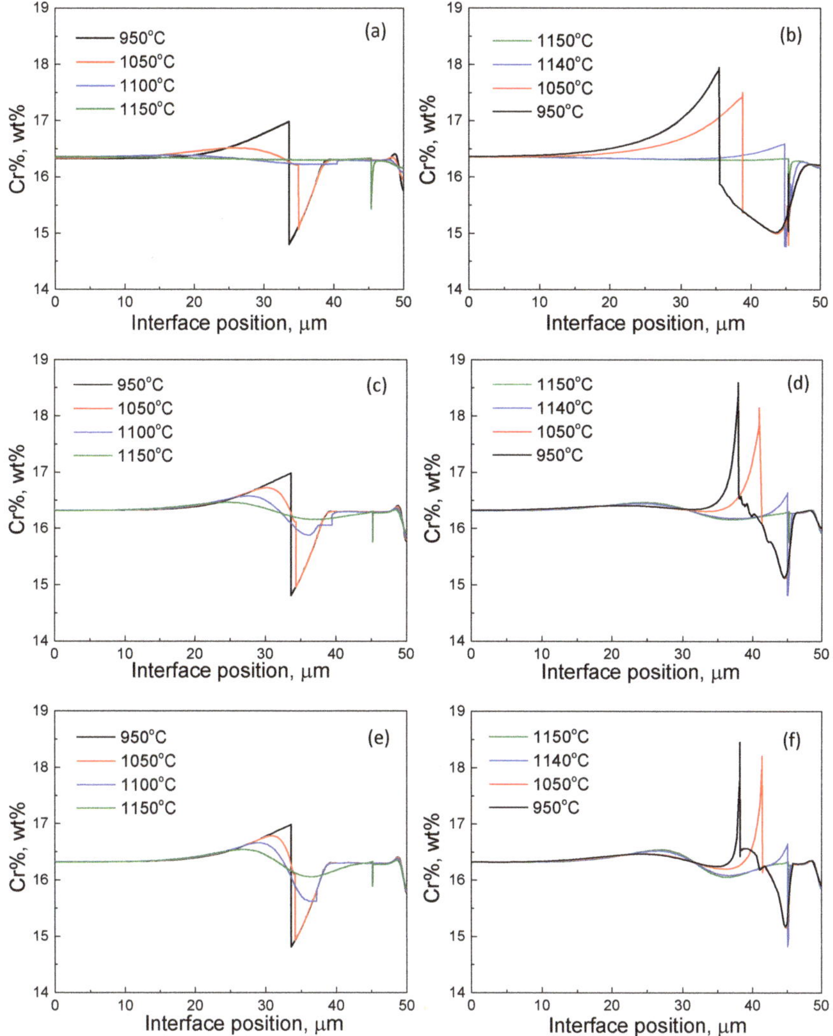

Figure 7. The evolution of the Cr profile during cyclic heat treatment at the RTC of (**a**,**b**) 10 °C/min; (**c**,**d**) 100 °C/min, and (**e**,**f**) 200 °C/min, where (**a**,**c**,**e**) and (**b**,**d**,**f**) correspond to the heating and cooling stage, respectively.

Simulation results from the PE condition are presented in Table 2. Under the PE condition, carbon diffusion plays a determining role for interface migration while the substitutional element Cr does not redistribute among ferrite and austenite at the interface. Therefore, the RTCs employed in this study have negligible effect on the transformation kinetics. Results from experiments and DICTRA simulation are all summarized in Table 2. It is seen that, when the RTC increases from 10 to 200 °C/min, the one-dimensional austenite fraction at the end of the cyclic heat treatment from PE simulation decreases marginally by 0.4%, in contrast to the noticeable decrease of 7% from LE simulation. From the above experimental study and DICTRA simulation, one could summarize that the transformation kinetics in type 430 stainless steel can be better captured by the simulation under the LE condition even though the Fe-C-N-Cr system is only a simplified representative of type 430 stainless steel.

Table 2. Measured and simulated austenite fraction by the end of cyclic heat treatment in type 430 stainless steel.

Stainless Steels	Value Type	10 °C/min	100 °C/min	200 °C/min
430	Measured [1]	23.7%	16.3%	16.7%
Fe-C-N-Cr system	simulated (LE) [2]	29.1%	24.3%	23.6%
	simulated (PE) [2]	28.4%	28.2%	28.0%

[1] area fraction; [2] one-dimensional fraction. LE: local equilibrium; PE: para equilibrium.

4.2. Mechanism for the Formation of Cr-Rich Precipitates at the Interphase Boundary in Type 430 Stainless Steel

Figure 8 shows the interruptedly quenched microstructure of the sample from the continuous cooling experiment, as shown in Figure 2b. Type 410S stainless steel is included here for comparison. The cooling rate employed, i.e., 30 °C/min was the same as the on-site measured value during the hot-rolling process. As the same as Figure 5, the observed martensite was transformed from the prior austenite by quenching. It is seen that, when samples were quenched at 850 °C, as shown in Figure 8a,c, ferrite and martensite were the only two constituent phases. After further slow cooling to 200 °C, the interphase precipitates as indicated by the arrow in Figure 8b were formed in type 430 stainless steel in contrast to its absence in type 410 stainless steel, as shown in Figure 8d, under the same heat treatment condition. Based on the calculated phase diagram in Figure 1, it is proposed that the Cr-rich precipitates at the interphase boundary were formed during the slow cooling process from 850 to 200 °C.

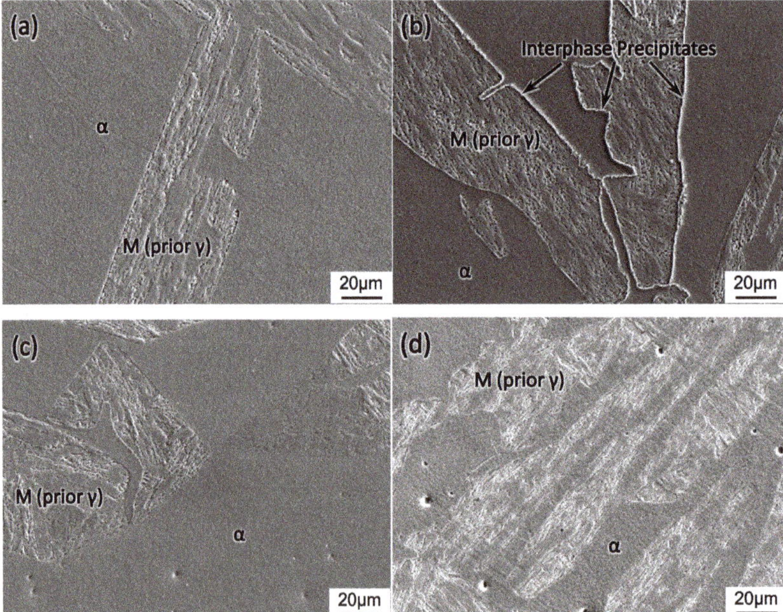

Figure 8. Optical micrographs showing the microstructure quenched from (**a**,**c**) 850 °C and (**b**,**d**) 200 °C, where (**a**,**b**) and (**c**,**d**) are from type 430 and 410S stainless steel, respectively.

The samples quenched at 850 °C from the continuous cooling experiment were subsequently re-examined by SEM with EDS to reveal the Cr profile across interphase boundaries. Figures 9 and 10 present the line scanning results at interphase boundaries in type 430 and 410S stainless steel, respectively. The line scanning was conducted at a sampling rate of 6 nm/point and under a magnification of ×20,000. The black rectangular data points in Figures 9 and 10 were the raw data from line scanning. Using the "adjacent-averaging method", where 50 neighboring data points included in the adjacent 0.3 μm length line were averaged to substitute the original data point, the Cr profiles were smoothed and more clearly presented in red lines. In type 430 stainless steel, as shown in Figure 9a,b, a substantial enrichment of Cr existed in the ferrite adjacent to the interphase boundary, i.e., 17.59% relative to 15.78% at the far-end of the ferrite matrix. While, in type 410S stainless steel, as shown in Figure 10a,b, the maximum Cr% in ferrite adjacent to the interphase boundary and at the far-end of the ferrite matrix was 12.77% and 12.52%, respectively. When ferrite is enclosed by austenite, soft impingement occurs. As illustrated in Figure 9c,d and Figure 10c,d, the average Cr% in ferrite enriched to 17.2% and 12.85% in type 430 and 410S stainless steel, respectively. Thus, the formation of Cr-rich precipitates at the interphase boundaries were facilitated by the segregated Cr in type 430 stainless steel. In type 410S stainless steel, the enrichment level, if represented by the difference of Cr% in the neighboring area of interphase boundaries from the far end of the ferrite region, was much lower, i.e., 0.25% in contrast to 1.8% in type 430 stainless steel.

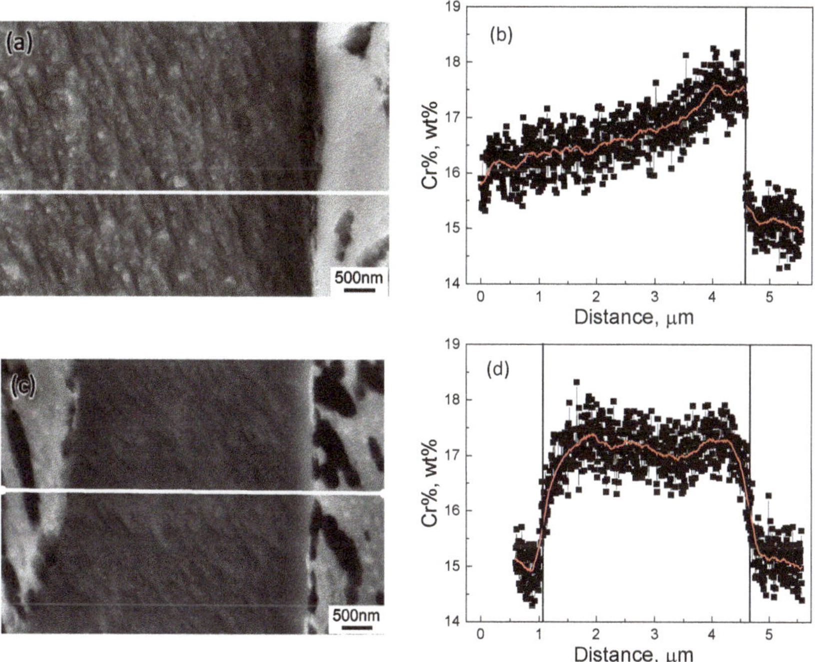

Figure 9. The Cr concentration profile across (**a**,**b**) single and (**c**,**d**) dual interphase boundaries in type 430 stainless steel interruptedly quenched at 850 °C.

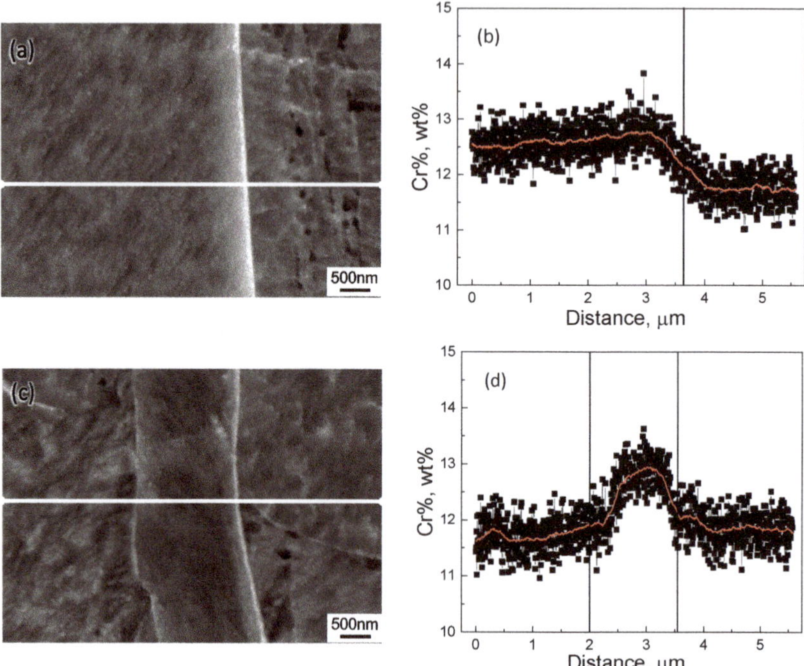

Figure 10. The Cr concentration profile across (**a**,**b**) single and (**c**,**d**) dual interphase boundaries in type 410S stainless steel interruptedly quenched at 850 °C.

In order to further interpret the formation of Cr enrichment, DICTRA simulation under the pre-determined LE condition in Section 4.1 is carried out where the Fe-C-N-Cr system was used as a representative of type 430 or 410S stainless steel as well. As shown in Figure 11, when the temperature decreases from 1200 to 900 °C, the interface is migrating toward the ferrite region and partitioning of Cr from austenite to ferrite can be seen. Further temperature decreases led to the backward migration of the interface and a switch of transformation kinetics to NPLE mode where a Cr spike exists in front of the interface. There are two interesting characteristics in this simulation. Firstly, the interface velocity during earlier austenite formation or the later austenite-to-ferrite transformation is much faster in type 410S stainless steel, possibly due to a large driving force as suggested by the phase diagram in Figure 1. Secondly, by the end of the simulation, a substantial Cr enrichment remains at the ferrite side of the interphase boundary in type 430 stainless steel. Compared with the line scanning results in Figures 9 and 10, an astonishing agreement has been achieved in terms of not only the shape of the Cr profile but also the Cr% in the adjacent region of the interphase boundary. Therefore, the experiment and simulation results have strongly supported the correlation between the formation of Cr-rich precipitates at the prior austenite/ferrite interphase boundary and the austenite–ferrite transformation kinetics.

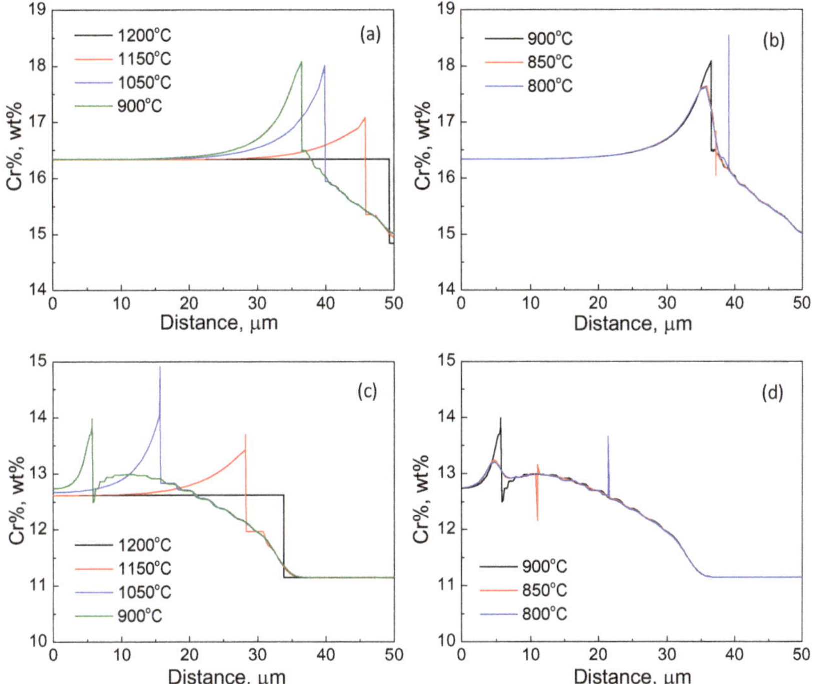

Figure 11. The evolution of the Cr profile during continuous cooling for type (**a**,**b**) 430; (**c**,**d**) 410S stainless steel.

5. Conclusions

The formation of Cr-rich precipitates at the interphase boundary in type 430 stainless steel, which not only induces intergranular corrosion and intergranular stress corrosion cracking but also significantly deteriorates the ductility and toughness, was investigated from the perspective of austenite–ferrite transformation kinetics. The following conclusions were drawn from this work.

1) The microstructure from cyclic transformation was largely affected by the rate of temperature change, which is in well accordance with the DICTRA simulation of austenite–ferrite transformation under the LE condition.
2) In contrast to type 410S stainless steel, a noticeable enrichment of Cr adjacent to the interphase boundary which facilitated the formation of interphase Cr-rich precipitates in type 430 stainless steel was revealed by EDS analysis and interpreted by DICTRA simulation under the LE condition. This has provided solid evidence for the correlation between the formation of interphase Cr-rich precipitates and austenite–ferrite transformation kinetics.

Author Contributions: Conceptualization, T.J. and J.S.; methodology, R.N. and T.J.; data curation, investigation and formal analysis, R.N. and T.J.; validation, H.W.; supervision, T.J.; writing—original draft preparation, review and editing, T.J.; resources, J.S. and Z.W.

Funding: This research was funded by the National Key R&D Program of China (2017YFB0304201) and the Fundamental Research Funds for the Central Universities (N180702012).

Conflicts of Interest: The authors declare no conflict of interest.

References

1. Hu, C.L.; Xia, S.; Li, H.; Liu, T.G.; Zhou, B.; Chen, W.J.; Wang, N. Improving the intergranular corrosion resistance of 304 stainless steel by grain boundary network control. *Corros. Sci.* **2011**, *53*, 1880–1886. [CrossRef]
2. Dayal, R.K.; Parvathavarthini, N.; Raj, B. Influence of metallurgical variables on sensitisation kinetics in austenitic stainless steels. *Int. Mater. Rev.* **2005**, *50*, 129–155. [CrossRef]
3. Du Toit, M.; Van Rooyen, G.T.; Smith, D. Heat-affected zone sensitization and stress corrosion cracking in 12% chromium type 1.4003 ferritic stainless steel. *Corrosion* **2007**, *63*, 395–404. [CrossRef]
4. Kim, J.K.; Lee, B.-J.; Lee, B.H.; Kim, Y.H.; Kim, K.Y. Intergranular segregation of Cr in Ti-stabilized low-Cr ferritic stainless steel. *Scr. Mater.* **2009**, *61*, 1133–1136. [CrossRef]
5. Kim, J.K.; Kim, Y.H.; Lee, B.H.; Kim, K.Y. New findings on intergranular corrosion mechanism of stabilized stainless steels. *Electrochim. Acta* **2011**, *56*, 1701–1710. [CrossRef]
6. Park, J.H.; Seo, H.S.; Kim, K.Y. Alloy design to prevent intergranular corrosion of low-Cr ferritic stainless steel with weak carbide formers. *J. Electrochem. Soc.* **2015**, *162*, C412–C418. [CrossRef]
7. Sedriks, A.J. *Corrosion of Stainless Steels*, 2nd ed.; Wiley-Interscience: New York, NY, USA, 1996; p. 110.
8. Pardo, A.; Merino, M.C.; Coy, A.E.; Viejo, F.; Carboneras, M.; Arrabal, R. Influence of Ti, C and N concentration on the intergranular corrosion behavior of AISI 316Ti and 321 stainless steel. *Acta Mater.* **2007**, *55*, 2239–2251. [CrossRef]
9. Yin, Y.F.; Faulkner, R.G.; Moreton, P.; Armson, I.; Coyle, P. Grain boundary chromium depletion in austenitic alloys. *J. Mater. Sci.* **2010**, *45*, 5872–5882. [CrossRef]
10. Kim, J.K.; Kim, Y.H.; Uhm, S.H.; Lee, J.S.; Kim, K.Y. Intergranular corrosion of Ti-stabilized 11 wt% Cr ferritic stainless steel for automotive exhaust systems. *Corros. Sci.* **2009**, *51*, 2716–2723. [CrossRef]
11. Lv, J.L.; Liang, T.X.; Dong, L.M.; Wang, C. Influence of sensitization on microstructure and passive property of AISI 2205 duplex stainless steel. *Corros. Sci.* **2016**, *104*, 144–151.
12. Shankar, P.; Shaikh, H.; Sivakumar, S.; Venugopal, S.; Sundararaman, D.; Khatak, H.S. Effect of thermal aging on the room temperature tensile properties of AISI type 316LN stainless steel. *J. Nucl. Mater.* **1999**, *264*, 29–34. [CrossRef]
13. Ghosh, S.; Kain, V.; Ray, A.; Roy, H.; Sivaprasad, S.; Tarafder, S.; Ray, K.K. Deterioration in fracture toughness of 304LN austenitic stainless steel due to sensitization. *Metall. Mater. Trans. A* **2009**, *40*, 2938–2949. [CrossRef]
14. Hilders, O.A.; Santana, M.G. Toughness and fractography of austenitic type 304 stainless steel with sensitization treatments at 973 K. *Metallography* **1988**, *21*, 151–164. [CrossRef]
15. Kumar, S.; Shahi, A.S. Studies on metallurgical and impact toughness behavior of variably sensitized weld metal and heat affected zone of AISI 304L welds. *Mater. Des.* **2016**, *89*, 399–412. [CrossRef]
16. Gilmour, J.B.; Purdy, G.R.; Kirkaldy, J.S. Thermodynamics controlling the proeutectoid ferrite transformations in Fe-C-Mn alloys. *Metall. Trans.* **1972**, *3*, 1455–1464. [CrossRef]
17. Coates, D.E. Diffusional growth limitation and hardenability. *Metall. Trans.* **1973**, *4*, 2313–2325. [CrossRef]
18. Borgenstam, A.; Engstrom, A.; Hoglund, L.; Ågren, J. DICTRA, a tool for simulation of diffusional transformations in alloys. *J. Phase Equilib.* **2000**, *21*, 269–280. [CrossRef]
19. Saha, A.; Ghosh, G.; Olson, G.B. An assessment of interfacial dissipation effects at reconstructive ferrite–austenite interfaces. *Acta Mater.* **2005**, *53*, 141–149. [CrossRef]
20. Dmitrieva, O.; Ponge, D.; Inden, G.; Millan, J.; Choi, P.; Sietsma, J.; Raabe, D. Chemical gradients across phase boundaries between martensite and austenite in steel studied by atom probe tomography and simulation. *Acta Mater.* **2011**, *59*, 364–374. [CrossRef]

Article

Enhanced Corrosion Resistance of SA106B Low-Carbon Steel Fabricated by Rotationally Accelerated Shot Peening

Chaonan Lei [1], Xudong Chen [1], Yusheng Li [2], Yuefeng Chen [1] and Bin Yang [1,*]

[1] Collaborative Innovation Center of Steel Technology, University of Science and Technology Beijing, Beijing 100083, China
[2] Nano and Heterogeneous Materials Center, School of Materials Science and Engineering, Nanjing University of Science and Technology, Nanjing 210094, China
* Correspondence: byang@ustb.edu.cn; Tel.: +86-6233-3351

Received: 29 May 2019; Accepted: 2 August 2019; Published: 8 August 2019

Abstract: The corrosion resistance of a SA106B carbon steel with a gradient nanostructure fabricated by rotationally accelerated shot peening (RASP) for 5, 10, 15 and 20 min was investigated. Electrochemical tests were carried out in the 0.05 M H_2SO_4 + 0.05 M Na_2SO_4 and 0.2 M NaCl + 0.05 M Na_2SO_4 solutions. The experimental results showed that the sample RASP-processed for 5 min exhibited the best corrosion resistance among them. TEM analysis confirmed that the cementite dissolution and formation of nanograins, which improved the corrosion resistance of the steel. Prominent micro-cracks and holes were produced in the samples when the RASP was processed for more than 5 min, resulting in the decrease of corrosion resistance.

Keywords: carbon steel; rotationally accelerated shot peening; nanocrystalline; corrosion resistance

---

1. Introduction

Nanocrystalline (NC) metallic materials have a high-strength without sacrificing toughness and ductility [1–5]. In general, a NC metallic layer with a certain thickness has often been utilized to improve the environmental service behavior [6,7]. The rotationally accelerated shot peening (RASP) technique can produce nanograins due to the generation of defects and interfaces (grain boundaries), the increasing of polycrystalline free energy, and the inducing of grain refinement, through the application of high strains at high strain rates [8–11]. The material exhibits unique mechanical and corrosion properties on account of strengthening the metal surface with the high density nanocrystals and interfaces.

Recently, some researchers have studied the corrosion mechanism of the surface nanocrystallization material. For instance, Balusamy et al. [12] studied whether, or not, the surface changes had a direct impact on the electrochemical activity of the metallic materials. The grain boundaries and the matrix structure formed numerous tiny electrochemical cells. Ye et al. [13] proposed that nanocrystalline 309 stainless steel (SS) by a DC magnetron sputtering exhibited different corrosion resistance changes in different solutions. Huang et al. [14] reported that the surface mechanical attrition treatment induced grain refinement and dislocations, had positive effects on the corrosion behavior of the Ti-25Nb-3Mo-3Zr-2Sn alloy, annealing experimental results further indicated that the improved corrosion resistance was mainly due to the grain refinement. Li et al. [15] reported that the effect of the grain size on the corrosion resistance of a NC low-carbon steel fabricated through an ultrasonic shot peening technique in a 0.05 M H_2SO_4 + 0.05 M Na_2SO_4 solution. When grain size was less than 35 nm, the corrosion rate increased as grain size decreased. They understood this was attributed to the increased number of the active sites caused by surface nanocrystallized low-carbon steel. Zhang et al. [16] found that in a saturated $Ca(OH)_2$ solution with and without Cl^-, the micro-cracks in the surface layer of the supersonic fine-particles (about 3 μm) bombarding the low-carbon steel did not degrade its passivity properties and pitting resistance. The interesting work of Chen et al. [17] proposed that stainless steel having underwent the RASP process showed a good corrosion resistance. However, the corrosion mechanisms of the nano-carbon steel and nano-stainless steel were very different. Normally, we think that surface nanocrystalline may worsen corrosion. In contrast, we found corrosion resistance performance of carbon steel increased after the RASP. The influence of RASP on the corrosion behavior of the carbon steel will be discussed below.

The RASP is a newly developed surface nanocrystallization technology for fabricating gradient structure with the grain size varying from nanometer to micrometer without changing the overall chemical compositions of a carbon steel. SA106B low-carbon steel after the RASP has excellent mechanical and corrosion properties, making them ideal for the secondary loop in a nuclear power plant to reduce the flow-accelerated corrosion. In this work, the surface nanocrystallization SA106B carbon steel was fabricated by RASP. The influence of the RASP on the corrosion behavior of the carbon steel has been investigated.

2. Experimental Procedures

2.1. Material

The material in this work is SA106B carbon steel, and its chemical composition is given in Table 1. The specimens were cut from the Jiangsu Chengde steel tube share Co., Ltd in China. Mainly, its metallographic structure consists of ferrite and pearlite, as shown in Figure 1, and its mean grain size is 30 μm. The projectile used for the shot peening was made of GCr15 balls with a diameter of 0.8 mm, and the shot velocity was 80 m/s. The process was carried out for 5, 10, 15 and 20 min, periods. The samples were rotated during the shot peening process. For the RASP-processed for 20 min SA106B carbon steel specimen, the average hardness at the top surface is 260 HV (Figure 2). The hardness decreased gradually from the surface to the inner, and finally approached a stable value (175 HV) at a depth of 120 μm. Furthermore, according to the well-known Hall–Petch equation, the changes in hardness are inversely proportional to the variation in the size of grains formed by RASP [18]. Therefore, the smaller the grain size, the higher the hardness will be.

Table 1. Chemical composition of the SA106B carbon steel (wt. %).

C	Mn	Si	S	P	Cr	Mo	V	Fe
0.21	0.53	0.26	0.008	0.011	0.02	0.01	0.01	Balance

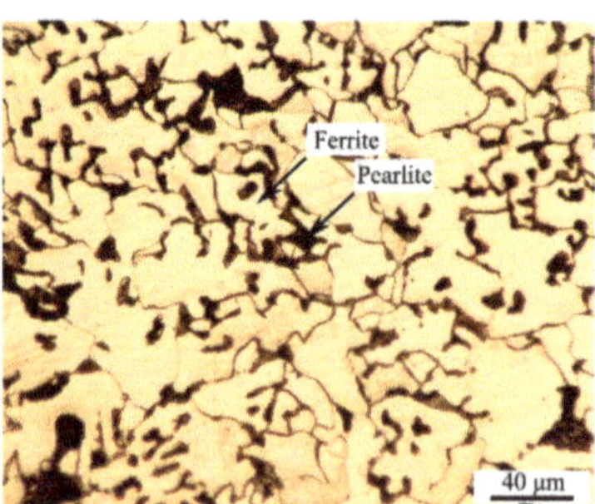

Figure 1. Optical microscope of SA106B low-carbon steel before the rotationally accelerated shot peening (RASP).

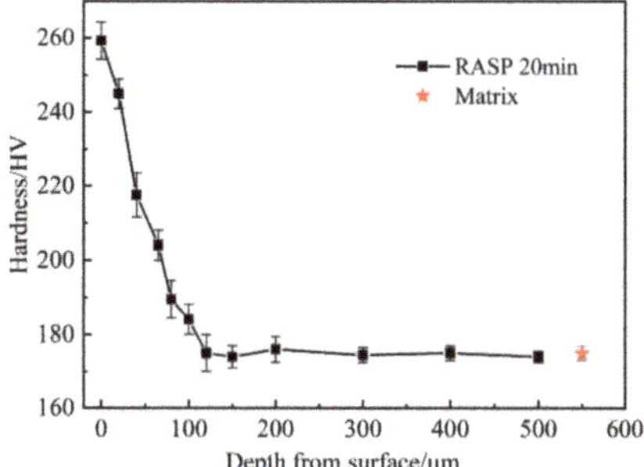

Figure 2. Hardness gradient as a function of depth for the RASP-processed for 20 min sample.

2.2. Electrochemical Measures

The specimens with a size of 10 mm × 10 mm × 5 mm were used in the electrochemical tests. They were spot-welded to the back of the samples with copper lines for the electrical contact. Prior to each electrochemical experiment, the exposed surface (with an area of 1 cm^2) of the specimens were wet ground with a series of SiC papers ranging from #600 to #3000, cleaned thoroughly with alcohol, acetone, and deionized water in sequence, and then dried quickly with flowing compressed hot air. The samples were polished first and then immersed in the 0.05 M H_2SO_4 + 0.05 M Na_2SO_4 and 0.2 M NaCl + 0.05 M Na_2SO_4 solutions, respectively. The electrochemical measurements were performed at room temperature. The potentiodynamic polarization and electrochemical impedance spectroscopy (EIS) experiments were conducted by using a three-electrode system. A platinum sheet was used as the auxiliary electrode. A saturated calomel electrode (SCE) as the reference electrode, and a specimen as the work electrode. The samples were washed with distilled water and immediately dried to analyze the corrosion morphology after the electrochemical measurements. To confirm the data reproducibility, the polarization tests were performed at least three times. The EIS tests were carried out at the open circuit potential (OCP) with a sinusoidal potential amplitude of 10 mV, running from 100 kHz to 10 MHz

3. Results and Discussion

3.1. Microstructure Characterizations

Figure 3 shows the SEM image of the RASP-processed sample in cross-section. According to the grain deformation degree, we found that the depth of the deformation layer of the carbon steel processed 20 min by the RASP was about 120 μm. It was shown that the top surface of the sample produced a distinct plastic deformation and the grain within the layer was remarkably refined. Black and white arrows indicate the treated surface and deformed region of the sample, respectively. It shows that the RASP successfully produced a steel with a structural gradient. Some bending striations were found near the surface. Duan et al. [19,20] also found these deformation characteristics under high strain, and two competing factors were suggested. One was the work-hardening effect through the increase of strain. The other was the thermal-softening effect caused by adiabatic temperature rise. In this study, severe plastic deformation (SPD) was introduced into the surface, resulting in the enhancement of the lattice distortion and micro-strain. This is a reason why the bending striations were formed.

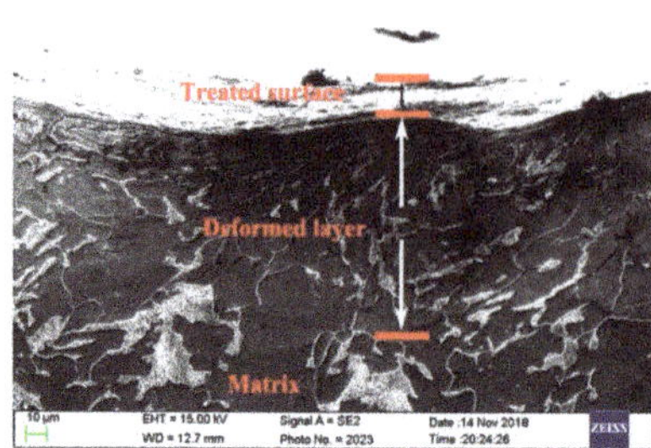

Figure 3. SEM image of the RASP-processed for 20 min sample.

Figure 4 shows the SEM images of the different RASP-processed samples in cross-sections. One can see that with the prolongation of the RASP time, the thickness of the deformed layer at the top surface increased. With a peening time of 5 min, the sample had a harder surface without defects. However, prominent micro-cracks and holes were produced easily in the samples when the RASP was processed for more than 5 min, as shown in Figure 4 by the red arrows.

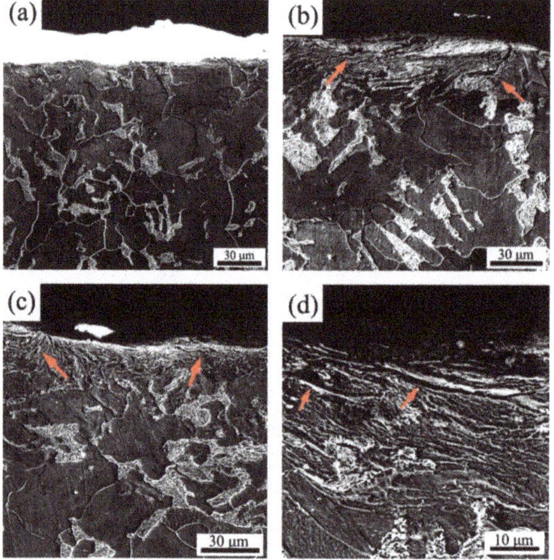

Figure 4. SEM images of the RASP-processed samples: (**a**) 5 min, (**b**) 10 min, (**c**) 15 min, (**d**) 20 min.

3.2. Potentiodynamic Polarization Results

The potentiodynamic polarization curves of the RASP-processed SA106B low-carbon steel with various duration times are shown in Figure 5. The potentiodynamic polarization test has usually been used to characterize the corrosion properties of materials [21]. The corrosion current density (I_{corr}) can be used to calculate the corrosion rate of metals. The I_{corr} is determined from the Tafel plot by extrapolating the linear portion of the polarization curve near corrosion potential (E_{corr}) [22]. As illustrated by the curve in Figure 5, all the potentiodynamic polarization curves show similar characteristics. When the polarization potential was higher than the corrosion potential, the current of the steel increased sharply with the increase of the polarization potential, indicating that the low-carbon steel exhibited the activation state in the solution. These results were consistent under two different electrolyte solutions, as indicated in Tables 2 and 3. The I_{corr} values decreased after the RASP, indicating the corrosion resistances for the samples were improved. It can also be seen that the I_{corr} increased slightly at a longer RASP time, the increasing corrosion rate can be attributed to the microscopic changes of the samples.

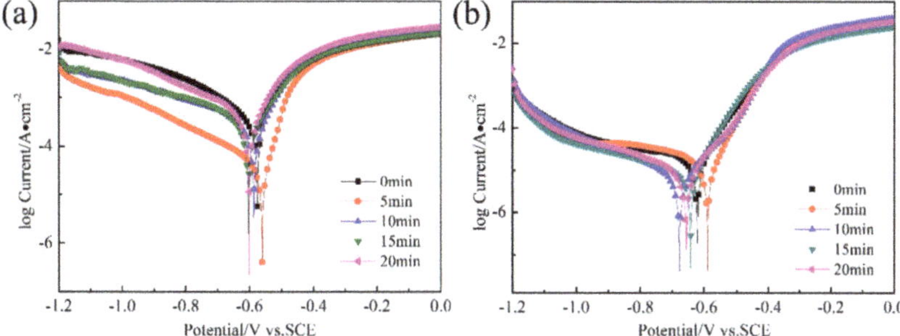

Figure 5. Potentiodynamic polarization curves of SA106B low-carbon steel for various RASP-processed periods ranging from 0 min to 20 min: (**a**) 0.05 M H_2SO_4 + 0.05 M Na_2SO_4 solution; (**b**) 0.2 M NaCl + 0.05 M Na_2SO_4 solution at room temperature.

Table 2. Electrochemical parameters of SA106B low-carbon steel for various RASP-processed periods ranging from 0 min to 20 min in the 0.05 M H_2SO_4 + 0.05 M Na_2SO_4 solution.

RASP Time (min)	E_{corr} (mV_{SCE})	I_{corr} ($\mu A/cm^2$)
0	−576 (±10)	490 (±11)
5	−560 (±8)	44 (±6)
10	−586 (±9)	227 (±8)
15	−601 (±12)	295 (±15)
20	−603 (±5)	380 (±26)

Table 3. Electrochemical parameters of SA106B low-carbon steel for various RASP-processed periods ranging from 0 min to 20 min in the 0.2 M NaCl + 0.05 M Na_2SO_4 solution.

RASP Time (min)	E_{corr} (mV_{SCE})	I_{corr} ($\mu A/cm^2$)
0	−620 (±11)	19.80 (±1.79)
5	−588 (±15)	7.78 (±0.37)
10	−676 (±12)	8.50 (±0.29)
15	−641 (±9)	8.90 (±0.31)
20	−645 (±12)	9.55 (±0.43)

The carbon steel which differs from a passive material, like a stainless steel, is a material with active dissolution behavior [23,24]. When we estimate its corrosion resistance, the primary factor parameter is the I_{corr}, followed by the E_{corr}. The corrosion resistance of the 5-min RASP-processed steel exhibited the best corrosion resistance among all of the samples. The original carbon steel had the worst corrosion resistance.

3.3. EIS Study

In studying corrosion and passivation processes, electrochemical impedance is a powerful tool to provide more information about the electrochemical processes. The impedance responses of these systems are given in Figures 6 and 7 in Nyquist and Bode formats, respectively. Figure 6 shows the effect of the RASP-processed time on the Nyquist plots of the SA106B low-carbon steel. One can see that the samples exhibited a high-frequency capacitive reactance arc and low-frequency inductance arc characteristics. At least two time constants are clearly observed in the Nyquist and Bode representations. With the increase of the RASP-processed time, the diameter of the semicircle decreased. The electrochemical impedance spectroscopy diagram shows that the larger the capacitive reactance arc, the better the corrosion resistance. We can conclude that the corrosion resistance of the samples improved after the RASP. The red circle with small capacitive reactance arc or inductive arc in the high-frequency part may be caused by the high-frequency phase shift. For the exact details, please refer to the specific introduction of Mansfeld [25]. A capacitive reactance generally above 10 kHz is essentially not a reflection of the electrochemical process.

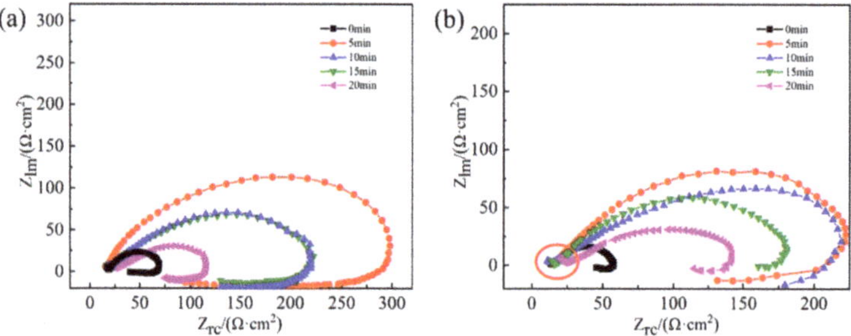

Figure 6. The effect of the RASP-processed time on the Nyquist plots of the SA106B low-carbon steel measured at open-circuit potential with a sinusoidal potential amplitude of 10 mV, running from 100 kHz to 10 MHz in the (**a**) 0.05 M H_2SO_4 + 0.05 M Na_2SO_4 solution and (**b**) 0.2 M NaCl + 0.05 M Na_2SO_4 solution at room temperature.

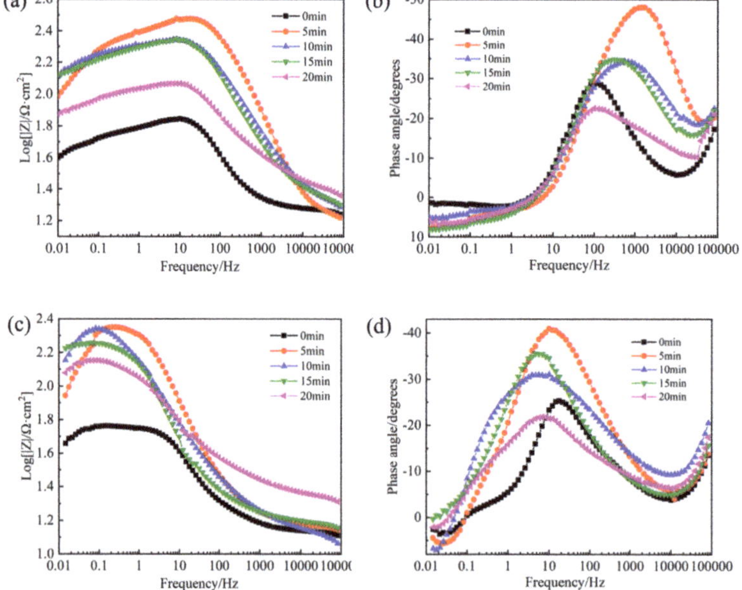

Figure 7. Bode plots of the RASP-processed SA106B specimens. (**a**) Frequency-impedance relation and (**b**) frequency-phase relation in the 0.05 M H_2SO_4 + 0.05 M Na_2SO_4 solution. (**c**) Frequency-impedance relation and (**d**) frequency-phase relation in the 0.2 M NaCl + 0.05 M Na_2SO_4 solution.

The fitted result for the impedance spectrum measured after potentiodynamic polarization in the 0.2 M NaCl + 0.05 M Na_2SO_4 solution with equivalent circuits in the inset of Figure 8a, as a representative example, is shown in Figure 8. The impedance data were fitted with ZSimpWin 3.50 version software (Ann Arbor, Michigan, MI, USA) using an equivalent circuit. It can be seen that the fitted and measured results match quite well in both Nyquist and Bode plots. The shrinkage of the Nyquist plots at the real parts are usually interpreted as the typical corrosive pitting appearing during the test [26]. It shows that intermediate products appear in the electrode reaction, producing a surface-adsorbing complex with the surface of the metal electrode. The oxidation film on the surface is slightly damaged and has not changed the original surface properties. Tables 4 and 5 list the fitted electrochemical parameters. R_s is the resistance of the solution affecting the charge transfer process, and R_{ct} is the charge transfer resistance of the surface electrode reaction of the low-carbon steel, which can reflect the surface of the electrode due to electricity generation. R_L is the inductance resistance and L is the inductance. C_1 is the interfacial capacitance and C_{ad} is the adsorption capacitance. Q is the constant phase element (CPE) which represents the capacitance of double layer. The CPE, which has non-integer power dependence on the frequency, is used to represent the capacitances of double layer to account for the deviation from the ideal capacitive behavior due to surface inhomogeneity, roughness and adsorption effects. The impedance of CPE is described by the expression

$$Z_{CPE} = Y_0^{-1}(jw)^{-\alpha}$$

where Y_0 is a proportional factor, j is the imaginary unit, w is the angular frequency, and α is the phase shift, which is a measure of the capacitance dispersion. For $\alpha = 0$, Q represents a resistance with $R = {Y_0}^{-1}$; for $\alpha = 1$, it represents a capacitance with $C = Y_0$; for $\alpha = 0.5$, it represents a Warburg element and for $\alpha = -1$, it represents an inductance with $L = {Y_0}^{-1}$. The R_{ct} value continuously decreased with the increase of the RASP-processed time (Tables 4 and 5), indicating that the corrosion resistance decreased, which may be related to the changes in the thickness, homogeneity, and composition of the oxidation film.

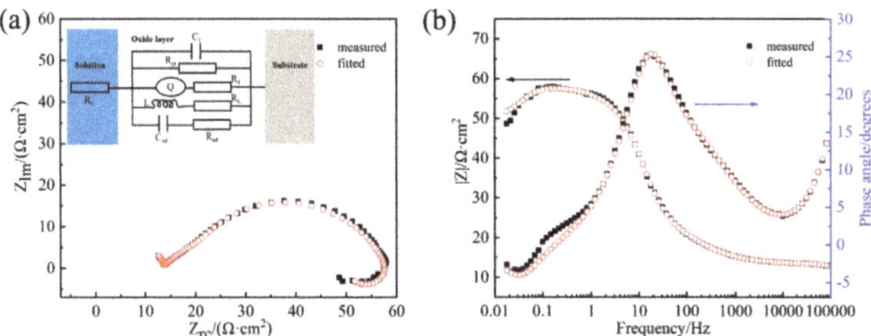

Figure 8. The fitted electrochemical impedance spectroscopy (EIS) of original specimen after potentiostatic polarization in the 0.2 M NaCl + 0.05 M Na_2SO_4 solution: (**a**) Nyquist plots and (**b**) Bode plots. The inset of (**a**) shows the equivalent electrical circuit for the specimen.

Table 4. Equivalent-circuit element values for EIS data corresponding to the SA106B low-carbon steel treated for different times in 0.05 M H_2SO_4 + 0.05 M Na_2SO_4 solution.

RASP Time (min)	R_s (Ω·cm²)	R_{ct} (Ω·cm²)	Q Parameters		C_1 (F/cm²)	C_{ad} (F/cm²)	R_1 (Ω·cm²)	R_{ad} (Ω·cm²)	R_L (Ω·cm²)	L (H·cm²)
			Y_0 (μF/cm²)	α						
0	4.1	63.4	20.83	0.91	5.59×10^{-8}	2.16×10^{-5}	33.3	45.1	168.2	84.3
5	8.3	282.1	2.73	0.92	1.20×10^{-7}	1.42×10^{-6}	12.7	248.1	320.3	398.4
10	10.3	202	11.63	0.79	1.23×10^{-7}	3.05×10^{-6}	15.8	222.2	534.6	718.6
15	10.2	201.7	12.93	0.83	1.15×10^{-7}	3.99×10^{-6}	16.8	164.4	416.6	483.2
20	11.0	101.9	16.75	0.78	8.55×10^{-8}	1.10×10^{-5}	23.9	95.5	231.4	381.5

Table 5. Equivalent-circuit element values for EIS data corresponding to the SA106B low-carbon steel treated for different times 0.2 M NaCl + 0.05 M Na_2SO_4 solution.

RASP Time (min)	R_s (Ω·cm²)	R_{ct} (Ω·cm²)	Q Parameters		C_1 (F/cm²)	C_{ad} (F/cm²)	R_1 (Ω·cm²)	R_{ad} (Ω·cm²)	R_L (Ω·cm²)	L (H·cm²)
			Y_0 (μF/cm²)	α						
0	1.8	56.9	1394	0.64	4.40×10^{-8}	0.000102	19.4	58.1	321.9	1656
5	8.2	222.3	684	0.67	1.65×10^{-7}	0.000107	7.2	141.2	186.5	867
10	7.0	208.7	837.6	0.61	2.64×10^{-7}	0.000200	7.9	154.8	386.0	1867
15	5.5	199.9	1171	0.58	8.59×10^{-8}	0.000142	12.2	74.4	1308.0	3746
20	3.5	172.1	1939	0.53	3.51×10^{-8}	0.000064	21.3	78.5	520.9	2847

3.4. SEM Photomicrographs

Figure 9 illustrates the dissolution morphologies of the SA106B low-carbon steel after the polarization tests. Some pitting pits were found in both ferrite and pearlite. The pearlite consists of ferrite and cementite phases. The cementite phases as the cathode have a higher potential compared to the ferrite phases, leading to the severe galvanic corrosion which can induce and worsen pitting [27,28]. With the prolongation of holding time, pits numbers gradually increased. Moreover, their size enlarges and the depth increases due to the micro-cracks effect and the roughness of carbon steel specimens treated by the peening. Thus, the corrosion resistance of the RASP-processed samples decreased. Among them, the sample RASP-processed for 5 min exhibited the best corrosion resistance. This is consistent with the results of the polarization curves and the impedance spectrum, as shown in Figures 5 and 6.

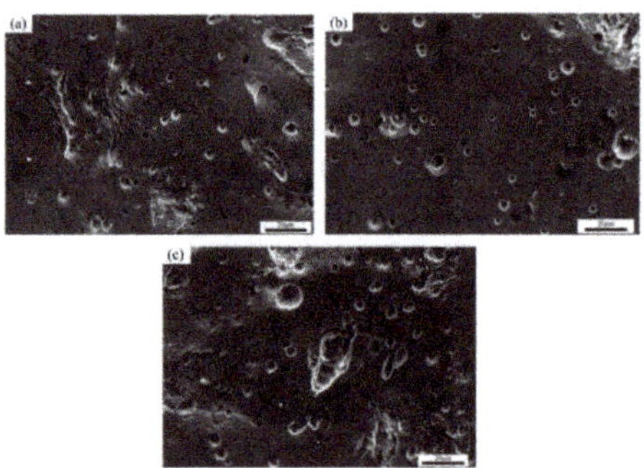

Figure 9. SEM micrographs for the samples RASP-processed for (**a**) 5 min, (**b**) 10 min and (**c**)15 min after potentiodynamic polarization in the 0.2 M NaCl + 0.05 M Na_2SO_4 solution.

3.5. TEM Analysis

TEM image of the SA106B carbon steel sample was depicted in Figure 10a. Figure 10a, b show TEM bright-field micrograph and selected-area electron diffraction (SAED) patterns, respectively, in the top surface layer of the sample RASP-processed for 5 min. One can note that the crystal grains have been refined into nanocrystal grains of relatively equiaxed and uniformly distributed compared with the irregular crystal grains of the matrix. The mean grain size is 25 nm (Figure 10c). Note that the SAED pattern of the sample consists of rings. It is known that the more continuous the rings, the smaller the grain sizes within the selected field of the view and the more uniform the distribution of grains [29]. The result demonstrates that the grain size of the steel can be markedly reduced when they are RASP-processed for 5 min. It is noteworthy that some spots spreading along the rings (Figure 10b) were found, indicating that there are also some coarse grains in the samples.

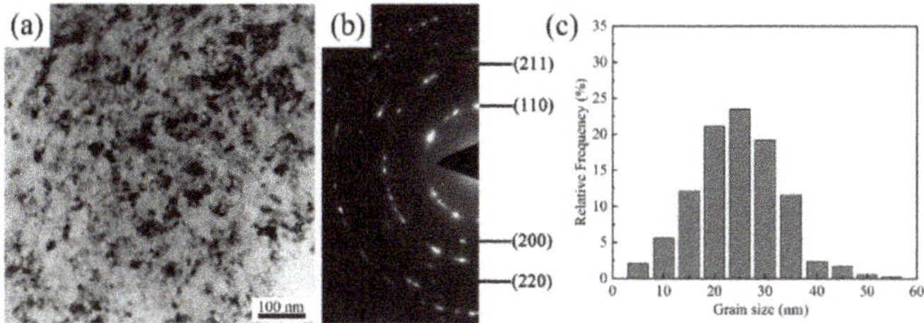

Figure 10. (**a**) TEM image of the top surface layer of the sample RASP-processed for 5 min; (**b**) the corresponding selected-area electron diffraction (SAED) pattern in (**a**); (**c**) grain size distribution in the top surface layer.

Figure 11 shows typical TEM images of the SA106B low-carbon steel, (a) RASP-processed for 0 min, (b) RASP-processed for 5 min and (c) shows the corresponding SAED pattern. In the carbon steel specimens untreated by the peening, one can see that the cementite is parallel, evenly distributed and the width of lamellar spacing is about 200 nm, as shown in Figure 11a. As shown in Figure 11b, the main result concerns the cementite phase which underwent a dissolution, at least partial, during SPD. The parallel distributions were changed severely. The diffraction spot, as shown in Figure 11c, confirmed the formation of supersaturated carbon atom ferrite. It is plausible that because the dislocation activity and cementite dissolution occurred simultaneously during plastic deformation, the carbon atoms can be dragged out of the cementite by mobile dislocations [30]. One study [31] has shown that cementite dissolution takes place through a global mechanism involving the whole volume of each individual lamellae, resulting in a carbon concentration gradient from cementite to ferrite. Indeed, as cementite partly dissolves, a large amount of carbon atoms is released, and must therefore be partitioned in the ferrite. Owing to the importance of the dissolution, the average carbon content in the ferrite may reach 1–2 at.%. After the cementite dissolution, a solid solution formed, which increased the electrode potential of the matrix [32]. Thus, the number of micro batteries reduced and the corrosion film was more stable and the surface layer of the sample was flattened.

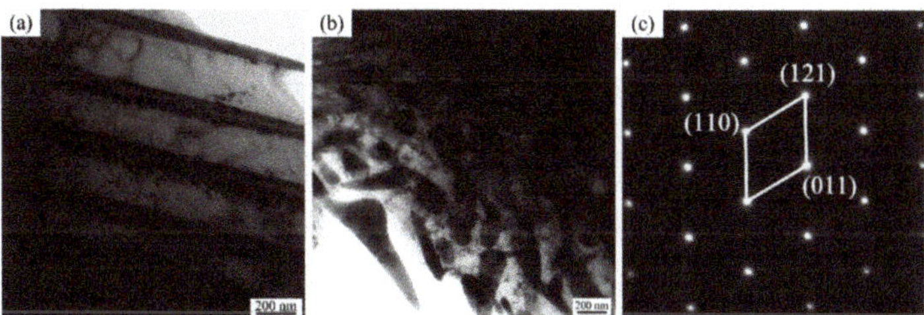

Figure 11. TEM images of the SA106B low-carbon steel. (**a**) RASP-processed for 0 min; (**b**) RASP-processed for 5 min; (**c**) The corresponding selected-area electron diffraction (SAED) pattern in (b).

4. Conclusions

In the present work, the influence of corrosion resistance of a SA106B low-carbon steel with gradient nanostructure produced by RASP was investigated. The main conclusions can be drawn as follows:

(1) The RASP-processed SA106B carbon steel surface formed a deformation layer with a depth about 120 μm. The mean grain size is 25 nm in the top surface layer.
(2) The sample RASP-processed for 5 min exhibited the best corrosion resistance due to the cementite dissolution and formation of nanograins.

(3) Prominent micro-cracks and holes were produced in the steel when the RASP was processed more than 5 min, resulting in the decrease of corrosion resistance.

Author Contributions: Writing-original draft preparation, figures and data curation, C.L.; data analysis, C.L., X.C., Y.C. and B.Y.; literature search, X.C.; study design, Y.L. and B.Y.; Writing—review and editing, B.Y.

Funding: This research was founded by the Beijing Municipal Natural Science Foundation under No 2162026 and the Beijing Municipal Science and Technology Commission under No Z181100005218005.

Conflicts of Interest: The authors declare no conflict of interest.

References

1. Moering, J.; Ma, X.; Chen, G. The role of shear strain on texture and microstructural gradients in low carbon steel processed by surface mechanical attritiontreatment. *Scr. Mater.* **2015**, *108*, 100–103. [CrossRef]
2. Zhu, Y.T.; Lowe, T.C.; Langdon, T.G. Performance and applications of nanostructured materials produced by severe plastic deformation. *Scr. Mater.* **2004**, *51*, 826–830. [CrossRef]
3. Fang, T.H.; Li, W.L.; Tao, N.R. Revealing extraordinary intrinsic tensile plasticity in gradient nano-grained copper. *Science* **2011**, *331*, 1587–1590. [CrossRef] [PubMed]
4. Mishra, R.; Balasubramaniam, R. Effect of nanocrystalline grain size on the electrochemical and corrosion behavior of nickel. *J. Corros. Sci.* **2004**, *46*, 3019–3029. [CrossRef]
5. González, J.; Peral, L.-B.; Colombo, C.; Fernández-Pariente, I. A study on the microstructural evolution of a low alloy steel by different shot peening treatments. *Metals* **2018**, *8*, 187. [CrossRef]
6. Segal, V.M. Severe plastic deformation: simple shear versus pure shear. *Mater. Sci. Eng. A* **2002**, *338*, 331–344. [CrossRef]
7. Lu, K.; Lu, J. Nanostructured surface layer on metallic materials induced by surface mechanical attrition treatment. *Mater. Sci. Eng. A* **2004**, *375*, 38–45. [CrossRef]
8. He, J.W.; Ma, S.N.; Ba, D.M. Investigation of surface nanocrystallization by pre-forceing rolling technology. *Adv. Mater. Res.* **2013**, *750*, 1963–1966. [CrossRef]
9. Sun, J.C.; Sheng, G.R.; Wang, Y.T. Surface self-nanocrystallization on industrial pure iron by high energy shot peening. (In Chinese). *Heat Treat. Met.* **2010**, *35*, 38–41.
10. Wang, X.; Li, Y.S.; Zhang, Q. Gradient structured copper by rotationally accelerated shot peening. *J. Mater. Sci. Technol.* **2016**, *7*, 758–761. [CrossRef]
11. Li, J.; Soh, A.K. Enhanced ductility of surface nano-crystallized materials by modulating grain size gradient. Modell. Simul. *Mater. Sci. Eng.* **2012**, *20*, 085002.
12. Balusamy, T.; Sankara Narayanan, T.S.N.; Ravichandran, K. Influence of surface mechanical attrition treatment (SMAT) on the corrosion behaviour of AISI 304 stainless steel. *Corros. Sci.* **2013**, *74*, 332–344. [CrossRef]
13. Ye, W.; Li, Y.; Wang, F. Effects of nanocrystallization on the corrosion behavior of 309 stainless steel. *Electrochim. Acta* **2006**, *51*, 4426–4432. [CrossRef]
14. Huang, R.; Han, Y. The effect of SMAT-induced grain refinement and dislocations on the corrosion behavior of Ti-25Nb-3Mo-3Zr-2Sn alloy. *Mater. Sci. Eng.* **2013**, *33*, 2353–2359. [CrossRef] [PubMed]
15. Li, Y.; Wang, F.; Liu, G. Grain size effect on the electrochemical corrosion behavior of surface nanocrystallized low-carbon steel. *Corros. Sci.* **2004**, *60*, 891–896. [CrossRef]
16. Zhang, L.; Ma, A.; Jiang, J. Electrochemical corrosion properties of the surface layer produced by supersonic fine-particles bombarding on low-carbon steel. *Surf. Coat. Technol.* **2013**, *232*, 412–418. [CrossRef]
17. Chen, X.D.; Li, Y.S.; Zhu, Y.T. Improved corrosion resistance of 316LN stainless steel performed by rotationally accelerated shot peening. *Appl. Surf. Sci.* **2019**, *481*, 1305–1312. [CrossRef]
18. Chabok, A.; Dehghani, K. Formation of nanograin in IF steels by friction stir processing. *Mater. Sci. Eng. A* **2010**, *528*, 309–313. [CrossRef]
19. Duan, C.; Wang, M. A review of microstructural evolution in the adiabatic shear bands induced by high speed machining. *Acta Metal.* **2013**, *26*, 97–112. [CrossRef]
20. Duan, Z.Q.; Li, S.X.; Huang, D.W. Microstructures and adiabatic shear bands formed by ballistic impact in steels and tungsten alloy. *Fatigue. Fract. Eng. Mater. Struct.* **2003**, *26*, 1119–1126. [CrossRef]
21. Chen, Y.F.; Chen, X.D.; Dai, X. Effect of spinodal decomposition on the pitting corrosion resistance of Z3CN20.09M duplex stainless steel. *Mater. Corros.* **2017**, *69*, 527–535. [CrossRef]

22. Yu, Y.C. *Electrochemistry basics tutorial*, 2nd ed.; Beijing Chemical Industry Press: Beijing, China, 2019; pp. 118–120.
23. Guo, H.X.; Lu, B.T.; Luo, J.L. Interaction of mechanical and electrochemical factors in erosion-corrosion of carbon steel. *Electrochim. Acta.* **2005**, *51*, 315–323. [CrossRef]
24. Kolawole, S.K.; Kolawole, F.O.; Enegela, O.P. Pitting corrosion of a low carbon steel in corrosive environments: experiments and models. *Adv. Mater. Res.* **2015**, *1132*, 349–365. [CrossRef]
25. Mansfeld, F.; Kending, M.W.; Lorenz, W.J. Corrosion inhibition in neutral, aerated media. *J. Electrochem. Soc.* **1985**, *132*, 290–296. [CrossRef]
26. Li, Y.R.; Lin, W.M.; Wei, Y.H. Electrochemical corrosion behavior of mechanical attrition treated surface layer with nanocrystallines on Cu-10Ni alloy. *Corros. Sci. Prot. Technol.* **2012**, *24*, 397–400.
27. Clover, D.; Kinsella, B.; Pejcic, B. The influence of microstructure on the corrosion rate of various carbon steels. *J. Appl. Electrochem.* **2005**, *35*, 139–149. [CrossRef]
28. Simoes, A.M.; Bastos, A.C.; Ferreira, M.G. Use of SVET and SECM to study the galvanic corrosion of an iron-zinc cell. *Corros. Sci.* **2007**, *49*, 726–739. [CrossRef]
29. Wu, X.; Tao, N.; Hong, Y. Microstructure and evolution of mechanically-induced ultrafine grain in surface layer of AL-alloy subjected to USSP. *Acta Mater.* **2002**, *50*, 2075–2084. [CrossRef]
30. Bang, C.W.; Seol, J.B.; Yang, Y.S. Atomically resolved cementite dissolution governed by the strain state in pearlite steel wires. *Scr. Mater.* **2015**, *108*, 151–155. [CrossRef]
31. Danoix, F.; Julien, D.; Sauvage, X.; Copreaux, J. Direct evidence of cementite dissolution in drawn pearlitic steels observed by tomographic atom probe. *Mater. Sci. Eng. A* **1998**, *250*, 8–13. [CrossRef]
32. Murayama, M.; Horita, Z.; Hono, K. Microstructure of two-phase Al-1.7 at% Cu alloy deformed by equal-channel angular pressing. *J. Acta Mater.* **2001**, *49*, 21–29. [CrossRef]

Article

Fatigue Behavior of Metastable Austenitic Stainless Steels in LCF, HCF and VHCF Regimes at Ambient and Elevated Temperatures †

Marek Smaga [1,*], Annika Boemke [1], Tobias Daniel [1], Robert Skorupski [2], Andreas Sorich [2] and Tilmann Beck [1]

1 Institute of Materials Science and Engineering, TU Kaiserslautern, 67663 Kaiserslautern, Germany; boemke@mv.uni-kl.de (A.B.); daniel@mv.uni-kl.de (T.D.); beck@mv.uni-kl.de (T.B.)

2 Former employee of the Institute of Materials Science and Engineering, TU Kaiserslautern, 67663 Kaiserslautern, Germany; skorupski@mv.uni-kl.de (R.S.); sorich@mv.uni-kl.de (A.S.)

* Correspondence: smaga@mv.uni-kl.de; Tel.: +49-631-205-2762

† On the occasion of his 70th birthday, this work is dedicated to our mentor, colleague and friend Prof. Dr.-Ing. habil. Dietmar Eifler.

Received: 31 May 2019; Accepted: 19 June 2019; Published: 21 June 2019

Abstract: Corrosion resistance has been the main scope of the development in high-alloyed low carbon austenitic stainless steels. However, the chemical composition influences not only the passivity but also significantly affects their metastability and, consequently, the transformation as well as the cyclic deformation behavior. In technical applications, the austenitic stainless steels undergo fatigue in low cycle fatigue (LCF), high cycle fatigue (HCF), and very high cycle fatigue (VHCF) regime at room and elevated temperatures. In this context, the paper focuses on fatigue and transformation behavior at ambient temperature and 300 °C of two batches of metastable austenitic stainless steel AISI 347 in the whole fatigue regime from LCF to VHCF. Fatigue tests were performed on two types of testing machines: (i) servohydraulic and (ii) ultrasonic with frequencies: at (i) 0.01 Hz (LCF), 5 and 20 Hz (HCF) and 980 Hz (VHCF); and at (ii) with 20 kHz (VHCF). The results show the significant influence of chemical composition and temperature of deformation induced α'-martensite formation and cyclic deformation behavior. Furthermore, a "true" fatigue limit of investigated metastable austenitic stainless steel AISI 347 was identified including the VHCF regime at ambient temperature and elevated temperatures.

Keywords: austenitic stainless steel; metastability; LCF; HCF; VHCF; ambient and elevated temperatures

1. Introduction

In the early stages of the development of stainless steels, the passivity of the material and, therefore, its "stainlessness" was the main scope of development [1,2]. Chemical passivity of steels in many environmental conditions is achieved by alloying at least about 11 wt% chromium to the base material. Due to an excellent combination of mechanical and technological properties, as well as corrosion resistance, austenitic stainless steels are the most prevalent group of stainless steels—widely used for components in nuclear power and chemical plants as well as a great variety of industrial, architectural and biological applications [3–6]. Since the chromium contents of typical austenitic stainless steels exceed 16 wt%, their equilibrium microstructure at room temperature would be fully ferritic, if no other austenite stabilizing alloying elements were added to the material. Elements most often used to obtain an austenitic microstructure are nickel, manganese, carbon and nitrogen. Because carbon has a very high affinity with chromium, chromium carbides are prone to develop particularly at high temperatures. This can reduce the chromium solid solution content in the austenitic matrix to less than 11 wt% and, consequently, lead to localized loss of stainlessness, especially at, or close to, the grain boundaries, resulting—under certain conditions—in intercrystalline corrosion. Therefore, in order to assure the corrosion resistivity of austenitic stainless steels, a very low carbon content and/or alloying with elements of a higher affinity with carbon, (e.g. niobium or titanium) are required [4]. In some cases, the ferromagnetic body centered cubic (bcc) δ-ferrite can be obtained in the paramagnetic face centered cubic (fcc) γ-austenitic microstructure directly after the manufacturing or melting process [7–10]. Indeed, δ-ferrite is a stable phase and its volume fraction does not change during mechanical loading. Generally, the influence of the chemical composition of Cr-Ni stainless steels on the initial microstructure at ambient temperature (AT) obtained after solution annealing heat treatment can be roughly estimated from the Cr und Ni content using the Maurer diagram [1]. Furthermore, the Schaeffler diagram provides more detailed information based on Cr and Ni equivalents taking other alloying elements besides Cr and Ni into account and is generally used for determining welding microstructures in Cr-Ni stainless steels [11].

The change in chemical composition not only influences the passivity of austenitic stainless steel but also significantly affects the metastability of austenite [12–16], meaning it is susceptible to undergoing phase transformations. Hence, the paramagnetic γ-austenite can transform by plastic deformation to a more stable microstructure, i.e. paramagnetic hexagonal close-packed (hcp) ε-martensite and/or body cubic centered (bcc) α´-martensite [17–24]. Scheil, one of the earliest researchers in this field, investigated the γ-austenite to α´-martensite transformation by measuring magnetic properties [25]. The change in magnetic properties by phase transformation can be used for non-destructive detection of α´-martensite as well as for monitoring fatigue processes in metastable austenites [26]. However, δ-ferrite, in the initial state of metastable austenite, has to be clearly separated from deformation-induced α´-martensite because both phases are ferromagnetic [7–10]. Deformation-induced phase transformations in metastable austenite affect significantly the fatigue process and are influenced by chemical composition [12–16], temperature [14,27–30] (encompassed by stacking fault energy (SFE) [31–39]), as well as grain size [32,40–43] strain amplitude [29,44–47] strain rate [48–50] and strain/stress state [51]. Additionally, SFE in fcc materials influences the slip character of dislocation, which has been summarized in the cyclic deformation map as a function of SFE and plastic strain amplitude given in [52–54]. For low SFE, independent of strain (stress) amplitude, the planar slip character occurs but for materials with higher SFE, a dependency of strain (stress) amplitude on dislocation slip character exists. High strain (stress) amplitude in the low cycle fatigue (LCF) regime leads to the development of the cell dislocation structure, which correlates with the wavy slip character of dislocation. The decrease of strain (stress) amplitude results in the formation of persistent slip bands in the high cycle fatigue (HCF) regime. Between the LCF and HCF regimes and lower and higher SFE, mixed dislocation structures can be observed [52–54]. A cyclic dislocation structure map of fcc materials with different SFE is not currently available for the very high cycle fatigue (VHCF) regime. The relationship between dislocation slip character and SFE described above was found for stable fcc materials [52–54]. In the case of metastable fcc austenite, SFE influences not only the dislocation slip character but also the deformation-induced phase transformation in ε and/or α´-martensite [17–24] as well as the formation of twins [55–57]. Figure 1 summarizes the microstructural changes in metastable austenitic Cr-Ni stainless steels due to cyclic mechanical loading, supported by TEM micrographs, which influence the cyclic deformation behavior of metastable austenitic stainless steels and consequently fatigue life. In this context, the present paper focuses on microstructural changes in metastable austenitic stainless steels, their cyclic deformation and transformation behavior at ambient and elevated temperatures in the whole fatigue regime from LCF to VHCF.

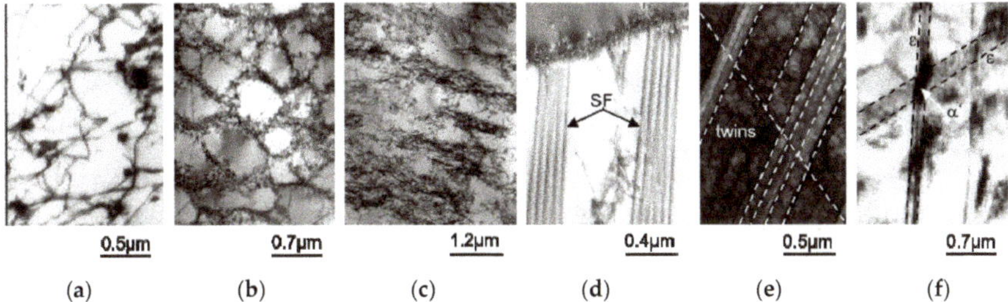

Figure 1. TEM micrographs with deformation and transformation microstructures of cyclically loaded metastable austenitic stainless steels: (**a**) typical dislocation density of the initial state after solution annealing, AISI 348 [58]; (**b**) cell dislocation structure (wavy slip character) at N_f in AISI 348 fatigued with σ_a = 280 MPa, R = −1 at AT with f = 5 Hz [58]; (**c**) dislocation accumulation in one direction (planar slip character) at N_f in AISI 348 fatigued with σ_a = 260 MPa, R = −1 at AT with f = 5 Hz [58]; (**d**) stacking faults (SF) at N_f in AISI 304 fatigued with $\varepsilon_{a,t}$ = 0.325%, R = -1 at AT with f = 5 Hz [58]; (**e**) twins in AISI 347 fatigued with σ_a = 160 MPa, R = −1 at T = 300 °C with f = 980 Hz at N = 5 × 10^8, (**f**) ε- and α´-martensite at N_f in AISI 321 fatigued with σ_a = 330 MPa, R = −1 at AT with f = 5 Hz [59].

2. Materials

The investigated material was metastable austenitic stainless steel AISI 347 (X10CrNiNb1810, 1.4550) in two batches (A and B). The chemical composition of both batches is given in Table 1, which corresponds to German and international standards [60,61]. However, it is important to note that these standards do not take into consideration the metastability of the austenitic microstructure, but focus on the stainlessness of the steel. It is known that the same type of austenitic stainless steel can exist in significantly different states of metastability [16] up to a fully stable state. The material's metastability can be characterized by experimentally estimated equations, which describe the martensitic start (M_S) or deformation-induced transformation temperature (M_d) as well as the stacking fault energy (SFE), according to the chemical composition of austenitic stainless steels. In the literature, several equations for M_S [13], M_{d30} [14] and SFE [34] are given. Table 1 gives examples of these metastability parameters according to the following equations:

$$M_{d30,Angel} \text{ in } ^\circ C = 413 - 462 \cdot (C+N) - 13.7 \cdot Cr - 9.5 \cdot Ni - 8.1 \cdot Mn - 18.5 \cdot Mo - 9.2 \cdot Si \quad (1)$$

$$M_{S,Eichelmann} \text{ in } ^\circ C = 1350 - 1665 \cdot (C+N) - 42 \cdot Cr - 61 \cdot Ni - 33 \cdot Mn - 28 \cdot Si \quad (2)$$

$$SFE \text{ in } mJ/m^2 = 25.7 + 2 \cdot Ni - 0.9 \cdot Cr - 1.2 \cdot Mn + 410 \cdot C - 77 \cdot N - 13 \cdot Si \quad (3)$$

The investigated austenitic stainless steel batches are in a metastable state (M_{d30} is in the range of ambient temperature). M_{d30} is the temperature at which 50 vol% of α´-martensite is developed by 30% of true plastic deformation [14] and was introduced for the comparison of the metastability of austenitic stainless steels. However, the deformation-induced phase transformation from γ-austenite in to α´-martensite can also take place at higher temperatures than M_{d30} [46,58,59,62,63]. The amount of deformation induced α´-martensite depends: (i) on the initial conditions, given by production process, such as chemical composition and initial microstructure e.g. the grain size of austenite, dislocation arrangements/density, precipitations and (ii) on loading parameters, like deformation temperature, amount of plastic deformation, as well as the stress and strain state/rate. Therefore, determining the true M_d-temperature above which no α´-martensite formation takes place is not practically possible. In order to take into account both aspects specified above (i and ii) on the susceptibility of forming α´-martensite in metastable austenite, a method based on dynamically applied local plastic deformation and magnetic measurement was developed [16]. The parameter established by this method as I_ξ correlates very well with the grade of α´-martensite formation during cyclic loading and allows to distinguish the susceptibility of austenitic stainless steels, which have the same chemical composition and grain size, to undergo phase transformation. The I_ξ parameter of the two investigated batches of AISI 347 is given in Table 1. Table 2 summarizes the mechanical properties of the investigated material at ambient temperature (AT) and T = 300 °C as well as α′-martensite content after specimen failure.

Table 1. Chemical composition in weight % and metastability parameters.

AISI 347	C	N	Nb	Cr	Ni	Mn	Mo	Si	M_{d30} °C	M_S °C	SFE mJ/m²	I_ξ FE%
Batch A	0.040	0.007	0.62	17.6	10.6	1.83	0.29	0.41	25	−189	39	0.25
Batch B	0.024	0.019	0.41	17.3	9.3	1.55	0.19	0.63	46	−81	27	0.66

Table 2. Mechanical properties and α′-martensite content after specimen failure (ξ).

AISI 347	$R_{p0.2}$ in MPa	UTS in MPa	A in %	ξ in FE%
Batch A, AT	242	569	66	4.41
Batch A, 300 °C	180	357	36	0.00
Batch B, AT	220	621	51	33.22
Batch B, 300 °C	155	428	37	0.00

Shown in Figure 2 are light and electron microscopy micrographs of longitudinal sections of the initial microstructure of the investigated materials in the solution-annealed state and after plastic deformation. Specimens from batch A were extracted from an original nuclear power plant surge line pipe with an outside diameter of 333 mm and a wall thickness of 36 mm. The pipe was manufactured seamless, drilled from the inside, turned from the outside and delivered in a solution-annealed state (1050 °C / 10 min / H_20), such that in the initial state no α′-martensite was detected. Note that the surge line pipe was investigated in the as-manufactured condition and has not been previously used in a nuclear power plant. The material of batch B was provided as rolled bars with a diameter of 25 mm in a solution-annealed state. A fully austenitic microstructure was obtained by additional annealing at 1050 °C for 35 minutes and quenching in helium atmosphere.

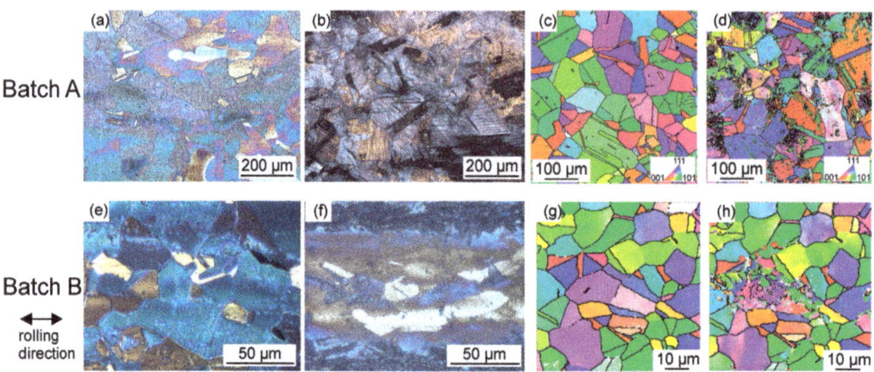

Figure 2. Optical micrographs of longitudinal sections using Bloech & Wedl I etching (**a**,**e**) initial state, (**b**,**f**) after plastic deformation as well as EBSD grain maps of (**c**,**g**) initial state and (**d**,**h**) after plastic deformation.

In Figure 2a the optical micrograph of the initial state of batch A is shown. This micrograph was taken after color etching using a Bloech & Wedl I etching agent, which is able to visualize local inhomogeneities in chemical composition [15]. A homogeneous distribution of the Cr and Ni content was detected in the microstructures of batch A. Furthermore, plastic deformation of batch A led to the homogeneous development of α′-martensite in the austenite matrix during plastic deformation (see Figure 2b,d). An electron backscatter diffraction (EBSD) micrograph of the initial state of both batches (Figure 2c,g) revealed a one-phase microstructure with annealing twins, typical for austenitic stainless steels. The same etching and observation techniques were used for characterizing the microstructure of batch B. In the case of batch B, the etching agent revealed local chemical inhomogeneities caused by slight variations of the Cr and Ni content as blue and brown bands (Figure 2e), which could not be removed during solution annealing. The blue band correlates with a lower Ni and higher Cr content, while the brown bands indicate higher Ni and lower Cr contents [15]. The band structure with a corresponding local metastability of austenitic microstructure led to a pronounced α´-martensite formation in regions with higher Cr content (Figure 2f). The local chemically induced band structure could not be observed in scanning electron micrographs using EBSD technique (Figure 2h). Instead, the EBSD image shows a homogeneous crystallographic microstructure in both cases, whereas the deformation-induced α´-martensite formation after plastic deformation can be clearly detected by EBSD images (Figure 2d,h). The comparison of EBSD grain maps (Figure 2c,g) of both batches presented a clear difference in the grain size. Quantitative evaluation yielded a mean grain size of 120 μm for batch A and of 17 μm for batch B, respectively.

3. Methods

To investigate the fatigue behavior of metastable austenitic stainless steels at ambient and elevated temperatures from the LCF to VHCF regime, servohydraulic and ultrasonic fatigue systems were used. The test temperature of T = 300 °C was achieved using an inductive heating system and control based on measurement by a type-K ribbon thermocouple in the center of the gauge length. The isothermal LCF and HCF tests at ambient temperature (AT) and T = 300 °C were performed with an MTS 100 kN servohydraulic testing system using load frequency of f = 0.01 Hz (LCF), 5 Hz and 20 Hz (HCF), see Figure 3a and b. The VHCF tests at T = 300 °C were performed with f = 980 Hz at an MTS 1 kHz servohydraulic testing system (Figure 3c). The VHCF tests at AT were realized in an ultrasonic fatigue testing system developed and built up at the authors' institute [64,65] (Figure 3d) with an operating frequency f = 20 kHz. In order to characterize the cyclic deformation behavior in the LCF and HCF regime, respectively, an extensometer (AT and T = 300 °C) and thermocouples (only AT) were used. The area of each hysteresis loop describes the cyclic plastic strain energy dissipated per unit volume during a given loading cycle, which is mainly dissipated into heat and, hence, results in a change in specimen temperature [66,67]. Temperature was measured with one thermocouple in the middle of the gauge length (T_1) and two reference thermocouples at the elastically loaded specimen shafts (T_2, T_3). The temperature change induced by cyclic plastic deformation was calculated according to:

$$\Delta T = T_1 - 0.5 \cdot (T_2 + T_3) \quad (4)$$

The in situ detection of fatigue induced α´-martensite formation was done by magnetic Feritscope™ (FISCHER, Windsor, CT, USA) measurements at AT (see Figure 3a). Due to the higher permeability of ferromagnetic ferrite compared to paramagnetic austenite, the response of the material to magnetic induction increases with the ferrite content. Using a non-destructive magnetic method, the Feritscope™ measures the relative permeability of a material in the alternating magnetic field of its probe. This provides a ferrite content signal (FE%), which is also influenced by the curvature of the specimen's surface and stress-strain state. Furthermore, to determine the ferromagnetic α´-martensite content in vol%, the Feritscope™ signal (FE%) needs to be multiplied by a factor of 1.7 [68]. Because the calibration factor was determined only for α´-martensite contents below 55 FE%, in the following diagrams the magnetically determined α´-martensite is indicated as ξ in FE% without calculating the vol% of α´-martensite. Furthermore, within one load cycle, the magnetic properties of α´-martensite are influenced by stress/strain due to the Villari effect (inverse magnetostriction), which describes a change of the magnetic susceptibility of ferromagnetic material due to mechanical stresses [26]. Therefore, an arithmetic mean value per load cycle of the measured Feritscope™ signal is given in the diagrams. For specimens loaded in LCF, HCF and VHCF tests at T = 300 °C and in VHCF tests at AT, ex situ

Feritscope™ measurements were performed. The fatigue tests at AT and T = 300 °C were total strain controlled in LCF regime and stress controlled in the HCF regime. The VHCF tests at T = 300 °C were stress controlled and the ultrasonic VHCF tests at AT performed in displacement-control. All tests were performed in symmetric push-pull conditions with load ratio R = −1.

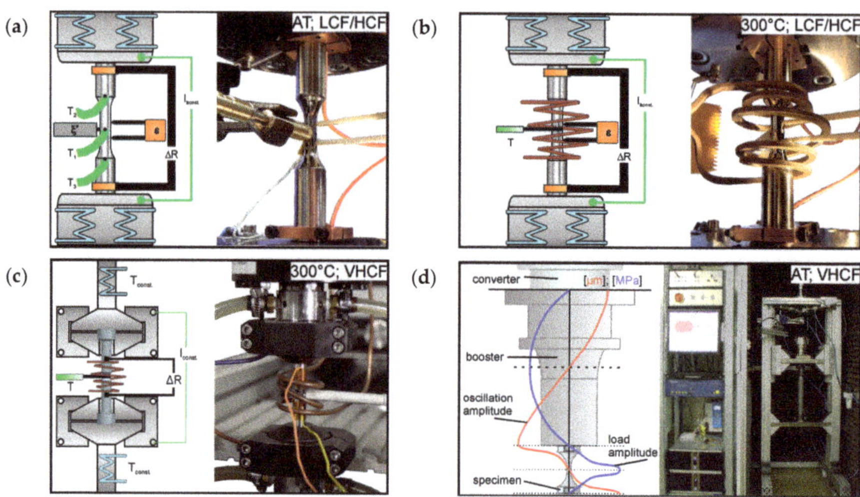

Figure 3. Schematic representation of the experimental setup for fatigue tests in the LCF and HCF regime at (**a**) AT and (**b**) at T = 300 °C, as well as in the VHCF regime at (**c**) T = 300 °C and (**d**) at AT.

VHCF testing of metastable austenitic stainless steels at ambient temperature using an ultrasonic fatigue system is more challenging than for stable metallic materials because of transient material behavior and significant self-heating of specimens. Due to the formation of high strength α′-martensite in softer austenite, less damping of the oscillation amplitude takes place, resulting in higher displacement amplitude. Therefore, an unstable displacement amplitude occurred. A representative pulse sequence from the VHCF fatigue test on a fully austenitic microstructure in its initial state and after cyclic loading up to $N \sim 2 \times 10^6$ clearly shows the challenge in performing the fatigue test with constant load amplitude (Figure 4). The first pulses were characterized by an oscillation plateau with a constant amplitude level during each pulse (Figure 4a). These results underline the necessity of a correct specimen design using FEM simulation to ensure fully reversed loading conditions with the maximum stress amplitude in the center of the gauge length and the maximum oscillation amplitude at the specimen´s end (see Figure 3d) [65]. However, due to fatigue-induced α′-martensite formation, the initial amplitude plateaus changed to pulses with an increasing displacement level (Figure 4b). This had to be leveled out by appropriate parameter adjustments during phase transformation. Further details can be found in further papers (see [65,69]).

Besides adjusting the proportional, integral and derivative (PID) parameters during VHCF tests, the deformation-induced specimen temperature increase has to be limited. Figure 5a shows the development of displacement for a pulse/pause ratio of 0.5 s/2.5 s, which is typically used for stable metallic materials [64], and which results in an effective load frequency f_{eff} = 3300 Hz. The progress of the specimen´s temperature is also plotted (Figure 5). It can be clearly seen that a pulse/pause ratio of 0.5 s/2.5 s cannot be used for fatigue testing of metastable austenite at ambient temperature because within two pulses the temperature increased to 200 °C (Figure 5a). To keep the specimen´s temperature below 50 °C, a change of temperature below 25 K and therefore a pulse/pause ratio of 0.06 s/2.94 s had to be used (Figure 5b). This led to an effective frequency of f_{eff} = 400 Hz. Theoretically, to achieve the limit of the number of cycles $N_l = 2 \times 10^9$ for a fatigue test with f_{eff} = 400 Hz, about 58 days would be required. In reality, with the development of α′-martensite, cyclic hardening of materials takes place, which reduces cyclic plasticity. Consequently, the development of the specimen's temperature decreases, and fatigue testing could be performed with a higher pulse/pause ratio with f_{eff} = 1650 Hz. The adjustment of PID parameters was also less critical given that saturation of α'-martensite was achieved in the cycle regime of N~10^8 (see Figure 10).

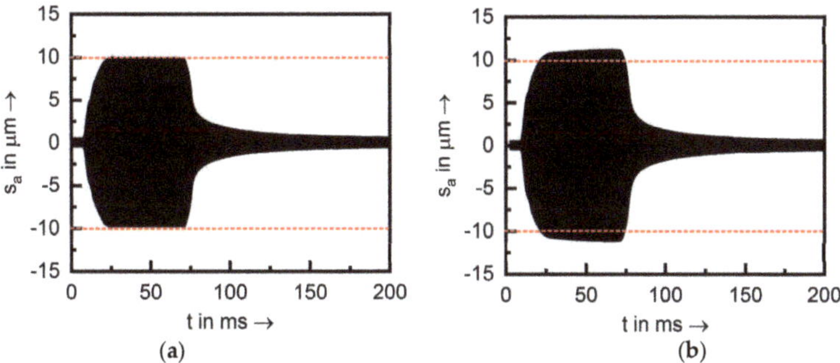

Figure 4. Displacement amplitude of batch A during VHCF testing at (**a**) the beginning of the fatigue test and (**b**) after fatigue-induced α'-martensite formation occurred without parameter adjustments.

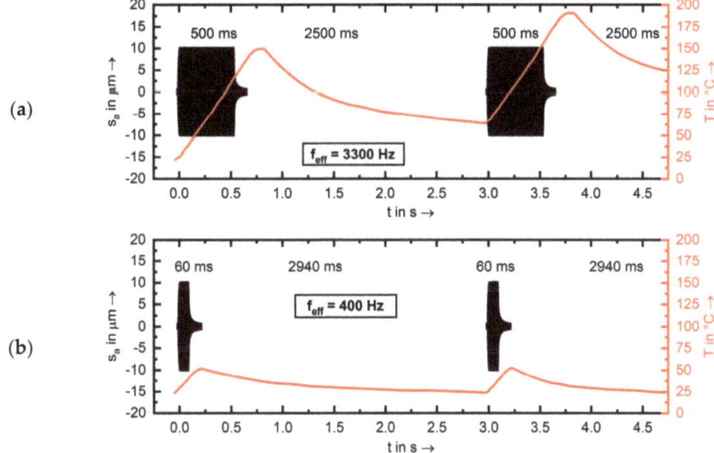

Figure 5. Development of the specimen's temperature using a pulse/pause ratio of (**a**) 0.5 s/2.5 s and (**b**) 0.06 s/2.94 s for AISI 347 batch A.

4. Results

Fatigue tests were performed with specimens from AISI 347 batch A taken from an original surge line pipe. Specimens from rolled bars from AISI 347 batch B represented the more metastable material and were used for selected fatigue tests to show the influence of metastability on the deformation and phase transformation behavior in the LCF regime as well as for characterizing HCF behavior at ambient temperature.

4.1. LCF Behavior

4.1.1. Ambient Temperature

Single step (constant amplitude) tests under total-strain control were performed in the LCF regime with total strain amplitudes 0.6% ≤ $\varepsilon_{a,t}$ ≤ 1.6%, load frequency f = 0.01 Hz and triangular waveform. Figure 6 presents the development of stress amplitude (σ_a), the change in specimen temperature (ΔT) and α'-martensite formation (ξ) for batch A and, additionally, the results from the test with total strain amplitude $\varepsilon_{a,t}$ = 1.0% for batch B. The cyclic deformation behavior of the investigated steel at ambient temperature was fundamentally determined by deformation-induced austenite-α'-martensite transformation. After a load-dependent number of cycles N, the formation of α'-martensite started and increased continuously with increasing cycle number until specimen failure (Figure 6c). The σ_a,N-curves (Figure 6a) illustrate the associated cyclic hardening processes, which led to a maximum stress amplitude for $\varepsilon_{a,t} \geq 1.2\%$ in the range of the tensile strength σ_f = 569 MPa (batch A) of the solution-annealed material. Cyclic hardening was detected by temperature measurement, temperature increases in total-strain controlled fatigue tests being indicative of increases of plastic strain energy. Because of very low load frequency, the maximal change in the specimen temperature was below 3 K, which corresponds with an absolute temperature of about 28 °C and consequently bears no significant influence on the formation of α'-martensite. Figure 6c shows the development of α'-martensite content during the fatigue tests discussed above. After an incubation period, which depended on the total strain amplitude, the α'-martensite volume fraction increased continuously with the number of cycles. With increasing $\varepsilon_{a,t}$, the onset of α'-martensite formation was shifted to lower N, and a higher volume fraction of α'-martensite was measured at specimen failure. To characterize the influence of metastability of nominally the same type of austenitic stainless AISI 347 steel, specimens from batches B were cyclically loaded with total strain amplitude $\varepsilon_{a,t}$ = 1%. Both batches A and B showed a continuous cyclic hardening, however, with different gradients after the first ten cycles. In the cycle range 10 < N < 100, batch A showed only a slight increase of stress amplitude, while batch B underwent a larger increase of σ_a, which correlated directly with a significant development of α'-martensite (Figure 6c) from 1 FE% to 41 FE%. As mentioned in the methods section above, ferromagnetic α'-martensite was measured in situ during fatigue testing using a Feritscope™ sensor, for which readings above 60 FE% could be inaccurate due to lack in instrument linearity and calibration difficulties at high α'-martensite contents. For this reason, measured data in the range above 60 FE% were plotted as dashed lines in Figure 6c. After a high rate of σ_a increase between N = 10 and N = 100, batch B showed a decrease in the stress amplitude rate $d\sigma_a/dN$. However, further cyclic hardening took place which led to specimen failure at σ_a = 760 MPa, which is 120 MPa higher than the tensile strength of this material in its as-received state (Table 2). Batch A also showed evidence of a correlation between the development of α'-martensite and an increase in stress amplitude. However, the start of α'-martensite formation occurred at a higher number of cycles in comparison to batch B and the α'-martensite content was found to be, at the same number of cycles, significantly lower with the maximum value at specimen failure ξ = 30 FE%. The results of these tests clearly depicted how substantial the differences can be between the formation of deformation-induced α'-martensite in the same type of austenitic stainless steel due to differences in chemical composition and grain size and consequently the material's metastability. It should be noted that both batches had a chemical composition within the range given in international standards for the investigated steel type.

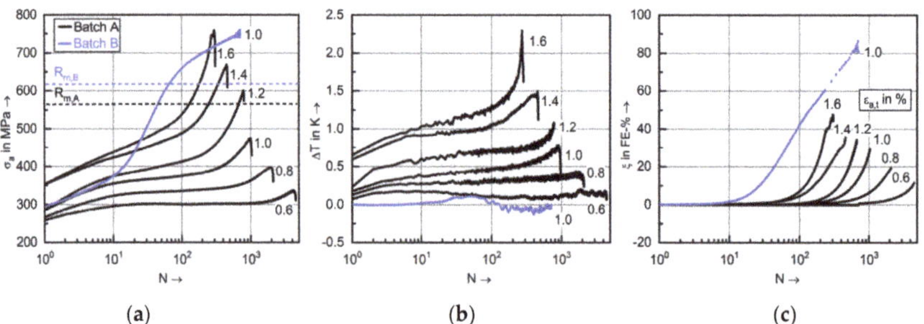

Figure 6. Development of (**a**) stress amplitude, (**b**) change in temperature and (**c**) α'-martensite versus N during total-strain controlled LCF test of batch A and B.

4.1.2. Elevated Temperature T = 300 °C

Single step total-strain control fatigue tests in the LCF regime at 300 °C were performed with total strain amplitudes $0.8\% \leq \varepsilon_{a,t} \leq 1.6\%$, a load frequency f = 0.01 Hz and triangular waveform. Figure 7 presents the development of the stress amplitude σ_a versus the number of cycles N. At 300 °C the cyclic deformation behavior of the investigated steel was characterized by initial cyclic hardening, followed by slight cyclic softening before the final stress amplitude drop associated with macro crack propagation. The microstructural changes due to cyclic plastic deformation are described elsewhere [46,63]. At these load parameters, no α′-martensite formation occurred and, compared to ambient temperature, the stress amplitude was significantly lower (compared Figure 6a with Figure 7).

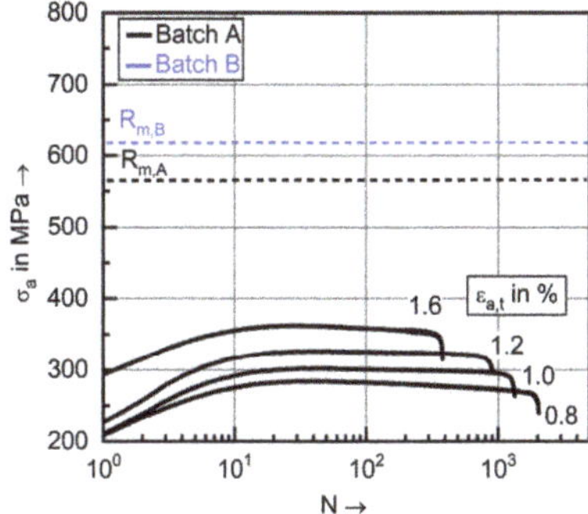

Figure 7. Development of stress amplitude σ_a versus load cycles N during total-strain controlled LCF test at T = 300 °C of batch A.

4.2. HCF Behavior

For fatigue tests in the HCF regime, stress-controlled single step tests were performed. Typically, in order to obtain the limited number of cycles N_l in the range of 10^6–10^7 cycles, test frequencies in the range 5 Hz ≤ f ≤ 20 Hz are used for metallic materials. For $N_l = 2 \times 10^6$ with a load frequency of f = 5 Hz, about 5 days, for $N_l = 2 \times 10^7$ with a load frequency f = 20 Hz, about 12 days would be required. Usually, higher test frequencies can be used for HCF tests at ambient temperature (AT) of metallic materials but when investigating the fatigue behavior of metastable austenitic stainless steels, self-heating of the specimens has to be taken into account. Even if external cooling of the specimen is undertaken, significant increases in specimen temperature cannot be suppressed [70,71]. Hence, in this study, test frequencies for fatigue tests at nominally ambient temperature were f = 5 Hz (batch B) and for T = 300 °C f = 20 Hz (batch A).

4.2.1. At Nominally Ambient Temperature

To characterize the cyclic deformation and transformation behavior during cyclic loading in the HCF regime, specimens taken from batch B were cyclically loaded in stress-controlled single step tests with stress amplitudes in the range 225 MPa ≤ σ_a ≤ 280 MPa. In Figure 8a the resulting developments of plastic strain amplitude $\varepsilon_{a,p}$ are plotted against the number of cycles N. The $\varepsilon_{a,p}$,N curves illustrate pronounced cyclic hardening after a short period of initial cyclic softening. Cyclic hardening is caused by an increase in dislocation and stacking fault density, as well as by formation of deformation-induced α′-martensite (Figure 8b). A positive influence of α′-martensite formation on fatigue life was observed, because austenite-α′-martensite transformation causes cyclic hardening, which significantly reduces plastic strain amplitude. Similarly, total-strain controlled fatigue tests in the LCF regime (Figure 6), after an incubation period dependent on stress amplitude, resulted in a continuously increasing volume fraction of α′-martensite over the number of cycles. With increasing stress amplitude, the onset of α′-martensite formation was shifted to lower N and higher α′-martensite fractions were measured at specimen failure. This development of α´-martensite in stress-controlled tests was comparable to the results given in [16] as well as for total-strain controlled fatigue tests shown in Figure 6c. However, with 1.7 FE% < ξ < 8.8 FE% the maximum values of α´-martensite at specimen failure/ultimate number of cycles were significantly lower compared to total strain-controlled LCF tests. The change in the specimen temperature during the fatigue tests is shown in Figure 8c. As in the LCF tests, changes in temperature were used to detect cyclic deformation behavior. However, in the stress-controlled fatigue tests, an increase in temperature correlated with an increase of the σ-ε hysteresis loop area, indicating cyclic softening, whereas a decrease in temperature correlated with a decrease of the σ-ε hysteresis loop area, and characterized cyclic hardening. Besides using ΔT to characterize cyclic deformation behavior, information about the general amount of self-heating is provided (Figure 8c). In consideration of Equation (4), a maximum specimen temperature of 102 °C was achieved while the value of ΔT signal shows 53 K. As is known, temperature has a significant influence on deformation-induced α´-martensite formation and consequently fatigue life. In literature [5,72–79] the fatigue life of metastable austenitic stainless steels is given as S-N curves obtained by fatigue testing with load frequencies up to 150 Hz. Obviously, specimen temperatures in HCF tests with frequency higher as 2 Hz were higher than ambient temperature; however, the information about the specimen temperatures is not at all times given. To obtain HCF results at true AT or at temperatures around 25 °C–30 °C, fatigue tests with significantly lower frequency, e.g. f = 0.2 Hz, have to be performed. This requires 115 days of test duration to achieve $N_L = 2 \times 10^6$, i.e. a duration that is impractical for systematic investigation of HCF behavior due to high resource expenses. In conclusion, to obtain representative results in the HCF regime in a reasonable time, it is necessary to perform fatigue tests of metastable austenite with variable test frequencies during single step fatigue tests, which effectively suppress self-heating of e.g. $\Delta T_{max} < 10$ K. Therefore, it has to be noted that the presented results (and many others given in literature [5,72–79], especially regarding the development of α´-martensite and its influence on fatigue life) are influenced by specimen temperature, which needs to be taken into consideration.

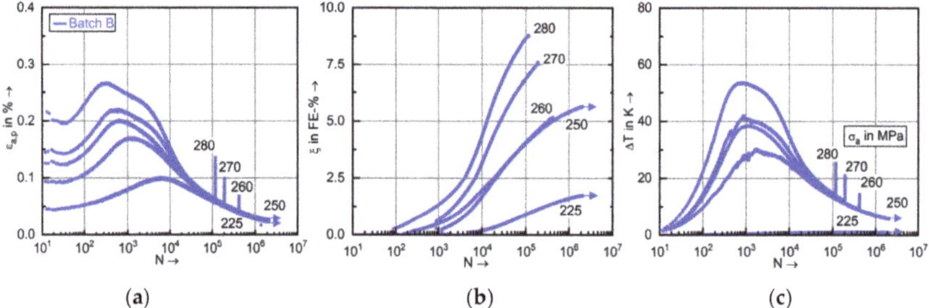

Figure 8. Development of (**a**) plastic strain amplitude $\varepsilon_{a,p}$, (**b**) α′-martensite and (**c**) change in the specimen's temperature ΔT versus load cycles N during HCF tests at AT with f = 5 Hz of batch B.

4.2.2. Elevated Temperature T = 300 °C

Figure 9 shows the development of plastic strain amplitude during fatigue loading of HCF specimens from batch A at 300 °C with a frequency f = 20 Hz. As these fatigue tests were performed at elevated temperature, a higher test frequency can be used to achieve a reasonable short testing time. At the beginning of the fatigue tests, cyclic softening was detected for all stress amplitudes followed by a saturation state for stress amplitudes 165 MPa ≤ σ_a ≤ 200 MPa and finally crack growth up to specimen failure. Inversely, the stress amplitude of 160 MPa, after initial softening up to about 2×10^4 cycles, led to cyclic hardening, which reduced plastic strain amplitude down to zero. Consequently, macroscopic elastic cyclic loading occurred and at this stress amplitude the limit number of cycles $N_l = 2 \times 10^7$ was achieved without failure. Ex situ magnetic Feritscope™ measurements quantified 0.13 FE% for this specimen and 0.00 FE% for all loadings > 160 MPa. That is, for only a small load amplitude at 300 °C, a change in the cyclic deformation behavior occurred from softening/saturation to cyclic hardening.

This change in cyclic deformation character was also observed during total strain-controlled fatigue tests and detected in situ using electromagnetic induced ultrasonic signals in the specimen [46,47,63]. At 300 °C, the investigated batch A showed no α′-martensite formation during monotonic tensile loading (see Table 2) as well as in LCF [46,47,63] and HCF regimes at "higher" loading amplitudes (see Figures 7 and 9). As mentioned in the introduction, the cyclic deformation behavior of metastable austenitic steels depends on load parameters such as stress amplitude, frequency, and temperature, and results in different microstructural changes from planar dislocation slip, over wavy slip, until the development of stacking faults, twins, as well as ε and α′-martensite (see. Figure 1). As the volume fraction of α′-martensite measured by Feritscope™ is relatively low (ξ = 0.13 FE%) further microstructural changes have to play a role in the cyclic hardening of metastable austenitic steel at low stress amplitude in fatigue testing at 300 °C, which are still not entirely investigated and explained.

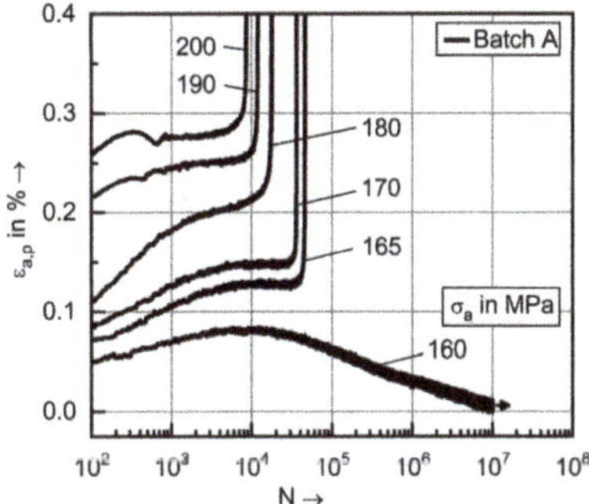

Figure 9. Development of plastic strain amplitude $\varepsilon_{a,p}$ versus load cycles N in HCF tests at T = 300 °C with f = 20 Hz.

4.3. VHCF Behavior

VHCF behavior was investigated at AT using an ultrasonic fatigue system (Figure 3d) at stress amplitudes 180 MPa ≤ σ_a ≤ 283 MPa, and a load frequency f = 20 kHz, while at T = 300 °C a servohydraulic test system was used and stress amplitudes 120 MPa ≤ σ_a ≤ 190 MPa were applied at a load frequency f = 980 Hz. All VHCF tests were performed on specimens from batch A.

4.3.1. Ambient Temperature

As σ-ε hysteresis measurements are (to date) not realizable during ultrasonic fatigue testing, and due to the macroscopically elastic behavior in the lower HCF/VHCF regime, the cyclic deformation behavior cannot be characterized using conventional data like $\varepsilon_{a,p}$, N curves. However, in situ measurement of further physical data, such as changes in the specimen temperature before and after a pulse sequence, dissipated energy and generator power, as well as "quasi in situ" measurement of magnetic properties, can be used to describe the cyclic deformation and transformation behavior of metastable austenites during fatigue testing in the VHCF regime. Figure 10 shows the aforementioned data and, in addition, the displacement amplitude as well as load frequency during an ultrasonic fatigue test with a stress amplitude of 250 MPa. After an approximately stationary phase up to N = 10^6, cyclic hardening occurred which could be detected by a strong increase in α′-martensite content. As mentioned before, and illustrated in Figure 4, α′-martensite formation promotes transient material behavior. To keep the displacement amplitude close to a constant value of 9.5 μm, corresponding to a stress amplitude of 250 MPa, the PID parameters of the ultrasonic testing system were stepwise readjusted. The cyclic hardening of the material resulted in lower self-heating illustrated by a decreasing change in temperature. This aspect enabled longer pulse and shorter pause times which resulted in a higher effective frequency. As described in Section 3, a stepwise increase of the pulse pause ratio up to 0.72 s/0.8 s, which represents an effective frequency f_{eff} = 1650 Hz, led to achieving the ultimate number of cycles within 25 days. After N = 1 × 10^8 load cycles, a lower α′-martensite formation rate dξ/dN occurred and until the limiting number of cycles a saturation state with a stabilized α′-martensite content of ξ = 2.2 FE% was reached. At the same time, temperature as well as displacement amplitude remained constant. In this phase, further adjustments of the displacement control were not necessary. The S-N curve resulting from single step tests is given in Figure 12c. Cyclic plastic deformation of metastable austenite at ambient temperature led to significant changes in phase distribution from single-phase austenitic to two-phase austenite/α′-martensite microstructures. At ambient temperature, and at all load amplitudes σ_a > 240 MPa, formation of α′-martensite in the range 0.3 FE% ≤ ξ ≤ 2.3 FE% took place. However, at smaller stress amplitudes σ_a < 240 MPa no deformation induced α′-martensite was measured. Fatigue failure only occurred in the HCF regime and no specimen failed in the VHCF regime beyond N = 10^7 load cycles. Accordingly, a true fatigue limit exists for metastable austenite [80,81].

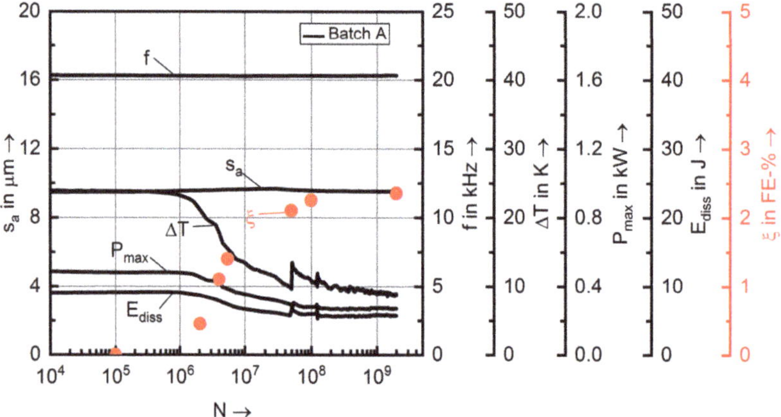

Figure 10. Displacement amplitude s_a, frequency f, temperature change ΔT, power P, dissipated energy E_{dis} and α′-martensite content ξ of batch A during VHCF test at σ_a = 250 MPa.

4.3.2. Elevated Temperature T = 300 °C

High precision stress-strain hysteresis measurement suitable to quantify microplastic deformations at low load amplitudes is, as with ultrasonic fatigue, impossible for tests using servohydraulic systems with a load frequency f = 980 Hz. The S-N Woehler curve resulting from constant amplitude fatigue tests at this frequency at 300 °C is given in Figure 12c. Similar to the HCF behavior at 300 °C at low stress amplitude, a very low volume fraction of α′-martensite was detected in the specimen which achieved the limit number of cycles N_l = 5 × 10^8. Since the Ferritscope™ sensor cannot be used in situ at this testing temperature, the kinetics of α′-martensite development was analyzed by an interrupted VHCF test with a stress amplitude of 160 MPa. At defined load cycles, ex situ Feritscope™ measurements were taken at the specimen surface at ambient temperature. Figure 11 shows no α′-martensite up to N = 1 × 10^7. The onset of α′-martensite formation occurred between N = 1 × 10^7 and N = 1 × 10^8. At N = 1 × 10^8 0.12 FE% and at N = 5 × 10^8 0.13 FE% were measured without specimen failure. As the measured volume fraction of α′-martensite was very low, further microstructural changes had to play a role in the cyclic hardening of metastable austenitic steel in the VHCF regime at 300 °C.

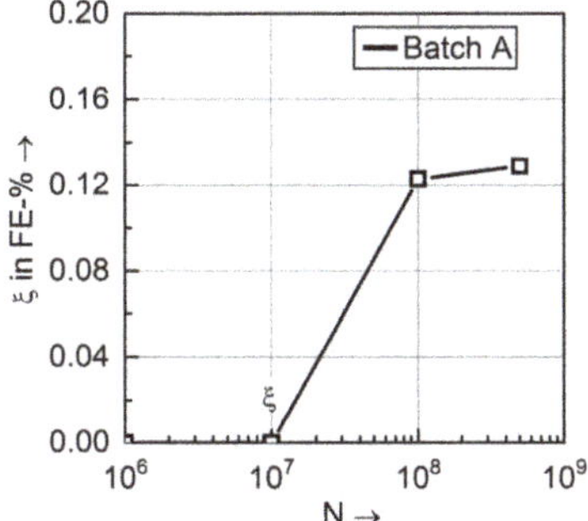

Figure 11. Development of α′-martensite during the interrupted VHCF test at σ_a = 160 MPa, f = 980 Hz and T = 300 °C.

5. Summary

Figure 12 summarizes the results from fatigue tests on AISI 347 in LCF, HCF and VHCF regimes at AT and 300 °C in the form of S-N curves. The influences of metastability of two different batches of AISI 347 on cyclic deformation behavior are clearly seen in LCF tests at ambient temperature (see Figure 6). The determined total strain-controlled fatigue life in the LCF regime at AT and 300 °C is similar (Figure 12a). However, significantly different cyclic deformation behavior was observed. At AT, cyclic hardening due to α′-martensite formation took place and the transformation from a single-phase austenitic to two-phase austenite/α′-martensite microstructure developed. The results epitomize a "dynamical composite material" with changing volume fraction and distribution of "reinforcements" during cyclic loading. At 300 °C in the LCF regime, no α′-martensite was measured and cyclic deformation behavior showed slight initial cyclic hardening followed by saturation/softening. Despite significantly different cyclic deformation behavior and resulting stress amplitudes in AT and T = 300 °C, respectively, similar fatigue lifetimes were estimated. However, the specimens loaded at AT achieved stresses higher than the ultimate tensile strength. A positive influence of α′-martensite formation on fatigue life was observed in stress-controlled fatigue tests in the HCF regime (Figure 12b) where austenite-α′-martensite transformation caused hardening, which significantly reduced the plastic strain amplitude and led to increased fatigue life. The α′-martensite formation occurred in all fatigue tests at AT. At 300 °C, albeit specimens loaded with low stress amplitude showed a very small volume fraction of α′-martensite. These specimens achieved the limit number of cycles without failure. A comparison of the S-N curves from tests at AT with T = 300 °C clearly showed a decrease in fatigue strength with an increase of temperature (Figure 12b). Similar behavior was observed for VHCF tests at AT and 300 °C (Figure 12c). At AT, higher fatigue strength existed, which correlated with the development of α′-martensite at each load amplitude. At 300 °C, only specimens with the formation of a low volume fraction of α′-martensite achieved the limit of the number of cycles without failure.

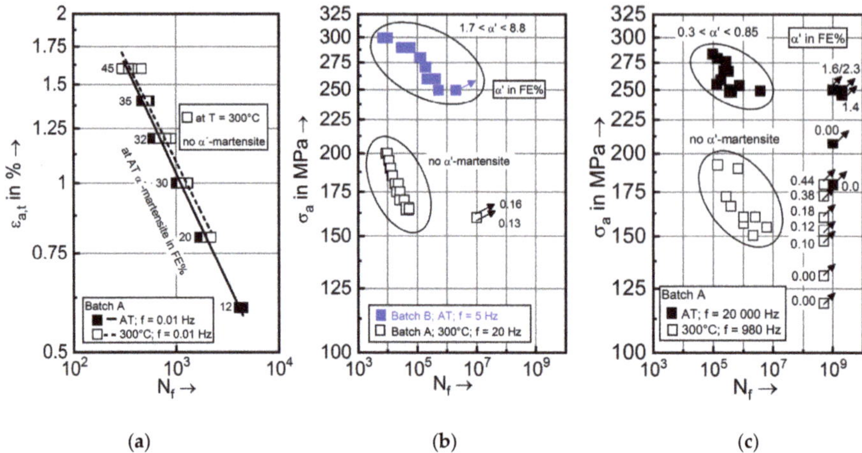

(a) (b) (c)

Figure 12. (**a**) $\varepsilon_{a,t}$ –N curve in LCF regime of batch A at AT and T = 300 °C, (**b**) S-N curve in HCF regime of batch A at T = 300 °C and batch B at AT, (**c**) S-N curve in VHCF regime of batch A at AT and T = 300 °C.

6. Conclusions

The present study characterized fatigue behavior of metastable austenitic stainless steels in the LCF, HCF and VHCF regimes at two temperatures, i.e. ambient temperature and 300 °C. Based on the acquired results, the following conclusions were drawn:

- The metastability of austenitic Cr-Ni steels can, even within the specifications given in international standards, e.g. for AISI 347 stainless steel, strongly vary due to relatively small changes in chemical composition.
- Different metastability of austenitic stainless steel affects the cyclic deformation and, consequently, the fatigue lifetime behavior.
- This is caused by the fact that during fatigue, various deformation (planar or wavy slip character, formation of stacking faults, twinning) and transformation ($\gamma \rightarrow \varepsilon$, $\gamma \rightarrow \varepsilon \rightarrow \alpha'$, $\gamma \rightarrow \alpha'$) mechanisms with different peculiarity take place.
- The investigated metastable austenitic stainless steel AISI 347 showed a true fatigue limit in the VHCF regime at AT and 300 °C.
- Further microstructural investigations have to be performed with focus on the development of deformation/transformation mechanism maps dependent on metastability parameters, e.g. stacking fault energy and loading parameters, in the LCF, HCF and VHCF regimes in accordance with deformation maps known in the literature [52–54] for stable materials.

Author Contributions: M.S. planned the experimental design and wrote the majority of the present paper; A.B., T.D., R.S. and A.S. realized experimental investigations and analysis of the given results; M.S. and T.B. supervision the work.

Funding: This research was funded by Federal Ministry for Economic Affairs and Energy (BMWi), Germany and the German Research Foundation (DFG).

Acknowledgments: The authors thank BMWi as well as the DFG for financial support as part of the CRC 926 project number 172116086 "Microscale Morphology of Component Surfaces".

Conflicts of Interest: The authors declare no conflict of interest.

References

1. Strauß, B.; Maurer, E. Die hochlegierten Chrom-Nickel-Stähle als nichtrostende Stähle. *Kruppsche Monatshefte* **1920**, *1*, 129–146.
2. Lo, K.H.; Shek, C.H.; Lai, J.K.L. Recent developments in stainless steels. *Mater. Sci. Eng.* **2009**, *R65*, 39–104. [CrossRef]
3. Leuk Lai, J.K.; Lo, K.H.; Shek, C.H. *Stainless Steels: An Introduction and Their Recent Developments*; Bentham Science Publishers: Sharjah, United Arab Emirates, 2012.
4. Marshall, P. Austenitic stainless steels: Microstructure and Mechanical Properties. Springer Science & Business Media: Berlin, Germany, 1984.
5. Chopra, O.K.; Shack, W.J. Effect of LWR coolant environments on the fatigue life of reactor materials. In *Final Report, NUREG/CR-6909*; Argon National Laboratory: Lemont, IL, USA, 2007.
6. Deutsche Edelstahlwerke Rost-, säure- und hitzebeständige Stähle. 2010. Available online: https://www.dew-stahl.com/fileadmin/files/dew-stahl.com/documents/Publikationen/Broschueren/014_DEW_RSH_D.pdf (accessed on 30 May 2019).
7. Hull, F.C. Delta ferrite and martensite formation in stainless steels. *Weld. J. Res. Suppl.* **1973**, *179*, 193–203.
8. Padilha, A.F.; Martorano, M.A. Predicting delta ferrite content in stainless steel castings. *ISIJ INT* **2012**, *52*, 1054–1065.
9. Padilha, A.F.; Tavares, C.F.; Martorano, M.A. Delta ferrite formation in austenitic stainless steel castings. *Mater. Sci. Forum* **2013**, *730–732*, 733–738. [CrossRef]
10. Graham, C.D.; Loren, B.E. Delta ferrite is ubiquitous in type 304 stainless steel: Consequences for magnetic characterization. *J. Magn. Magn. Mater.* **2018**, *458*, 15–18. [CrossRef]
11. Schäffler, A.L. Constitution diagram for stainless steel weld metal. *Met. Prog.* **1949**, *56*, 608–680B.
12. Becker, H.; Brandis, H.; Küppers, W. Zur Verfestigung instabil austenitischer nichtrostender Stähle und ihre Auswirkung auf das Umformverhalten von Feinblechen. *Thyssen Edelst. Techn. Berichte* **1986**, *12*, 35–54.
13. Eichelman, G.H.; Hull, F.C. The effect of composition in the temperature of spontaneous transformation of austenite to martensite in 18-8 type stainless steel. *Trans. ASM* **1952**, *45*, 77–104.
14. Angel, T. Formation of martensite in austenitic stainless steels—Effects of deformation, temperature and composition. *J. Iron. Steel Inst.* **1954**, *177*, 165–174.
15. Man, J.; Smaga, M.; Kuběna, I.; Eifler, D.; Polák, J. Effect of metallurgical variables on the austenite stability in fatigued AISI 304 type steels. *Eng. Fract. Mech.* **2017**, *185*, 139–159. [CrossRef]
16. Smaga, M.; Boemke, A.; Daniel, T.; Klein, M.W. Metastability and fatigue behavior of austenitic stainless steels. *MATEC Web Conf.* **2018**, *165*, 04010. [CrossRef]
17. Olson, G.B.; Cohen, M. A general mechanism of martensitic nucleation: Part I. General concepts and the FCC → HCP transformation. *Metall. Trans. A* **1976**, *7*, 1897–1904.
18. Olson, G.B.; Cohen, M. A general mechanism of martensitic nucleation: Part II. General concepts and the FCC → HCP transformation. *Metall. Trans. A* **1976**, *7*, 1905–1914.
19. Olson, G.B.; Cohen, M. A general mechanism of martensitic nucleation: Part III. General concepts and the FCC → HCP transformation. *Metall. Trans. A* **1976**, *7*, 1915–1923. [CrossRef]
20. Schumann, H. Verformungsinduzierte Martensitbildung in metastabilen austenitischen Stählen. *Krist. Tech.* **1975**, *10*, 401–411. [CrossRef]
21. Sinclair, C.W.; Hoagland, R.G. A molecular dynamics study of the fcc -> bcc transformation at fault intersections. *Acta Mater.* **2008**, *56*, 4160–4171. [CrossRef]
22. Shimizu, K.; Tanaka, Y. The $\gamma \rightarrow \varepsilon \rightarrow \alpha'$ martensitic transformations in an Fe–Mn–C Alloy. *Trans. Jpn. Inst. Met.* **1978**, *19*, 685–693. [CrossRef]
23. Spencer, K.; Embury, J.D.; Conlon, K.T.; Véron, M.; Bréchet, Y. Strengthening via the formation of strain-induced martensite in stainless steels. *Mater. Sci. Eng. A* **2004**, *387–389*, 873–881. [CrossRef]
24. Venables, J.A. Martensite transformation in stainless steels. *Philo. Mag.* **1962**, *7*, 35–44. [CrossRef]
25. Scheil, E. Uber die Umwandlung des Austenits in Martensit in Eisen-Nickellegierungen unter Belastung. *Z. Anorg. Allg. Chem.* **1932**, *207*, 21–40. [CrossRef]
26. Smaga, M.; Eifler, D. Fatigue life calculation of metastable austenitic stainless steels on the basis of magnetic measurements. *Mater. Test.* **2009**, *51*, 370–375. [CrossRef]

27. Biermann, H.; Solarek, J.; Weidner, A. SEM Investigation of High-Alloyed Austenitic Stainless Cast Steels With Varying Austenite Stability at Room Temperature and 100 degrees C. *Steel Res. Int.* **2012**, *83*, 512–520. [CrossRef]
28. Hahnenberger, F.; Smaga, M.; Eifler, D. Influence of γ-α'-phase transformation in metastable austenitic steels on the mechanical behavior during tensile and fatigue loading at ambient and lower temperatures. *Adv. Eng. Mater.* **2012**, *14*, 853–858. [CrossRef]
29. Hahnenberger, F.; Smaga, M.; Eifler, D. Microstructural investigation of the fatigue behavior and phase transformation in metastable austenitic steels at ambient and lower temperatures. *Int J. Fatigue* **2014**, *69*, 36–48. [CrossRef]
30. Smaga, M.; Hahnenberger, F.; Sorich, A.; Eifler, D. Cyclic deformation behavior of austenitic steels in the temperature range -60 C $\leq$ T $\leq$ 550 °C. *Key Eng. Mater.* **2011**, *465*, 439–442. [CrossRef]
31. Curtze, S.; Kuokkala, V.T.; Oikari, A.; Talonen, J.; Hänninen, H. Thermodynamic modeling of the stacking fault energy of austenitic steels. *Acta Mater.* **2011**, *59*, 1068–1076. [CrossRef]
32. Jun, J.H.; Choi, C.-S. Variation of stacking fault energy with austenite grain size and its effect on the M_S temperature of $\gamma \rightarrow \varepsilon$ martensitic transformation in Fe–Mn alloy. *Mater. Sci. Eng. A* **1998**, *257*, 353–356. [CrossRef]
33. Llewellyn, D.T. Work hardening effects in austenitic stainless steels. *Mater. Sci. Technol.* **1997**, *13*, 389–400. [CrossRef]
34. Martin, S.; Fabrichnaya, O.; Rafaja, D. Prediction of the local deformation mechanisms in metastable austenitic steels from the local concentration of the main alloying elements. *Mater. Lett.* **2015**, *159*, 484–488. [CrossRef]
35. Otte, H.M. The transformation of stacking faults in austenite and its relation to martensite. *Acta Metall.* **1957**, *5*, 614–627. [CrossRef]
36. Rhodes, C.G.; Thompson, A.W. The composition dependence of stacking fault energy in austenitic stainless steels. *Metall. Mater. Trans. A* **1977**, *8*, 1901–1906. [CrossRef]
37. Rémy, L.; Pineau, A.; Thomas, B. Temperature dependence of stacking fault energy in close-packed metals and alloys. *Mater. Sci. Eng.* **1978**, *36*, 47–63. [CrossRef]
38. Shen, Y.F.; Li, X.X.; Sun, X.; Wang, Y.D.; Zuo, L. Twinning and martensite in a 304 austenitic stainless steel. *Mater. Sci. Eng A* **2012**, *552*, 514–522. [CrossRef]
39. Schramm, R.E.; Reed, R.P. Stacking-fault energies of 7 commercial austenitic stainless steels. *Metall. Trans. A* **1975**, *6*, 1345–1351. [CrossRef]
40. Basu, K.; Das, M.; Bhattacharjee, D.; Chakraborti, P.C. Effect of grain size on austenite stability and room temperature low cycle fatigue behavior of solution annealed AISI 316LN austenitic stainless steel. *Mater. Sci. Technol.* **2007**, *23*, 1278–1284. [CrossRef]
41. Gonzáles, J.L.; Aranda, R.; Jonapá, M. The influence of grain size on the kinetics of strain induced martensite in type 304 stainless steel. In *Applications of Stainless Steel*; Nordberg, H., Björklund, J., Eds.; ASM International: Stockholm, Sweden, 1992; pp. 1009–1016.
42. Shrinivas, V.; Varma, S.K.; Murr, L.E. Deformation-Induced martensitic characteristics in 304 stainless and 316 stainless steels during room temperature rolling. *Metall. Mater. Trans. A* **1995**, *26*, 661–671. [CrossRef]
43. Yang, H.S.; Bhadeshia, H.K.D.H. Austenite grain size and the martensite–Start temperature. *Scr. Mater.* **2009**, *60*, 493–495. [CrossRef]
44. Bayerlein, M.; Christ, H.J.; Mughrabi, H. Plasticity-induced martensitic transformation during cyclic deformation of AISI 304L stainless steel. *Mater. Sci. Eng. A* **1989**, *114*, L11–L16. [CrossRef]
45. Smaga, M.; Walther, F.; Eifler, D. Deformation-induced martensitic transformation in metastable austenitic steels. *Mater. Sci. Eng. A* **2008**, *483–484*, 394–397. [CrossRef]
46. Sorich, A.; Smaga, M.; Eifler, D. Fatigue monitoring of austenitic steels with electromagnetic acoustic transducers (EMATs). *Mater. Performance Charact.* **2014**, *4*, 263–274. [CrossRef]
47. Sorich, A.; Smaga, M.; Eifler, D. Influence of cyclic deformation induced phase transformation on the fatigue behavior of the austenitic steel X6CrNiNb1810. *Adv. Mater. Res.* **2014**, *891–892*, 1231–1236. [CrossRef]
48. Das, A.; Sivaprasad, S.; Ghosh, M.; Chakraborti, P.C.; Tarafder, S. Morphologies and characteristics of deformation induced martensite during tensile deformation of 304 LN stainless steel. *Mater. Sci. Eng. A* **2008**, *486*, 283–286. [CrossRef]

49. Murr, L.E.; Staudhammer, K.P.; Hecker, S.S. Effects of strain state and strain rate on deformation-induced transformation in 304 stainless steel: part ii. microstructural study. *Metall. Trans. A* **1982**, *13*, 627–635. [CrossRef]
50. Staudhammer, K.P.; Murr, L.E.; Hecker, S.S. Nucleation and evolution of strain-induced martensitic (bcc) embryos and substructure in stainless-steel—A transmission electron-microscope study. *Acta Metall.* **1983**, *31*, 267–274. [CrossRef]
51. Hecker, S.S.; Stout, M.G.; Staudhammer, K.P.; Smith, J.L. Effects of strain state and strain rate on deformation-induced transformation in 304 stainless steel: Part I. Magnetic measurements and mechanical behavior. *Metall. Trans. A* **1982**, *13*, 619–626. [CrossRef]
52. Klesnil, M.; Lukas, P. Fatigue of metallic materials. *Mater. Sci. Monogr.* **1980**, *7*.
53. Lukáš, P.; Klesnil, M. Dislocation structures in fatigued single crystals of Cu-Zn system. *Phys. Stat. Sol. (a)* **1971**, *5*, 247–258. [CrossRef]
54. Mughrabi, H. Fatigue, an everlasting materials problem—still en vogue. *Proc. Eng. 2* **2010**, *2*, 3–26. [CrossRef]
55. Karaman, I.; Sehitoglu, H.; Chumlyakov, Y.I.; Maier, H.J. The deformation of low-stacking-fault-energy austenitic steels. *JOM* **2002**, *7*, 31–37. [CrossRef]
56. Lee, T.-H.; Shin, E.; Oh, C.-S.; Ha, H.-Y.; Kim, S.-J. Correlation between stacking fault energy and deformation microstructure in high-interstitial-alloyed austenitic steels. *Acta Mater.* **2010**, *58*, 3173–3186. [CrossRef]
57. Xie, X.; Ninga, D.; Suna, J. Strain-controlled fatigue behavior of cold-drawn type 316 austenitic stainless steel at room temperature. *Mater. Charact.* **2016**, *120*, 195–202. [CrossRef]
58. Nebel, T. Verformungsverhalten und Mikrostruktur zyklisch beanspruchter metastabiler austenitischer Stähle. Ph.D. Thesis, Technical University of Kaiserslautern, Kaiserslautern, Germany, September 2002.
59. Bassler, H.J. Wechselverformungsverhalten und verformungsinduzierte Martensitbildung bei dem metastabilen austenitischen Stahl X6CrNiTi1810. Ph.D. Thesis, Technical University of Kaiserslautern, Kaiserslautern, Germany, December 1999.
60. *German version EN 10088-1:2014 Stainless steels – Part 1: List of stainless steels*; DIN Deutsches Institut für Normung e. V.: Berlin, Germany, 2014.
61. Ausschuss, K. *Sicherheitstechnische Regel des KTA – KTA 3204 Reaktordruckbehälter-Einbauten*; Carl Heymanns Verlag: Cologne, Germany, 2015.
62. Smaga, M.; Sorich, A.; Eifler, D.; Beck, T. Very high cycle fatigue behavior of metastable austenitic steel X6CrNi1810 at 300 °C. In Proceedings of the 7th Internaltional Conference on Very High Cycle Fatigue (VHCF-7), Dresden, Germany, 3–5 July 2017.
63. Sorich, A. Charakterisierung des Verformungs- und Umwandlungsverhaltens des zyklisch beanspruchten metastabilen austenitischen Stahls X6CrNiNb1810. Ph.D. Thesis, Technical University of Kaiserslautern, Kaiserslautern, Germany, July 2006.
64. Ritz, F.; Beck, T. Influence of mean stress and notches on the very high cycle fatigue behavior and crack initiation of a low-pressure steam turbine steel. *Fatigue Fract. Eng. Mater. Struct.* **2017**, *40*, 1762–1771. [CrossRef]
65. Daniel, T.; Boemke, A.; Smaga, M.; Beck, T. Investigations of very high cycle fatigue behavior of metastable austenitic steels using servohydraulic and ultrasonic testing systems. In Proceedings of the ASME 2018 Pressure Vessels and Piping Conference, Prague, Czech Republic, 15–20 July 2018.
66. Biallas, G.; Piotrowski, A.; Eifler, D. Cyclic stress-strain, stress-temperature and stress-electrical resistance response of NiCuMo alloyed sintered steel. *Fatigue Fract. Eng. Mat. Struct.* **1995**, *18*, 605–615. [CrossRef]
67. Piotrowski, A.; Eifler, D. Characterization of cyclic deformation behaviour by mechanical, thermometrical and electrical methods. *Materialwiss. Werkstofftech.* **1995**, *26*, 121–127.
68. Talonen, J.; Aspegren, P.; Hänninen, H. Comparison of different methods for measuring strain induced alpha '-martensite content in austenitic steels. *Mater. Sci. Technol.* **2004**, *20*, 1506–1512. [CrossRef]
69. Boemke, A.; Smaga, M.; Beck, T. Influence of surface morphology on the very high cycle fatigue behavior of metastable and stable austenitic Cr-Ni steels. *MATEC Web Conf.* **2018**, *165*, 20008. [CrossRef]
70. Nikitin, B.; Besel, M. Effect of low-frequency on fatigue behavior of austenitic steel AISI 304 at room temperature and 25 °C. *Int. J. Fatigue* **2008**, *30*, 2044–2049. [CrossRef]
71. Pessoa, D.F.; Kirchhoff, G.; Zimmermann, M. Influence of loading frequency and role of surface micro-defects on fatigue behavior of metastable austenitic stainless steel AISI 304. *Int. J. Fatigue* **2017**, *103*, 48–59. [CrossRef]

72. De Backer, F.; Schoss, V.; Maussner, G. Investigations on the evaluation of the residual fatigue life-time in austenitic stainless steels. *Nucl. Eng. Des.* **2001**, *206*, 201–219. [CrossRef]
73. Le Duff, J.-A.; Lefrançois, A.; Meyzaud, Y.; Vernot, J.-Ph.; Martin, D.; Mendez, J.; Lehericy, Y. High cycle thermal fatigue issues in PWR nuclear power plants, lifetime improvement of some austenitic stainless steel components. *Metall. Res. Technol.* **2007**, *104*, 156–162.
74. Laamouri, A.; Sidhom, H.; Braham, C. Evaluation of residual stress relaxation and its effect on fatigue strength of AISI 316L stainless steel ground surfaces: Experimental and numerical approaches. *Int J. Fatigue* **2013**, *48*, 109–121. [CrossRef]
75. Vincent, L.; Le Roux, J.-C.; Taheri, S. On the high cycle fatigue behavior of a type 304L stainless steel at room temperature. *Int. J. Fatigue* **2012**, *38*, 84–91. [CrossRef]
76. Puchi-Cabrera, E.S.; Staia, M.H.; Tova, C.; Ochoa-Pérez, E.A. High cycle fatigue behavior of 316L stainless steel. *Int. J. Fatigue* **2008**, *30*, 2140–2146. [CrossRef]
77. Roth, I.; Kübbeler, M.; Krupp, U.; Christ, H.-J.; Fritzen, C.-P. Crack initiation and short crack growth in metastable austenitic stainless steel in the high cycle fatigue regime. *Proc. Eng.* **2010**, *2*, 941–948. [CrossRef]
78. Soo, P.; Chow, J.G.Y. High cycle fatigue behavior of solution annealed and thermally aged Type 304 stainless steel. *J. Eng. Mater. Technol.* **1980**, *102*, 141–146. [CrossRef]
79. Herrera-Solaz, V.; Niffenegger, M. Application of hysteresis energy criterion in a microstructure-based model for fatigue crack initiation and evolution in austenitic stainless steel. *Int. J. Fatigue* **2017**, *100*, 84–93. [CrossRef]
80. Müller-Bollenhagen, C.; Zimmermann, M.; Christ, H.-J. Adjusting the very high cycle fatigue properties of a metastable austenitic stainless steel by means of the martensite content. *Proc. Eng.* **2010**, *2*, 1663–1672. [CrossRef]
81. Grigorescu, A.; Kolyshkin, A.; Zimmermann, M.; Christ, H.-J. Effect of martensite content and geometry of inclusions on the VHCF properties of predeformed metastable austenitic stainless steels. *Procedia Struct. Integrity* **2016**, *2*, 1093–1100. [CrossRef]

Article

The Effect of Rapid Heating and Fast Cooling on the Transformation Behavior and Mechanical Properties of an Advanced High Strength Steel (AHSS)

Juan Pablo Pedraza [1], Rafael Landa-Mejia [2], Omar García-Rincon [1] and C. Isaac Garcia [2,*]

[1] Ternium-Mexico, 66450 San Nicolas de los Garza, Nuevo Leon, Mexico; JPEDRAZA@ternium.com.mx (J.P.P.); ogarciar@ternium.com.mx (O.G.-R.)

[2] Ferrous Physical Metallurgy Group, Mechanical Engineering and Materials Science Department, University of Pittsburgh, Pittsburgh, PA 15261, USA; ral76@pitt.edu

* Correspondence: cigarcia@pitt.edu

Received: 11 March 2019; Accepted: 7 May 2019; Published: 10 May 2019

Abstract: The major goal of this work was to study the effect of rapid heating and fast cooling on the transformation behavior of 22MnB5 steel. The effect of the initial microstructure (ferrite + pearlite or fully spheroidized) on the transformation behavior of austenite (during intercritical and supercritical annealing) in terms of heating rates (2.5, 30 & 200 °C/s) and fast cooling, i.e., 300 °C/s rate, were studied. As expected, the kinetics of austenite nucleation and growth were strongly related to the heating rates. Similarly, the carbon content of the austenite was higher at lower intercritical annealing temperatures, particularly when slower heating rates were used. The supercritical temperatures used in this study were similar to those used during commercial hot stamping operations, i.e., 845 and 895 °C, respectively, followed by a fast cooling rate. The prior austenite grain size (PAGS) was not strongly influenced by the nature of the initial microstructure, heating rate, reheating temperatures (845 or 895 °C), at 30 s holding time. The decomposition of austenite using fast cooling rates was examined. The results showed that 100% martensite was not obtained. The observed low temperature transformation products consisted of mixtures of martensite-bainite plus undissolved Fe_3C carbides and small amounts of martensite-austenite (M-A). At higher supercritical temperatures, i.e., 1000 °C and 1050 °C, the final microstructure showed an increase in the volume fraction of martensite and a decrease in the volume fraction of bainite. The Fe_3C and the M-A microconstituent were not observed. The best combination of tensile properties was obtained on samples reheated in the lower temperature range (845 to 895 °C). Interestingly, when the samples where reheated at the higher temperature range (1000 to 1050 °C) and fast cooled, the results of the mechanical properties did not exhibit significantly higher strength levels independent of heating rate or initial microstructural condition. This can be attributed to the change in the microstructural balance %martensite+%bainite as the reheating temperature increases. The results of this study are presented and discussed.

Keywords: EBSD-IQ; fast heating rate; formation of austenite; initial microstructure; PAGS; transformation behavior; tensile properties

---

1. Introduction

The continuous demand of using AHSS and UHSS steels, particularly Press Hardenable Steels (PHS), during hot stamping operations requires a systematic and fundamental understanding of the physical metallurgy of these steels. According to recent report on the use of AHSS and UHSS in North America, light weight vehicles are predicted to increase from the current 332 pounds per vehicle to 483 pounds by the year 2025 [1]. To achieve the desired performance of AHSS many studies are being conducted by universities and steel companies around the world. For example, Figure 1 shows the different pathways currently being studied to increase the strength and ductility of AHSS [1]. In addition, this figure also presents actual coating data for AHSS GEN 1 and 3 steels. It seems that the major research trends to increase the performance of AHSS strength-ductility are strongly related with the ability to control composition, microstructure and dislocation motion.

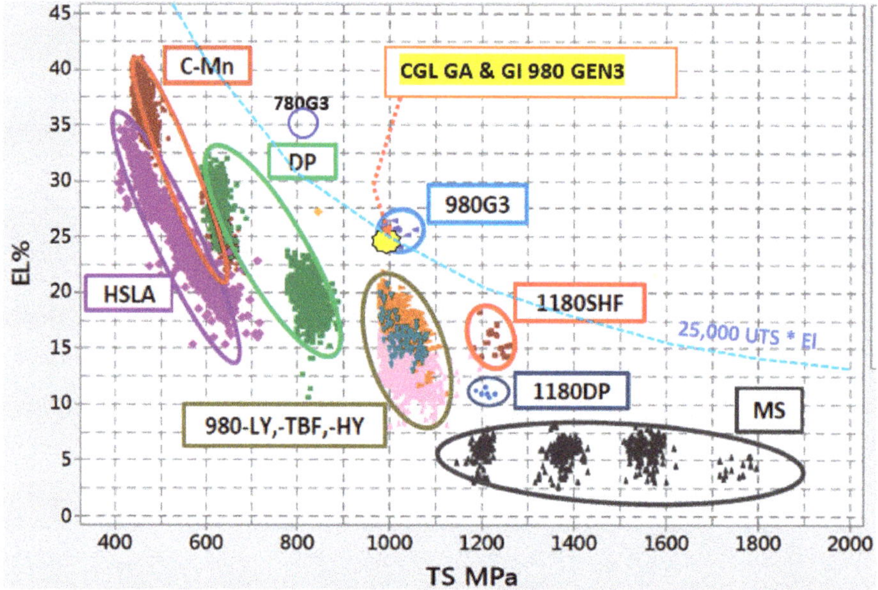

Figure 1. Different pathways to increase strength-ductility of AHSS [1].

Rapid Austenitization

According to Stahl Zentrum Stahl fur nachhaltige Mobilitat [2] hot stamped parts used in autos represent approximately 30% of all the body components. Mori et.al. [3], suggest that the use of high energy input during the process of hot stamping becomes an interesting proposition due to the effects of rapid heating on; (1) the transformation behavior of austenite; (2) oxidation behavior; (3) limited decarburization; (4) fine and/or coarser austenite grain sizes depending on reheating temperatures, holding times and heating rates; (5) localized non-uniform microstructures with substantial chemical heterogeneity; and (6) good hot workability. In summary, rapid heating and cooling during hot stamping seems to provide both a robust processing scheme and more cost effective approach than traditional reheating methods. Furthermore, the use of high input energy also opens the door to the potential development of new AHSS with higher strength and better total elongation.

Rapid heating technology is not new, the concept was initially developed to increase the mechanical properties of Armor plates for the US Army in 2010 [4]. Later on, the process was patented as a Micro-treatment of Iron-Based Alloy and Microstructure Resulting Therefrom [4]. It seems that by using rapid heating it is feasible to produce significant grain refinement of the decomposition products of austenite. There is substantial evidence that rapid hot forming based on high power and process integrated heating methods is an attractive proposition for the development of high strength steels with good formability [5–9]. Despite these claims, the state-of the-art in the development of AHSS steels using rapid heating seems to lack a better understanding on the effect of the starting microstructural condition and ultra-fast heating techniques. This is important because permits the better understanding of the formation and subsequent transformation behavior of austenite. To explore the effects of low and high input energy, starting microstructural condition, rapid cooling and resulting mechanical properties, the present study was conducted on a conventional 22MnB5 AHSS steel grade.

2. Experimental Procedure

The samples used in this investigation were sectioned from the hot band condition of a commercial strip of 22MnB5 steel coiled around 660 °C having a final thickness of 2.5 mm. The chemical composition of the steel is shown in Table 1.

Table 1. Chemical composition of 22MnB5 steel (Wt. %).

Element	C	Mn	Si	Ti	B	N	Cr
Wt. %	0.234	1.51	0.124	0.025	0.0022	0.0048	0.011

Other elements: Al, S, and P.

2.1. Microstructural Starting Condition

Prior to any experimental testing, coupons of approximately 25.4 cm × 25.4 cm × 2.5 mm in size were sectioned from the hot band condition. The initial microstructural condition in the hot band was ferrite + pearlite. A number of coupons were subjected to subcritical spheroidization treatments (675 °C for 10 h.). The two initial microstructural conditions are illustrated in Figure 2.

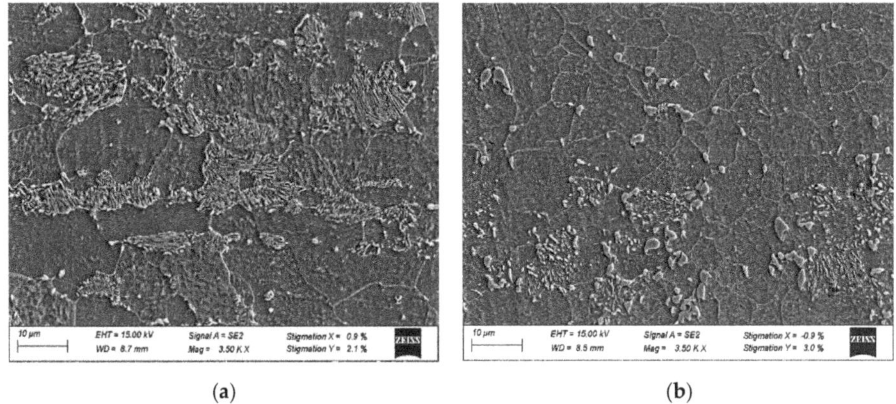

(a) (b)

Figure 2. SEM micrographs of initial microstructures (**a**) ferrite + pearlite and (**b**) spheroidized.

2.2. Heat Treatments

Samples having the initial microstructural conditions were subjected to intercritical and supercritical treatments using a continuous annealing line (CAL) induction 30 kW, 3-phase 480 V laboratory simulator with automated computer control for heating and cooling. This system is capable of reheating a sample 1.9 cm in thickness from RT (room temperature) to 1400 °C in 2.5 s and capable of rapid controlled cooling at 300 °C/s. The samples used in this study were 25.4 cm in length × 2.54 cm in width × 0.25 cm in thickness. A schematic of some of the heat treatments and cooling conditions used in this study are shown in Figure 3. The range of heat treatments studied included the temperatures of 732, 756, 772, 792, 845, 895, 1000 and 1050 °C, respectively.

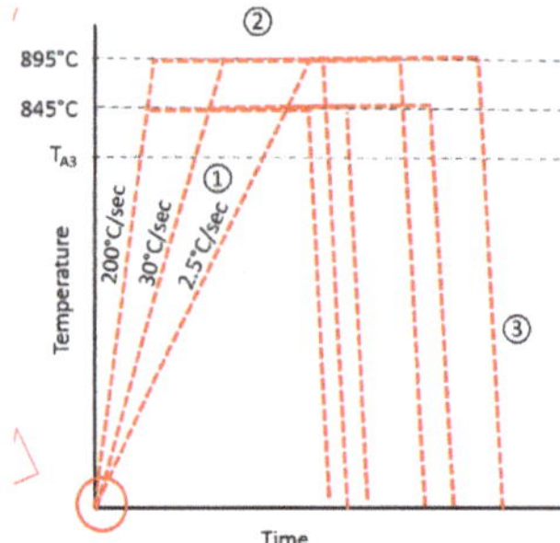

① **Heating at 2.5, 30 and 200°C/sec**

② **Austenitization for 30 seconds at 845 and 895°C**

③ **Rapid cooling (ice brine water solution) - cooling rate of 300°C/sec**

Figure 3. Schematic representation of the heat treatment cycles and photo of the laboratory CAL-induction simulator. T_{A3} is the transformation temperature.

2.3. Optical, Scanning Electron Microscopy and EDS—EBSD Analysis

The microstructural characterization of the samples in the starting and after heat treated were prepared for optical (OM), scanning electron microscopy (SEM) and EDS analysis by using standard metallographic procedures. Specimens for EBSD analysis were subjected to standard sample preparation followed by vibro-polishing in a VibroMet® polisher (Pace Technologies, Tucson, AZ, USA) for 3 h with 0.05 µm nanometer alumina suspension.

The SEM analysis was conducted on a ZEISS Sigma 500 VP scanning electron microscope (ZEISS, Pittsburgh, PA, USA) equipped with Oxford Aztec X-EDS (Oxford Instruments, Abingdon, UK) with an operating voltage between 10 kV and 20 kV. Electron Backscattered Diffraction (EBSD) and Image Quality (IQ) analyses were conducted on a FEI Scios FEG scanning electron microscope (University of Pittsburgh, Pittsburgh, PA, USA) equipped with an EBSD system. An accelerating voltage of 20kV with a beam current of 13 nA were used. The details of dwelling time, tilt angle, distance, scanned area and step size and the details for the EBSD-IQ phase-microstructural analysis are also described elsewhere [10]. The EBSD-IQ approach was used to measure the volume fraction of the microconstituents formed during intercritical and supercritical annealing treatments. An area of 100 µm × 100 µm with a step size of 0.2 µm was employed.

2.4. Mechanical Testing

Standard tensile coupons according to ASTM A8 were machined into sub-sized sheet tensile specimens with a gage length of 25 mm. These tensile samples were sectioned and machined from the heat treated samples. The tensile coupons were tested at RT following the standard described in the ASTM specification mentioned before.

3. Results and Discussion

3.1. Effect of Heating Rate on Transformation Temperatures (A_{C1} and A_{C3}) and the Nucleation of Austenite

It is well-known that the transformation temperatures, A_{C1} and A_{C3}, and the nucleation of austenite, are strongly affected by the heating rate. The effect of heating rates on the transformation temperatures can be calculated using a commercial thermodynamic software program J-MatPro (version 7 [11]), the results are presented in Figure 4. In this figure, according to the predictions of J-Mat Pro, it is shown that in order to obtain a fully homogeneous austenite transformation, the reheating temperature must be increased well above the A_{C3} temperature, especially at high heating rates. The formation of austenite is well-accepted to be a diffusion-controlled process, controlled by the slower diffusion process of interstitial elements, i.e., C, in austenite when compared to the diffusion rates in ferrite. Therefore, for heat treatments in the temperature range of 845 and 895 °C, or even at higher temperatures, i.e., 1000 °C, it would not be expected that austenite will transform into 100%

martensite during fast quenching. This view is independent of heating rates (2.5, 30 and 200 °C/s) prior to austenite transformation. For example, Figure 5 shows the EBSD-IQ results of ferrite-pearlite samples reheated at 1000 °C using 2.5 and 200 °C/s heating rates respectively, with a holding time of 30 s prior to rapid quenching. The results shown on this figure clearly support the theoretical predictions that a fully martensitic microstructure can't be obtained using slow or fast heating rates and short holding times at the supercritical temperatures. The resulting microstructure was a combination of martensite + bainite. This behavior can be explained by the effect of a heating rate and holding time on the nucleation and growth of austenite and the dissolution of Fe_3C carbides. Several studies [12–14] have indicated that during rapid reheating the classical view of phase transformations will deviate markedly from those observed during equilibrium conditions. That is, the kinetics of transformation will have a different behavior. For example, faster heating rates favors nucleation of austenite, while slower heating rates leads to a significant growth of austenite. That is, slower heating rates permit C diffusion through the austenite, enabling its growth. An additional effect is the holding time at a given temperature, fast heating rates and short holding times, promotes substantial local compositional differences in austenite.

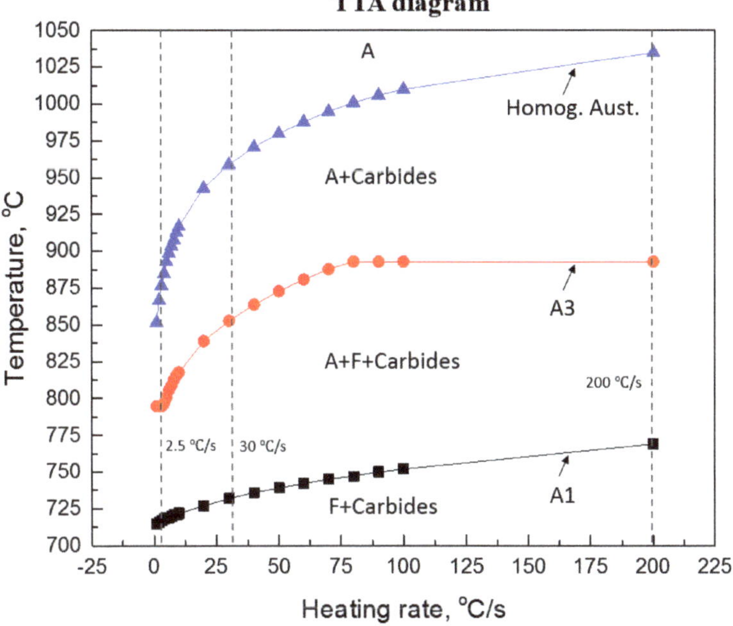

Figure 4. Shows the effect of heating rate on the transformation temperatures of 22MnB5 steel [11].

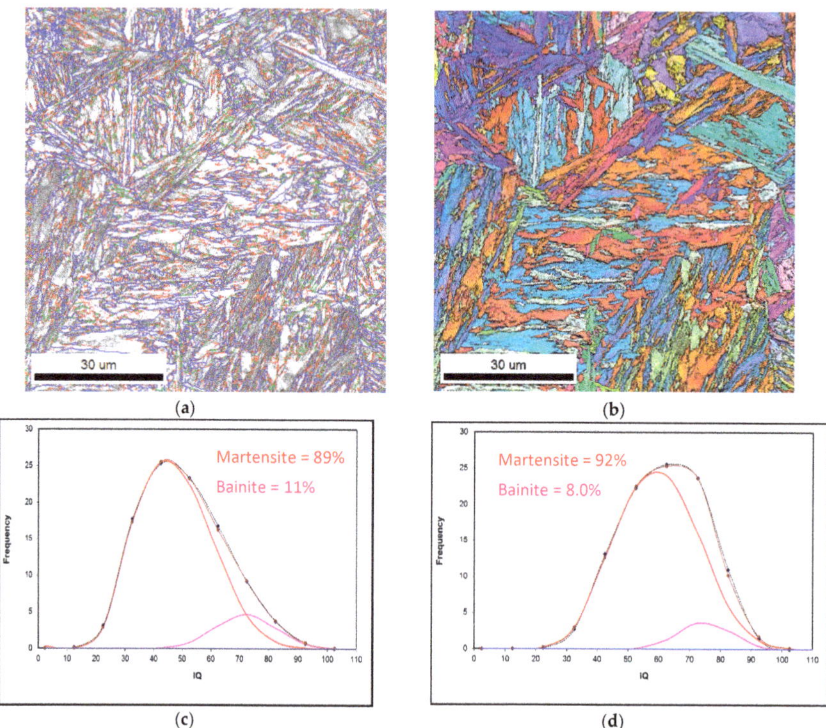

Figure 5. EBSD-IQ showing the microstructural balance of martensite and bainite after reheating a ferrite-pearlite microstructure at 1000 °C using heating rates of 2.5 °C/s and 200 °C/s and WQ. (**a**) and (**b**) represent the reconstructed EBSD-IQ microstructure and inverse pole figure and grain boundary character distribution for the sample reheated at 2.5 °C/s WQ sample. While (**c**) and (**d**) represent the EBSD-IQ analysis of the % final microstructure for 2.5 °C/s and 200 °C/s heating rates, respectively.

3.2. Formation of Austenite during Intercritical Reheating (Annealing)

The effect of the initial microstructure on the formation of austenite during intercritical reheating has been extensively studied. These studies provided a comprehensive view of the metallurgical reactions of austenite formation that take place during intercritical annealing; (1) nucleation and growth of austenite [15,16]; (2) the role of the initial microstructure [17]; (3) the incomplete dissolution of Fe_3C carbides [18,19]; (4) the non-uniformity of carbon content in intercritical austenite [20]; (5) partitioning of interstitial and substitutional solutes between $\alpha$and γphases [21]; and (6) the effect of heating rate [22].

Figure 6 shows the volume fraction of austenite formed during intercritical annealing (in the temperature range 732 to 792 °C) as function of heating rate and initial microstructure at very short holding times, i.e., 30 s. The results shown in this figure seem to indicate the influence of the initial microstructure and heating rate on the kinetics of transformation. That is, in a ferrite-pearlite microstructure, slower heating rate leads to higher nucleation and growth of austenite formation compared to faster heating rates for a given intercritical annealing temperature. When the initial microstructure is ferrite-100%spheroidized Fe_3C carbides, the formation of austenite doesn't have a similar dependence on the heating rate. This might be related to the fact that in a fully spheroidized microstructure, not all the Fe_3C particles nucleate austenite. The Fe_3C carbides located at the ferrite grain boundaries nucleate austenite preferentially, while those Fe_3C carbides located in the matrix do not contribute to the nucleation of austenite. These carbides dissolve and the carbon contributes to the growth of austenite. In summary, the kinetics of austenite formation in a ferrite-pearlite microstructure can proceed in one or two stages depending on the heating rate. The results shown in Figure 6 also seem to support the view that the formation of austenite from an initial ferrite-spheroidized Fe_3C microstructure tend to exhibit lower kinetics of austenite transformation compared to ferrite-pearlite or fully martensitic starting microstructures [19,23,24].

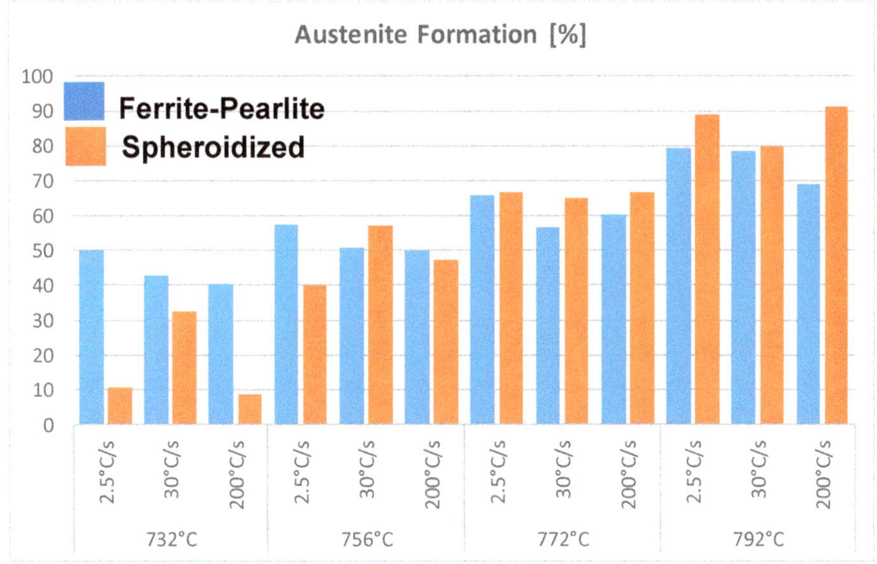

Figure 6. Formation of austenite during intercritical annealing as function of initial microstructural condition and heating rate.

It is well-accepted that the reheating temperature, holding time, and the effect of substitutional elements on the activity of carbon, controls the diffusion of carbon at the dissolving Fe_3C/γ interphase, hence the growth rate of austenite can be described by the equation shown below [25].

$$v = D\frac{dC}{dx}\left(\frac{1}{\Delta C^{\gamma\leftrightarrow\alpha}} + \frac{1}{\Delta C^{C\leftrightarrow\alpha}}\right) \quad (1)$$

where v is the velocity of the austenite phase boundary, D is the diffusion coefficient of C in austenite, dC/dx is the carbon concentration in the austenite matrix, $\Delta C^{\gamma\leftrightarrow\alpha}$ and $\Delta C^{C\leftrightarrow\gamma}$ are the differences in carbon concentration between austenite and ferrite and carbide and austenite, respectively. Mn segregation at the Fe_3C/γinterphase will decrease the diffusion of C through the austenite, hence decreasing the growth rate of austenite. The segregation of Mn in the lamellae pearlite and at the Fe_3C/αinterphase, i.e., spheroidized carbides is shown in Figure 7. This segregation affects the dissolution of Fe_3C and hence the kinetics of carbon diffusion in austenite during intercritical and supercritical heat treatments.

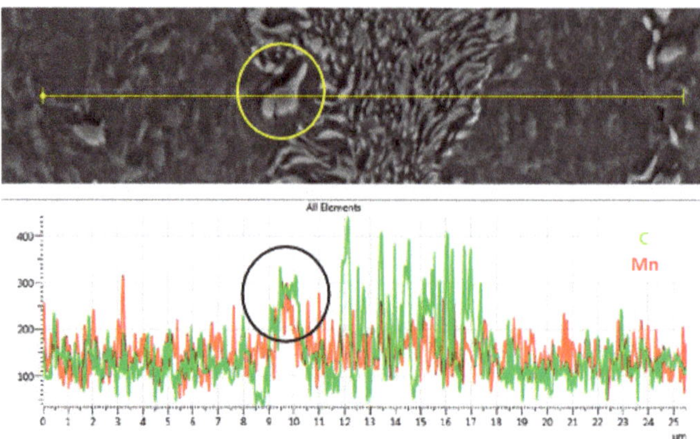

Figure 7. SEM-EDS line scan showing the segregation of C and Mn in the pearlite and Fe_3C carbides.

3.3. Formation of Austenite during Supercritical Reheating

Prior to studying the effect of reheating temperatures, heating rates and initial microstructure on the formation of austenite during supercritical annealing, the prior austenite grain size (PAGS) was determined. The results of the PAGS from different initial microstructural conditions and reheated at 845 °C and 895 °C, held 30 s at temperature and rapidly quenched in an ice brine solution are shown in Figure 8. The results show that the heating rate doesn't have a strong influence on the average PAGS. Meanwhile, as expected, the average PAGS value increases slightly with the reheating temperature (see table in Figure 8). The scale in the micros is (50 μm).

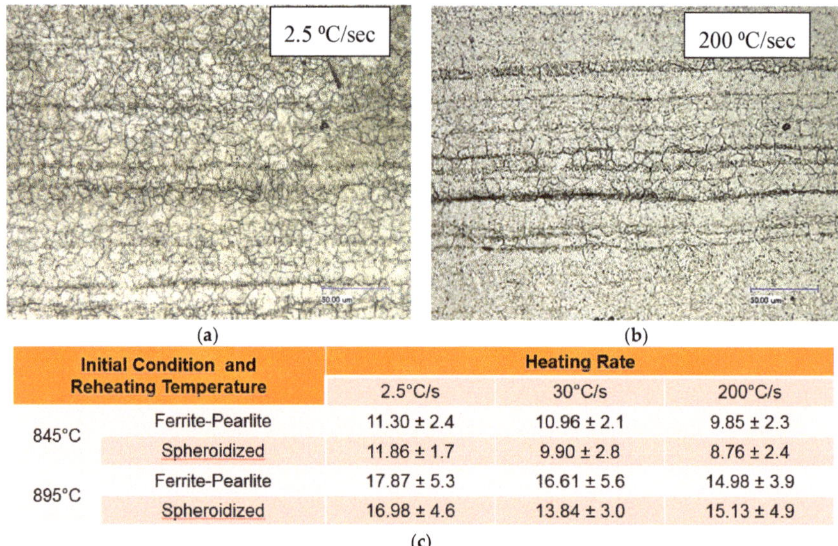

Initial Condition and Reheating Temperature		Heating Rate		
		2.5°C/s	30°C/s	200°C/s
845°C	Ferrite-Pearlite	11.30 ± 2.4	10.96 ± 2.1	9.85 ± 2.3
	Spheroidized	11.86 ± 1.7	9.90 ± 2.8	8.76 ± 2.4
895°C	Ferrite-Pearlite	17.87 ± 5.3	16.61 ± 5.6	14.98 ± 3.9
	Spheroidized	16.98 ± 4.6	13.84 ± 3.0	15.13 ± 4.9

(c)

Figure 8. Average PAGS in μm as function of heating rate, reheating temperature and initial microstructural condition; (**a**) OM of PAGS from 845 °C and (**b**) 895 °C. While (**c**) shows the PAGS values.

The decomposition products of austenite as function of initial microstructure, reheating temperature, heating rate and 30 s holding time prior to fast quenching are shown in Figure 9. The microstructural balance was obtained using the EBSD-IQ method described by Wu et al. [10], this method was also used in Figure 5. The results shown in Figure 9 indicate that the microstructures consisted of a mixture of martensite + bainite + undissolved Fe_3C carbides and small amount of martensite-austenite (MA) microconstituents. As expected, the amount of undissolved carbides and the MA seems to decrease as the reheating temperature increases. This is supported by the theoretical prediction (Figure 4) and the results presented in Figure 5. It is important to indicate that 100% martensite was not observed in any of the samples observed in this study.

Temperature (°C)	Initial Microstructure	Heating Rate	Martensite (%)	Bainite (%)	Fe_3C + MA (%)
845	Ferrite-Pearlite	2.5 °C/s	70.5	24	5.5
	Spheroidized	2.5 °C/s	79.2	19.25	1.55
845	Ferrite-Pearlite	30 °C/s	70.1	26.4	3.5
	Spheroidized	30 °C/s	75	22.9	2.1
845	Ferrite-Pearlite	200 °C/s	69.4	27.6	3.0
	Spheroidized	200 °C/s	71.7	25.8	2.5
895	Ferrite-Pearlite	2.5 °C/s	68.4	27.8	3.8
	Spheroidized	2.5 °C/s	73	25	2
895	Ferrite-Pearlite	30 °C/s	70.5	26.6	2.9
	Spheroidized	30 °C/s	71	27.29	1.71
895	Ferrite-Pearlite	200 °C/s	72	23.5	4.5
	Spheroidized	200 °C/s	71.6	24.9	3.5

(a)

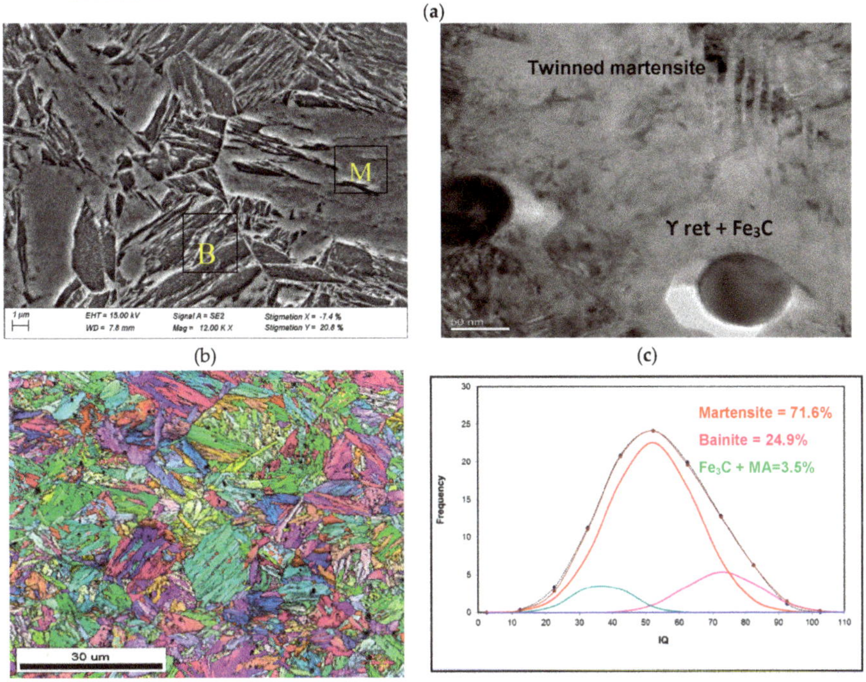

(b) (c)

(d) (e)

Figure 9. SEM-TEM and EBSD-IQ of WQ microstructure after reheating at 895 °C at a heating rate of 200 °C/s and fast quenched. Initial microstructure fully spheroidized; (**a**) Table of microstrucrual components based on the EBSD-IQ technique; (**b**) SEM micrograph showing martensite and bainite; (**c**) TEM micrograph showing undissolved Fe_3C carbides and retained γ at the carbide/matrix interface; (**d**) shows the inverse pole figure (IPF) and grain boundary character distribution, and (**e**) are the results from the EBSD-IQ analysis showing the percent of microstructural components.

3.4. Mechanical Properties

The tensile properties of a selected number of fully processed samples (ferrite-pearlite) from Figure 9 were tested and the resulting mechanical properties were evaluated, see Figure 10. As expected, the flow stress was continuous for all the samples tested. A comparison of the UTS shows that the samples reheated at 845 °C with a heating rate of 2.5 °C/s and those reheated using 30 °C/s exhibited a slightest difference in UTS value 1591 MPa versus 1648 MPa, respectively. This behavior can be explained by the increased amount of martensite + bainite and less Fe_3C + MA in the overall microstructure observed in the samples after reheating at 30 °C/s compared to those reheated using 2.5 °C/s. Interestingly reheating at higher supercritical temperatures, i.e., 1000 °C, did not increase the mechanical properties, as can be seen in Table 2. In this table, the average mechanical properties of the ferrite-pearlite and spheroidized samples reheated at 845 °C and WQ (water quenched) are also shown for comparison purposes. The YS and UTS for both starting conditions were very similar, the total elongation of the spheroidized samples was slightly lower compared to that of the ferrite-pearlite samples. The explanation for this behavior could be that the presence of some undissolved Fe_3C carbides acted as nucleation sites for the onset of the diffusive necking. Reheating at higher supercritical temperatures and fast cooling produces an increase in the percent of martensite, lower percent of bainite and the presence of Fe_3C + MA was not observed, compare Figures 5 and 9. These results are in strong agreement with the common knowledge that optimum microstructural combinations of martensite + bainite are stronger than 100% martensite.

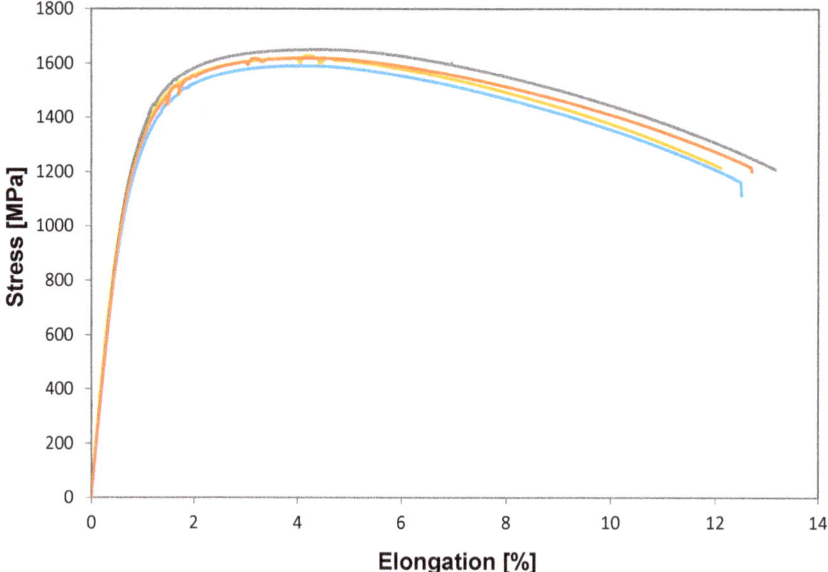

Figure 10. Flow behavior of ferrite-pearlite samples reheated at 845 and 895 °C using 2.5 and 30 °C/s respectively followed by rapid cooling (300 °C/s). All the results shown on this figure correspond to samples with an initial ferrite-pearlite microstructure.

Table 2. Average tensile properties of samples after reheating and WQ.

Temp. (°C)	Initial Microstructure	H.R. (°C/s)	UTS (MPa)	YS (MPa)	Elongation (%)
845	Ferrite-Pearlite	2.5	1591.3	1116.5	12.8
		30	1648.2	1120.2	13.5
		200	1620.0	1117.7	11.4
845	Spheroidized	2.5	1575.2	1106.1	11.9
		30	1637.3	1099.1	11.4
		200	1600.4	1115.7	10.6
1050	Ferrite-Pearlite	2.5	1566.0	1126.4	10.7
		30	1575.2	1072.2	10.4
		200	1606.7	1152.3	10.0

4. Conclusions

The results of this study clearly show the effect of the heating rate and initial microstructure on the nucleation and growth of austenite during intercritical and supercritical temperature. In ferrite-pearlite microstructures, slower heating rates, nucleation and growth reactions are preferred. Fast heating rates have a strong influence on the carbide dissolution during intercritical and supercritical annealing treatments. The transformation behavior of austenite after supercritical annealing at small ΔT (T-T_{Ac3}) and fast cooling rates always resulted in a multi-phase complex microstructure. At higher ΔT's a duplex microstructure with more martensite and less bainite was observed. The PAGS was not strongly influenced by fast heating rates and short holding times. The best combination of mechanical properties was obtained at small ΔT when the balance of martensite and bainite was optimum.

Author Contributions: All the authors contributed to this work. J.P.P. obtained his M.Sc. degree with this work; R.L.-M. conducted the EBSD-IQ and SEM analysis; O.G.-R. performed the mechanical testing; and C.I.G. was the academic advisor and wrote the manuscript.

Acknowledgments: The authors would like to thank TERNIUM-MEXICO for providing the commercial samples used on this study and the financial support. Also the authors would like to thank the members of the Ferrous Physical Metallurgy-MEMS Department at the University of Pittsburgh for their support.

Conflicts of Interest: The authors have not conflict of interest.

References

1. Goodwin, F.E.; Silva, E.A. Current Priorities in Developing, Forming and Joining of Advanced Galvanized Sheet Steels. In Proceedings of the GALVATECH 2017: 11th International Conference on Zinc and Zinc Alloy Coated Steel Sheet, Tokyo, Japan, 12–16 November 2017.
2. Stahl Zentrum Stahl fur Nachhaltige Mobilitat. Available online: http:/www.stahl-online.de/index.php/themen/stahanwendung/mobilitaet/ (accessed on 20 November 2018).
3. Mori, K.; Maeno, T.; Yamada, H.; Matsumoto, H. 1-Shot hot stamping of ultra-high strength steel parts consisting of resistance heating, forming, shearing and die quenching. *Int. J. Mach. Tools Manuf.* **2015**, *89*, 124–131. [CrossRef]
4. Cola, G.M. Flash Bainite: Room Temp Stamping 1500 to 1800 MPa Structural and Energy Absorbing Components to <2T Bend Radii and the Coil-Coil Production Equipment. In Proceedings of the AISTech 2018, Philadelphia, PA, USA, 7 May 2018; pp. 2731–2743.
5. Cola, G.M. Microtretament of Iron-Bases Alloy, Apparatur and Method Therefor, and Microstructure Resulting Therefrom. USA Patent 2010/0163140 A1, 1 July 2010.
6. Lolla, T.; Alexandrov, B.; Babu, S.; Cola, G. *Towards Understanding the Microstructure Developments during Flash Heating and Cooling of Steels*; The Ohio State University: Columbus, OH, USA, 2009.

7. HolzweiBig, M.J.; Lackmann, J.; Konrad, S.; Schaper, M.; Niendorf, T. Influence of short austenitization treatments on the mechanical properties of low alloy steels for hot forming applications. *Metall. Mater. Trans. A* **2015**, *46*, 3199–3207. [CrossRef]
8. Niesse, M.; Lackmann, J.; Frost, G.; Konrad, S.; Lambers, H.G. Warmformlinie und Verfahren zur Herstellung von Warmumgeformten Blechprodukten. Patent DE102014101539A1, 13 August 2015.
9. Kollec, R.; Veit, R.; Merklein, M.; Lechler, J.; Geiger, M. Investigation on induction heating for hot stamping of boron alloyed steels. *CIRP Ann.-Manuf. Technol.* **2009**, *58*, 275–278. [CrossRef]
10. Wu, J.; Wray, P.; Garcia, C.I.; Hua, M.; DeArdo, A.J. Image Quality Analysis: A new method of characterizing microstructures. *ISIJ Int.* **2005**, *45*, 254–262. [CrossRef]
11. Garcia, C.I. *JMatPro Software V7*; Sente Software Ltd.: Guildford, UK.
12. Lobbe, C.; Becker, C.; Tekkaya, A.E. Warm bending of microalloyed high-strength steel by local induction heating. In *Forming Technology Forum*; University of Twente: Enschede, The Netherlands, 2014; pp. 99–104.
13. Allwood, J.M.; Childs, T.H.C.; Clare, A.T.; De Silva, A.K.M.; Dhokia, V.; Hutchings, I.M.; Leach, R.K.; Leal-Ayala, D.R.; Lowth, S.; Majewski, C.E. Manufacturing at double the speed. *J. Mater. Process Technol.* **2016**, *229*, 729–757. [CrossRef]
14. Lobbe, C.; Hering, O.; Hiegemann, L.; Tekkaya, A.E. Setting Mechanical Properties of High Strength Steels for Rapid Hot Forming Process. *Materials* **2016**, *9*, 229. [CrossRef]
15. Roberts, G.A.; Mehl, R.F. The mechanism and the Rate of Austenite from Ferrite-Cementite Aggregates. *AMS Trans.* **1943**, *31*, 613–649.
16. Molinder, G. A Quantitative Study of the Formation of Austenite and the Solution of Cementite at Different Austenitizing Temperatures for a 1.27 wt % Carbon Steel. *Acta Metall.* **1956**, *4*, 565–571. [CrossRef]
17. Judd, R.R.; Paxton, H.W. Kinetics of Austenite Formation from a Spheroidized Ferrite-Carbide Aggregate. *Trans. TMS-AIME* **1968**, *242*, 206–215.
18. Speich, G.R.; Szirmae, A. Formation of Austenite from Ferrite and Ferrite-Carbide Aggregate. *Trans. TMS-AIME* **1969**, *245*, 1063–1074.
19. Garcia, C.I.; DeArdo, A.J. Formation of Austenite in 1.5 Pct. Mn Steels. *Metall. Trans. A Phys. Metall. Mater. Sci.* **1981**, *12*, 521–530. [CrossRef]
20. Speich, G.; Demarest, V.; Miller, R. Formation of austenite during intercritical annealing of dual-phase steels. *Metall. Mater. Trans. A* **1981**, *12*, 1419–1428. [CrossRef]
21. Azizi-Alizamini, H.; Militzer, M.; Poole, W.J. Austenite Formation in Plain Low-Carbon Steels. *Metall. Mater. Trans. A: Phys. Metall. Mater. Sci.* **2011**, *42*, 1544–1557. [CrossRef]
22. Dyachenko, S.S. The Austenite Formation in Fe-C Alloys. *Metallurgiya* **1982**, *128*.
23. Golovanenko, S.A.; Fonstein, N.M. Dual-Phase Low Alloyed Steels. *Metallurgiya* **1986**, *206*.
24. Law, N.C.; Edmonds, D.V. The Formation of Austenite in a Low-Alloy Steel. *Metall. Mater. Trans. A* **1980**, *11*, 33–46. [CrossRef]
25. Wycliffe, P.; Purdy, G.R.; Embury, J.D. Austenite Growth in the Intercritical Annealing of Ternary and Quaternary Dual-Phase Steels. In *Fundamentals of Dual-Phase Steels, Proceedings of the Symposium at the 110th AIME Annual Meeting, Chicago, IL, USA, 23–24 February 1981*; Metallurgical Society of AIME: Warrendale, PA, USA, 1981; pp. 59–83.

Article

Quenching and Partitioning of Multiphase Aluminum-Added Steels

Tuomo Nyyssönen [1,2,*], Olli Oja [3], Petri Jussila [3], Ari Saastamoinen [1], Mahesh Somani [4] and Pasi Peura [1]

1 Laboratory of Metals Technology, Faculty of Engineering and Natural Sciences, Materials Science, Tampere University, Korkeakoulunkatu 6, 33720 Tampere, Finland; ari.saastamoinen@tuni.fi (A.S.); pasi.peura@tuni.fi (P.P.)

2 Outotec Oyj, Pori Outotec Research Center, Kuparitie 10, 28330 Pori, Finland

3 Product Development, SSAB Europe Oy, Harvialantie 420, 13300 Hämeenlinna, Finland; olli.oja@ssab.com (O.O.); petri.jussila@ssab.com (P.J.)

4 Faculty of Technology, Materials and Mechanical Engineering, University of Oulu, 90014 Oulun Yliopisto, Finland; mahesh.somani@oulu.fi

* Correspondence: tuomo.nyyssonen@outotec.com or tuomo.nyyssonen@tut.fi; Tel.: +358-503-721-641

Received: 27 February 2019; Accepted: 19 March 2019; Published: 22 March 2019

Abstract: The quenching and partitioning response following intercritical annealing was investigated for three lean TRIP-type high-Al steel compositions. Depending on the intercritical austenite fraction following annealing, the steels assumed either a ferrite/martensite/retained austenite microstructure or a multiphase structure with ferritic, bainitic and martensitic constituents along with retained austenite. The amount of retained austenite was found to correlate with the initial quench temperature and, depending on the intercritical annealing condition prior to initial quenching, with the uniform and ultimate elongations measured in tensile testing.

Keywords: steel; martensite; austenite; quenching; partitioning; dilatometry

---

1. Introduction

Intercritical annealing of low-alloy steel occurs in the temperature regime between the *Ac1* and *Ac3* temperatures, where thermodynamic equilibrium corresponds to some mixture of the austenite and ferrite phases. The amount of austenite at the conclusion of intercritical annealing depends on several factors, such as the starting microstructure, heating rate [1] and, more importantly, the alloying contents of the steel. Upon quenching at higher than critical cooling rates, the intercritical austenite phase is transformed to martensite, resulting in low-alloy dual-phase steel consisting of martensitic islands in a matrix of ferritic grains.

If the quenching is interrupted somewhere between the martensite start temperature M_s and the temperature where the last of the austenite transforms (M_f), the martensitic transformation will be incomplete and the martensitic islands will include some untransformed metastable austenite. If the steel is then transitioned for a certain duration to a suitable intermediate holding temperature (also commonly referred as 'partitioning temperature'), this metastable austenite can be either partially or fully stabilized down to room temperature by enriching it with carbon partitioned from the carbon-supersaturated martensite. The resulting microstructure then would consist of martensitic islands interspersed with carbon-enriched retained austenite, suspended in the ferritic matrix. In effect, the intercritical austenite phase has undergone a heat treatment known as quenching and partitioning (Q&P), originally proposed in 2003 by Speer et al. [2].

The end goal of the Q&P treatment of intercritical austenite is to improve the ductility of a dual-phase microstructure via the strain-induced transformation of the retained austenite to martensite during deformation, essentially increasing the strain hardening capability of the steel. It was found by some of the present authors in a previous study [3] that austenite stabilized in this manner can increase ductility even for a low-carbon, high-aluminum steel. However, it was recently observed by Tan et al. [4] that the ability of the retained austenite to contribute to strain hardening depends on its localization, morphology and stability essentially with respect to the carbon-content. They found that if the deformation occurs primarily in the ferrite phase, the transformation of retained austenite to martensite remains limited, thus limiting the total elongation. A fine film-like morphology (resulting in high mechanical stability) and localization of the austenite away from ferrite-prior austenite interfaces will also promote this type of undesirable mechanical stability.

Successfully promoting quenching and partitioning in a steel imposes some requirements on alloying. First, the steel must have the ability to suppress or delay the formation of carbides or other precipitates during partitioning. This is typically achieved by adding a combination of ferrite-stabilizing elements such as aluminum [5], silicon [6] or chromium [7]. Second, the steel must contain a sufficient amount of carbon to enable stabilizing an optimum retained austenite fraction to achieve the desired increase in ductility. Third, the steel must be sufficiently hardenable in order to reach the initial quench temperature without allowing undesired ferritic or bainitic phase transformations during the initial cooling step.

Intercritical annealing opens up some interesting possibilities for controlling the condition of the austenite at the conclusion of annealing. The grain size and the relative fraction of the intercritical austenite can be controlled to a large extent by suitably varying the annealing parameters [1,8,9]. The growth of the austenite phase in low- and medium-carbon steels has been observed in various research works to be controlled essentially by the diffusion of carbon across the interphase boundary. Therefore, carbon can be expected to partition nearly completely to austenite [1,10], while the partitioning of heavier elements (such as silicon, aluminum or manganese) is usually limited and will depend on the annealing temperature and duration [11].

The conditioning of the austenite will drastically affect the quenching response, both in terms of martensite start temperature M_s as well as the hardenability. Assuming total partitioning of carbon from ferrite to austenite, a smaller initial fraction of austenite will greatly increase its carbon content. This will both decrease M_s and improve hardenability [12], at least up to the eutectoid composition. To illustrate the point further, three preliminary calculations were made with the thermodynamic and kinetic calculation software JMATPRO® [13] for two of the experimental steels studied in this work: Steel A and Steel B (see Table 1 for chemical composition). JMATPRO was used to calculate continuous cooling transformation (CCT) curves based on the model by Kirkaldy et al. [14] modified by Lee and Bhadeshia [15]. Figure 1 shows the CCT curves for the steels with modified carbon contents, from nominal composition to carbon contents corresponding to 50 and 25 vol % fraction of intercritical austenite at the conclusion of annealing (assuming full partitioning of carbon). In the figure, the 50 vol % fraction of austenite is indicated by an assumed carbon content of 0.32C for Steel A and 0.22C for Steel B, and 25 vol % austenite content is indicated by 0.64C for Steel A and 0.44C for Steel B. Dashed lines in the figure indicate continuous cooling curves corresponding to a linear cooling rate of 10 °C/s, which is a realistically achievable cooling rate in modern continuous annealing lines. Referring to Figure 1, the onset of bainite transformation is delayed due to an increase in carbon content at lower austenite fractions. At an austenite fraction of 25 vol %, bainite transformation is avoided completely at a cooling rate of 10 °C/s, based on the JMATPRO® calculations.

Table 1. Compositions of the experimental steel grades.

wt %	C	Mn	Al	Ni	Ti + Nb + V	Si + Cr + Mo	Balance
Steel A	0.16	2.08	1.03	0.044	0.008	0.689	Fe and traces of Cu, P, S
Steel B	0.11	2.65	0.81	0.041	0.035	0.978	Fe and traces of Cu, P, S
Steel C	0.16	1.62	1.31	0.041	0.023	0.498	Fe and traces of Cu, P, S

These calculations do not take into account prior austenite grain size, which has been observed to have a significant effect on the martensite transformation of intercritically annealed steel in a previous study [8]. Essentially, the prior austenite grain size potentially remains very small during intercritical annealing, further reducing the M_s temperature. Such an effect of fine austenite grain size on M_s has been exhaustively investigated, for instance, by Yang and Bhadeshia [16].

Austenite conditioning via intercritical annealing was shown to affect the quenching and partitioning response of a high-aluminum steel in a previous study [3]. In this case, the prior austenite grain size was of the order of 1.5 μm and the austenite had a high carbon content directly after intercritical annealing. This decreased M_s significantly beyond the value predicted by empirical equations and promoted the formation of martensite with a small packet size and coarse, irregularly-shaped lath morphology. This, in turn, resulted in blocky rather than film-like retained austenite formation, situated primarily at prior austenite and packet boundaries. The amount of retained austenite was found to correlate well with the initial quench temperature, as well as the uniform and ultimate elongations of the studied steels. The effect of intercritical annealing on M_s was also observed by Yi et al. [17], who showed that with a suitably high carbon and aluminum content in the steel, intercritical annealing can be used to lower the initial quench temperature QT to ambient temperature or lower. Q&P-aided dual-phase microstructures were also produced for a Fe-0.2C-2.0Mn-1.5Si (wt %) steel after intercritical annealing by Wang et al. [18].

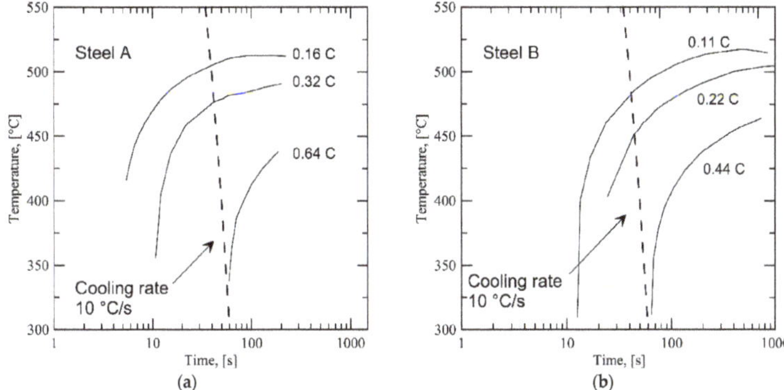

Figure 1. Partial CCT (continuous cooling transformation) curves showing the bainite start lines, calculated for (**a**) Steel A and (**b**) Steel B (see Table 1 for compositions), assuming full partitioning of carbon from ferrite to austenite during intercritical annealing. The carbon concentration for each annealing condition is indicated in wt % next to the corresponding bainite start line.

The above discussion should emphasize the point that the condition of the austenite at the conclusion of intercritical annealing should be known (at least at the level of prior austenite grain size and its carbon content), if any reliable estimation for Q&P heat treatment parameters should be made beforehand. Otherwise, it is necessary to determine M_s experimentally by suitable methods, such as dilatometry.

Despite several promising early results [3,4,17,18], the alloy compositions that have been investigated thus far are challenging to produce in an industrial setting with the current level of knowledge and technology. Furthermore, either the high silicon content of the steels makes hot dip galvanizing difficult [19] or the high aluminum or other alloying contents complicate the casting procedure.

With this in mind, the quenching and partitioning response was investigated for three experimental alloys following intercritical annealing: a conventional aluminum-alloyed TRIP-type steel and two novel lean-alloyed complex phase-type compositions with a mixture of aluminum, silicon and chromium elements as suppressors of carbide formation. The steels have been cast, rolled and could be hot dip galvanized in an industrial setting, at the current level of technology. The annealing temperatures and timeframes investigated here were also selected from the viewpoint of industrial relevance.

The phase transformation behavior during quenching of the steels was investigated via dilatometry and suitable quenching and partitioning parameters were determined based on the results. The effect of quench temperature on the retained austenite content and the morphology of the various phases in the final microstructures were investigated via optical microscopy, X-ray diffraction (XRD) and electron backscatter diffraction (EBSD). Preliminary evaluation of Q&P treated samples was made in respect of tensile properties.

2. Materials and Methods

The compositions of the investigated steels are shown in Table 1. The steels were supplied in a cold rolled, fully hard condition as 1.3-mm thick sheets.

Dilatometry experiments were conducted with a Gleeble 3800 thermomechanical simulator (Dynamic Systems Inc., Poestenkill, NY, USA) to determine the M_s temperatures of the steels after intercritical annealing. 10 mm × 60 mm specimens were cut from the sheets for the experiments. The specimens were subjected to a computer-controlled resistance heating followed by controlled cooling, with the aid of compressed argon gas below 400 °C down to room temperature. The temperature of each specimen was monitored with a K-type thermocouple, and the dilatation was measured in the transverse direction of the specimen at the thermocouple location using an extensometer fitted with quartz rods. The heat treatment parameters are shown in Table 2.

Table 2. Dilatometry parameters for the steels. *HR* = heating rate, *HT* = time at annealing temperature, *AT* = annealing temperature and *CR* = cooling rate.

HR (°C/s)	*HT* (min)	*AT* (°C)	*CR* (°C/s)
4	3	850	25

The transformation curves were extracted from the dilatometric data by line fitting over the linear thermal contraction portion of both the austenite and martensite phases. Martensite and austenite cooling contraction curves were extrapolated from the linear portions of the curve. The extent of the martensitic transformation was calculated by using the lever rule. A curve was fitted to this data corresponding to the Koistinen-Marburger [20] equation:

$$V_m = 1 - e^{-K(Ms-T)} \tag{1}$$

The martensite start temperature M_s and empirical fitting constant K were determined by curve fitting of Equation (1), using Trust-Region-Reflective Least Squares Algorithm in the Matlab® Curve Fitting Toolbox in Matlab® R2018b (Mathworks Inc., Natick, MA, USA). The transformation data up to a martensite fraction of 0.2 were excluded from the fitting, in order to minimize the effect of initial gradual martensite start on the fit. Two additional fits were also made, excluding data up to 0.4 and 0.6 martensite fraction. Of these three fits, the best fit according to the adjusted R-square statistics, when compared to the whole transformation data, was chosen as the optimum fitting solution. The M_s temperature was obtained directly from the fitting parameters for Equation (1). The fitting procedure is shown schematically in Figure 2.

The specimens were then sectioned in the transverse direction at the thermocouple location and mounted into cold setting resin for metallography. The specimens were ground and polished with colloidal silica Buehler Mastermet2 used in the final polishing step. The specimens were then treated with the color etching procedure proposed by LePera [21]. The polished and etched specimens were examined and micrographed with the Alicona InfiniteFocus G5 profilometer (Optimax IIM Ltd., Market Harborough, UK).

Typical quenching and partitioning parameters were designed for the steels based on the M_s values measured in the dilatometry experiments and the fitted martensitic transformation behavior, Figure 2. The parameters are shown in Table 3. Figure 3 shows a schematic diagram of the Q&P treatment cycle including linear heating and cooling rates.

Heat treatment specimens of 10 mm × 60 mm size were cut from the steel sheets. The Q&P treatments were conducted in the Gleeble 3800 simulator with the pocket jaw grips set at a free span distance of 35 mm. Two specimens were heat treated for each Q&P cycle: one specimen was later used for tensile testing and another for microstructural evaluation and retained austenite measurements using XRD. The test setup was essentially similar to the one employed in dilatometry, with the only difference that the extensometer was not used, if deemed unnecessary.

To determine if significant heating or cooling gradients exist in the specimens during the heat treatment cycles and to assess uniform temperature zone, three thermocouples were fitted on one of the Steel C specimens: one at the center, one 3 mm from center to the side and one 6 mm from center to the other side. It was found that at a distance of 3–6 mm from the controlling thermocouple, the temperature gradient remains below 3–5 °C at all stages of the heat treatment.

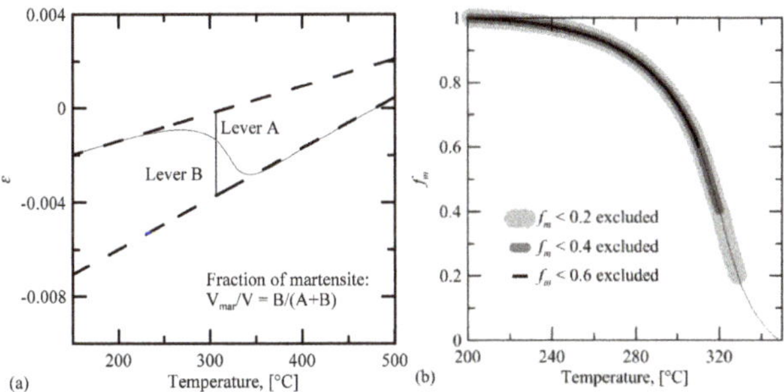

Figure 2. (**a**) The cooling contraction curve around the martensitic transformation. (**b**) The martensite volume fraction with respect to temperature extrapolated from the data shown in (**a**).

The retained austenite contents of the Q&P specimens were measured at room temperature using X-ray diffraction. A thickness of ~0.2 mm was ground from the specimen surface using a P800 SiC emery paper. The specimens were then ground with progressively finer papers, ending with the roughness P2000 in accord with standard polishing practice. The specimens were then electrolytically polished for 12 s at 40 V with the A2 electrolyte in a Lectropol-5 polisher (Struers Inc., Cleveland, OH, USA). The XRD analyses were conducted with the Panalytical Empyrean X-Ray diffractometer (Malvern Panalytical Ltd., Malvern, UK) using Co K_α-radiation (40° < 2θ < 102°, 40 kV, 45 mA). The site of the thermocouple location was used for centering the X-ray beam on the specimens. The peaks used in the analysis were (110), (200), (211) and (220) for martensite and (111), (200), (220) and (311) for austenite. The method for retained austenite calculation was the four-peak method described in SP-453 [22] (four peaks for both ferrite and austenite). The carbon content was estimated from the measured average austenite lattice parameter using Equation (2) [23]:

$$C_\gamma = (a_\gamma - 0.3555)/0.0044 \quad (2)$$

in which a_γ stands for the average austenite lattice parameter in Å.

After the XRD measurements, the specimens were sectioned in the transverse direction at the location of the thermocouple and prepared for electron backscatter diffraction (EBSD). The specimens were polished in a manner similar to the practice adopted for dilatometry specimens prepared for light optical microscopy. However, the specimens were not etched after polishing. The polished specimens were removed from the mounts, washed in ethanol and placed in a low pressure desiccator overnight to remove moisture. Prior to electron microscopy, the specimens were cleaned in an evacuated chamber fed with ionized oxygen plasma to remove any leftover organic residues on the surface.

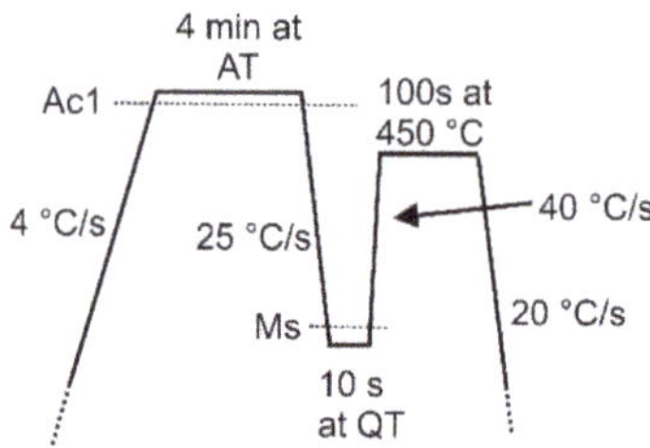

Figure 3. Schematic of the heat treatment cycle for the quenched and partitioned specimens.

Table 3. Annealing temperature (*AT*), initial quench temperature (*QT*), partitioning temperature (*PT*) and holding time at partitioning temperature (t_H) for the heat treated specimens.

Alloy	*AT* (°C)	*QT* (°C)	*PT* (°C)	t_H (s)
Steel A	850	125	450	100
-	-	150	-	-
-	-	175	-	-
Steel B	850	200	450	100
-	-	250	-	-
-	-	275	-	-
-	-	300	-	-
Steel C	850	75	450	100
-	-	100	-	-

The specimens were then subjected to electron backscatter diffraction (EBSD) studies. The scanning electron microscope (SEM) used was a Zeiss ULTRAPLUS UHR FEG-SEM system (ZEISS International, Oberkochen, Germany) fitted with a field emission gun (FEG) and an HKL Premium-F Channel EBSD system with a Nordlys F400 detector (Oxford Instruments plc, Abingdon, UK), which was used for phase contrast and orientation mapping. The parameters for EBSD analysis were 20 kV acceleration voltage, 14 mm working distance and a tilt angle of 70° with a step size of 0.05 µm.

Tensile testing was carried out using an Instron 8800 servohydraulic materials testing machine (Instron, Norwood, MA, USA). Non-standard tensile specimens were prepared by precision milling a 6 mm long, 3.5 mm wide gage area with 1 mm roundings into the center of each specimen. A cooling lubricant jet was used to reduce thermal effects on the specimen during milling. Each specimen was tested in tension to fracture at an engineering strain rate of 0.001 s^{-1}. Elongation was measured using a miniature axial extensometer. The total elongation A was recalculated to correspond to standard test geometry of a 120 mm × 20 mm gage section using the Oliver equation as implemented by ISO 2566/1 [24]:

$$A_2 = A_1 \times \left(\frac{k_1}{k_2}\right)^n \tag{3}$$

where A_2 is the calculated elongation value, A_1 is the known elongation value, k_1 and k_2 are the proportionality ratios of the two test pieces, and n is a material dependent constant. The standard adopts $n = 0.4$. The proportionality ratios k_1 and k_2 were calculated with the equation:

$$k_1 = \frac{w_s}{\sqrt{w_a t_a}} \tag{4a}$$

$$k_2 = \frac{l_s}{\sqrt{w_s t_a}} \tag{4b}$$

where w_s and l_s are the new width and length, respectively, and w_a, l_a and t_a are the measured width, length and thickness, respectively.

3. Results

3.1. Dilatometry Experiments

Table 4 shows the measured M_s temperatures for the steels estimated from dilatation curves based on dilatometer data. Figure 4 shows examples of the optical micrographs taken from the quenched dilatometry specimens. For Steel A and Steel C, the microstructures consist of martensite (including prior austenite) (lighter shade) and intercritical ferrite (darker shade). For Steel B, Le Pera etching produced an unexpected coloring result: the microstructure appears to consist of a primarily martensitic matrix (brown) with islands of intercritical ferrite (white).

Assuming that no ferrite transformation occurred during quenching, the intercritical austenite fraction can be estimated from the amount of the lighter phase (martensite) in the optical micrographs by image analysis [21]. The intercritical austenite contents could be estimated from the micrographs for Steel A and Steel C and they are shown in Table 4. In Steel B, the amount of intercritical austenite was judged to be approximately 65 vol %, although a reliable analysis is not possible due to the poor etching response. Assuming full partitioning of carbon, 65 vol % of austenite would correspond to an M_s temperature of 339 °C when calculated with the method proposed by Bhadeshia [25,26], which is in line with the measured M_s value shown in Table 4.

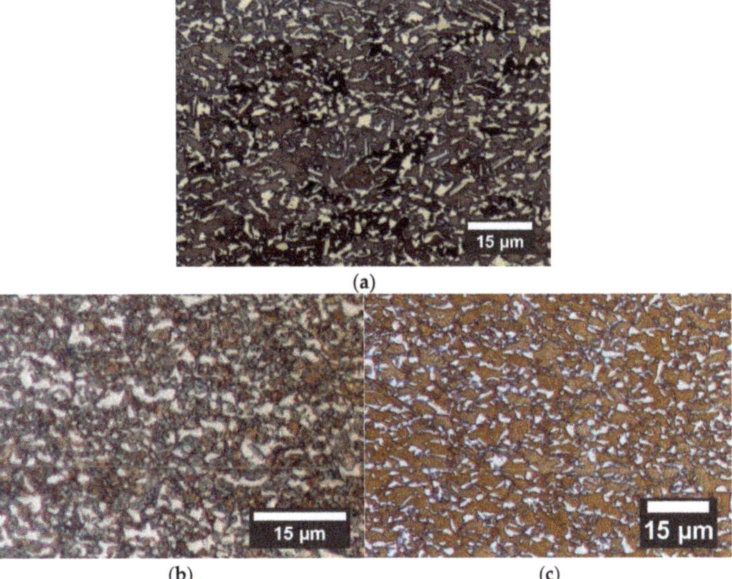

Figure 4. Typical examples of the optical micrographs of the dilatometry specimens for (**a**) Steel A, (**b**) Steel B and (**c**) Steel C.

Table 4. Experimentally determined M_s temperatures.

Alloy	*AT* (°C)	M_s (°C)	γ_{frac} (vol %)
Steel A	850	203	27 ± 1
Steel B	850	339	~65 (see text)
Steel C	850	116	23 ± 0.5

3.2. Q&P Experiments, Steel A

Figure 5 shows examples of the obtained XRD spectra for the quenched and partitioned specimens. There is clearly an appreciable amount of retained austenite following quenching and partitioning. The peaks corresponding to the ferrite phase coincide almost exactly among Steels A, B and C, while a slight scatter in the locations of the austenite peaks can be detected between the Steels. This is an indication that the lattice parameters of austenite in the Steels are affected by carbon partitioned from supersaturated martensite, each Steel having received a different amount of carbon in the austenite phase after quenching to the temperatures indicated in the figure.

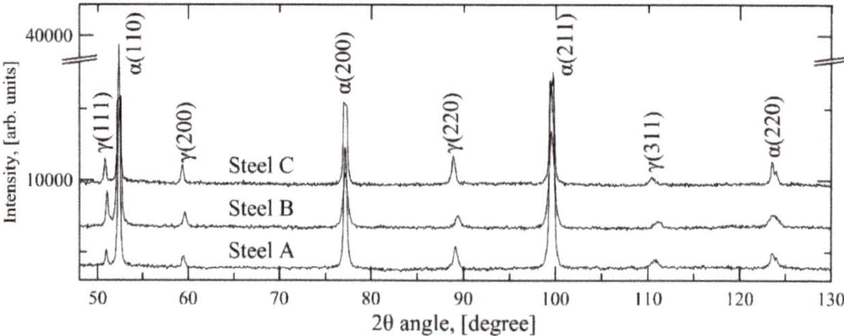

Figure 5. The observed XRD spectra for Steel A, quenched to 175 °C, Steel B, quenched to 300 °C and Steel C, quenched to 100 °C. All specimens were annealed at 850 °C for 4 min and partitioned at 450 °C for 100 s.

Figure 6 shows the measured retained austenite fraction and the calculated austenite carbon content with respect to the initial quench temperature for Steel A. As can be seen from the figure, the quench temperature correlates with the amount of retained austenite in the final microstructure. The carbon content of the retained austenite drops as the initial quench temperature is raised.

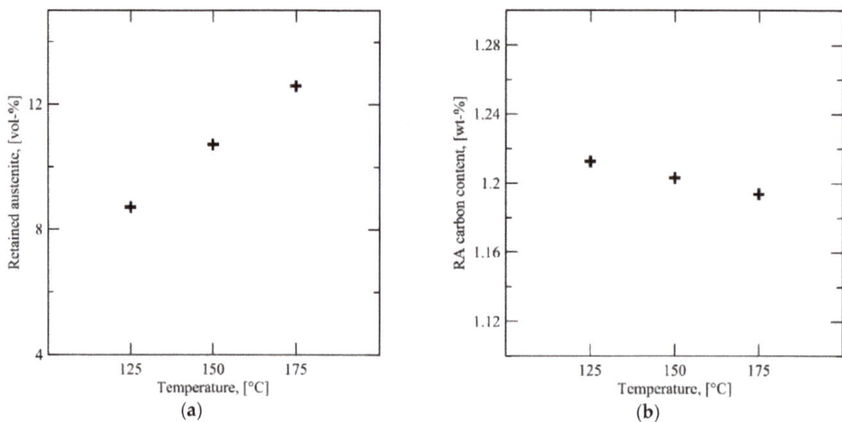

Figure 6. (**a**) The retained austenite fraction of Steel A with respect to initial quench temperature QT. (**b**) The carbon content of the retained austenite with respect to QT. Specimens annealed at 850 °C, quenched to QT and partitioned at 450 °C for 100 s.

Figure 7 shows the measured 0.5% proof strength (denoted R_{p05} in the Figures), ultimate tensile strength R_m, uniform elongation Ag and the total elongation A with respect to the initial quench temperature. Figure 6b shows that *Ag* correlates with the initial quench temperature. There is a slight inverse correlation with the total elongation. It is to be noted that the R_m practically stays at about 1000 MP irrespective of the quench temperature.

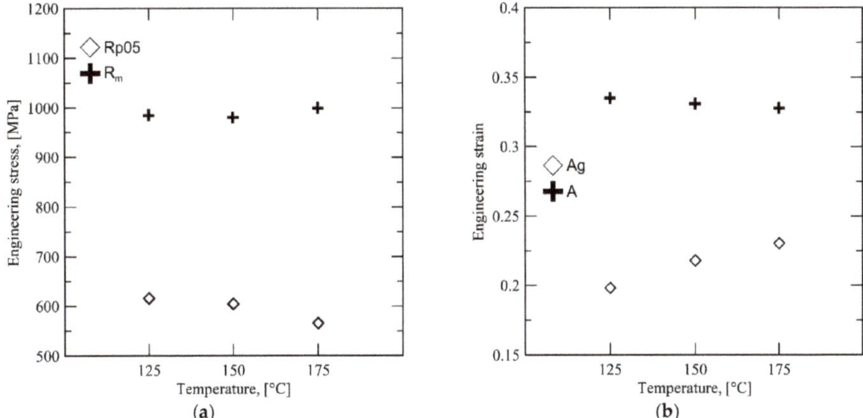

Figure 7. (**a**) The measured 0.5% proof strength and R_m values for Steel A with respect to *QT*. (**b**) The uniform elongation *Ag* and total elongation *A* for Steel A with respect to *QT*. Specimens annealed at 850 °C, quenched to *QT* and partitioned at 450 °C for 100 s.

3.3. Q&P Experiments, Steel B

Figures 8 and 9 show the results of the XRD measurements and tensile tests for Steel B similar to Figures 6 and 7. The behavior of Steel B differs from Steel A. Instead of a steady reduction in austenite fraction, there is an initially high amount (12 vol %) at *QT* = 300 °C followed by an appreciable drop at *QT* = 275 °C to a nearly stable austenite fraction (7–8 vol %) irrespective of further reduction in *QT* temperature down to 200 °C, Figure 8a. The corresponding average carbon content in retained austenite decreased from about 1.2% to 1% with an increase in *QT* from 200 to 300 °C, Figure 8b.

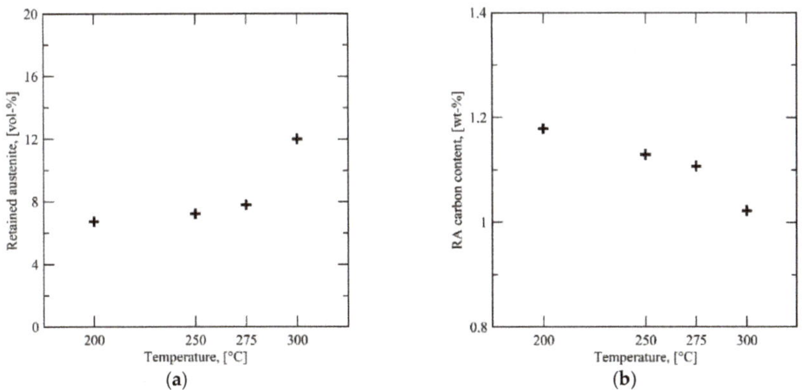

Figure 8. (**a**) The retained austenite fraction of Steel B with respect to initial quench temperature *QT*. (**b**) The carbon content of the retained austenite with respect to *QT*. Specimens annealed at 850 °C, quenched to *QT* and partitioned at 450 °C for 100 s.

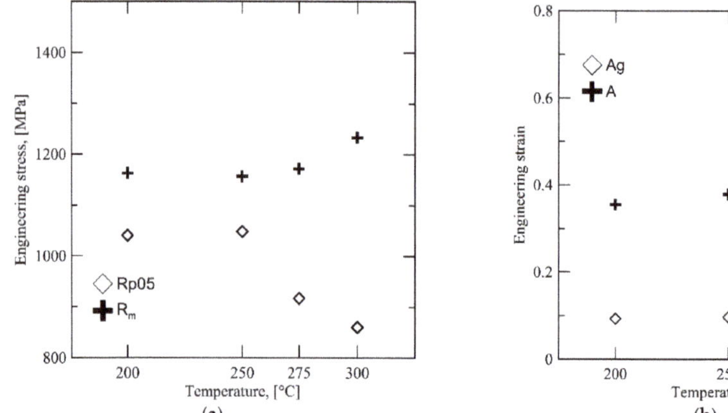

Figure 9. (**a**) The measured 0.5% proof strength and R_m values for Steel B with respect to *QT*. (**b**) The uniform elongation *Ag* and total elongation *A* for Steel B with respect to *QT*. Specimens annealed at 850 °C, quenched to *QT* and partitioned at 450 °C for 100 s.

Figure 9 shows the results of the tensile property characterization of the Steel B specimens. Interestingly, there is a significant drop in yield strength beyond the *QT* > 250 °C. The low yield strength at *QT* = 300 °C is, however, accompanied by a high R_m of 1250 MPa, even though the *Ag* and total elongation *A* seem somewhat insensitive to the *QT*.

3.4. Q&P Experiments, Steel C

The XRD measurements and tensile test results for Steel C are shown in Figures 10 and 11. The behavior of Steel C is very similar to Steel A, as the retained austenite fraction drops with decreasing *QT*. The higher retained austenite fraction of ≈11 vol % at *QT* = 100 °C corresponds to a significant increase in both uniform and total elongations, as well as a drop in yield strength.

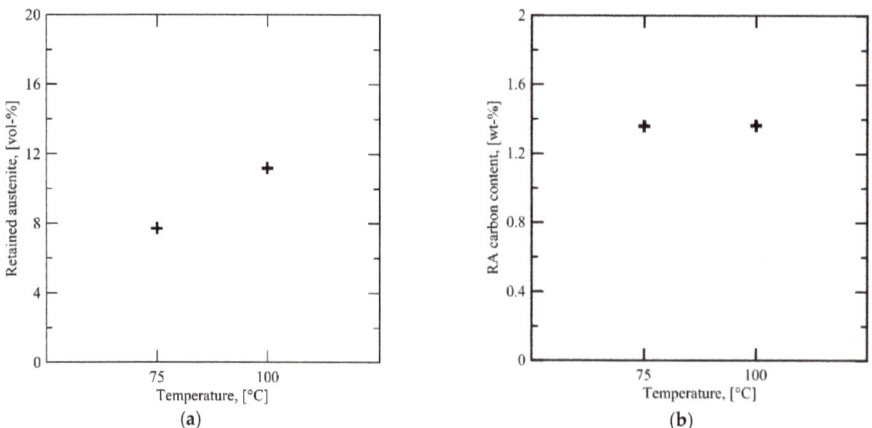

Figure 10. (**a**) The retained austenite fraction of Steel C with respect to initial quench temperature *QT*. (**b**) The carbon content of the retained austenite with respect to *QT*. Specimens annealed at 850 °C, quenched to *QT* and partitioned at 450 °C for 100 s.

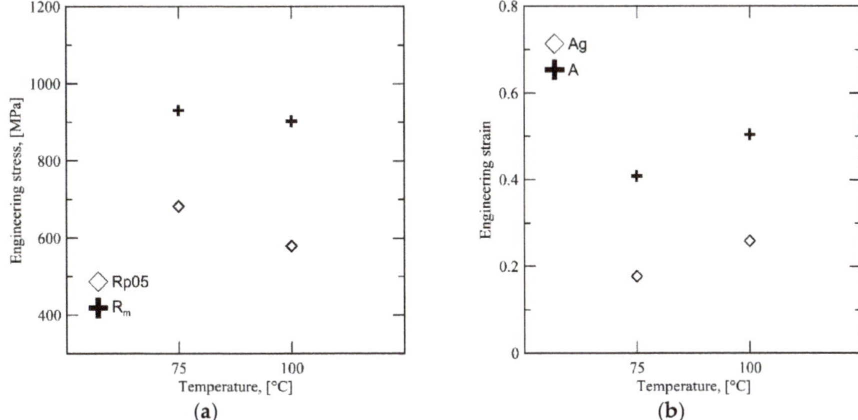

Figure 11. (**a**) R_{p05} and R_m and (**b**) uniform and total elongations for Steel C with respect to initial quench temperature QT. Specimens annealed at 850 °C, quenched to QT and partitioned at 450 °C for 100 s.

3.5. EBSD Measurements

Figure 12 shows representative results of the EBSD measurements as band contrast maps overlaid with austenite grains (shown in random coloring). Prior austenite and packet boundaries were determined using a previously developed iterative reconstruction algorithm [8,27] and are shown in red (packet boundaries) and black (prior austenite grain (PAG) boundaries and ferrite boundaries).

The observed microstructures in Figure 12a,c show that the martensitic transformation has been heterogeneous on a grain-by-grain basis for Steels A and C: untransformed, partially transformed and almost completely transformed austenite grains can be found in the microstructure. It should be noted that some austenite has probably transformed to martensite during EBSD specimen preparation, as the austenite fraction in EBSD measurements was much lower than in XRD.

Steel B does not exhibit a typical martensitic lath structure, although crystallographic analysis with the iterative method shows the presence of block- and packet-type subunits within prior austenite grains. This is an indication that some degree of bainite transformation has taken place either during the initial quenching or the partitioning stage of the heat treatment. In addition, several large, irregular-shaped intercritical ferrite grains are present in the microstructure.

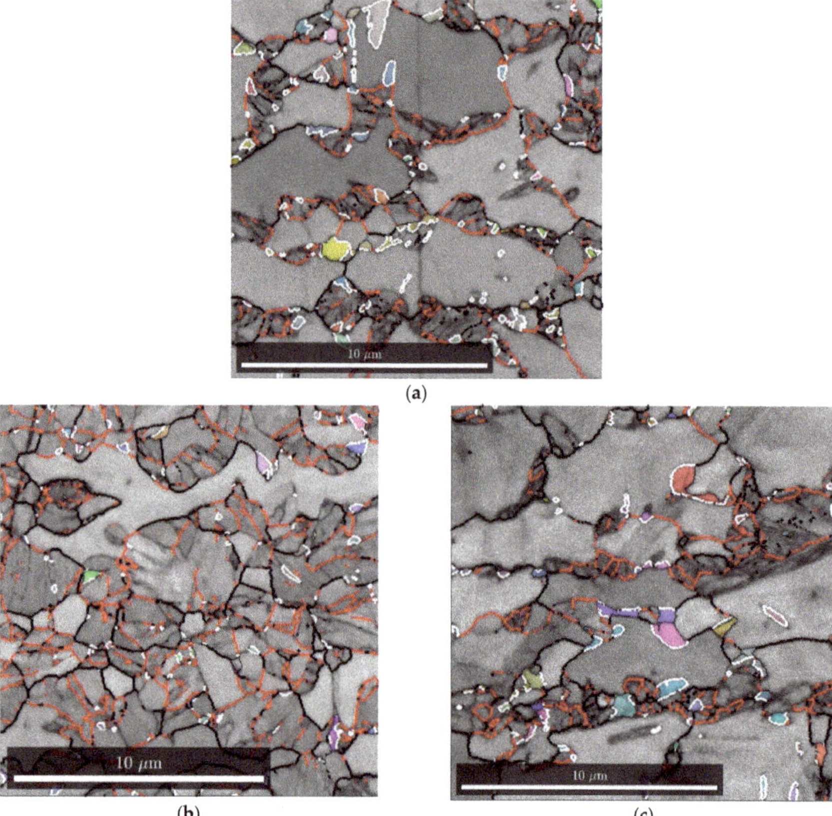

(a)

(b) (c)

Figure 12. Electron backscatter diffraction (EBSD) band contrast image overlaid with indexed retained austenite grains (random coloring). White boundaries indicate Kurdjumov–Sachs type relationship between γ and α. Black boundaries mark prior austenite grains and red boundaries indicate packet boundaries. (**a**) Steel A quenched to 150 °C, (**b**) Steel B quenched to 275 °C and (**c**) Steel C quenched to 100 °C. All specimens partitioned at 450 °C for 100 s.

4. Discussion

Based on the image analysis of the dilatometry specimens, there is an approximately 27 vol % intercritical austenite fraction in the microstructure of the Steel A after annealing. Assuming that all of the untransformed austenite that remains directly after the interrupted quenching is stabilized with carbon and is also retained at room temperature; 55 vol % is transformed at 175 °C, 63 vol % at 150 °C and 70 vol % at 125 °C. Fitting the Koistinen-Marburger equation to these values does not give a meaningful result, because the apparent martensitic transformation is too gradual with respect to temperature to obtain a good fit. Besides, martensite finish temperature M_f is an indistinct term. It is therefore probable that the martensitic transformation is not actually homogeneous in the microstructure and the degree of transformation varies from grain to grain. This conclusion is supported by the EBSD maps in Figure 11a, which shows a heterogenous martensitic transformation. The behavior of Steel C appears to follow a similar trend, based on the image analysis, XRD and EBSD results.

The behavior of Steel B differs from that of Steels A and C. Instead of a steady reduction in austenite fraction, there is an initially high amount at $QT = 300$ °C followed by an appreciable drop at $QT = 275$ °C to a nearly stable austenite fraction irrespective of further QT temperature reduction.

As shown by Figure 4b, there is much more austenite in the microstructure of Steel B after intercritical annealing compared to Steels A and C. This has two consequences—the average carbon content of the austenite is significantly lower (assuming total partitioning of carbon) and austenite grain size is higher. Both factors lower the critical driving force necessary for martensite nucleation, contributing to the rapid formation of martensite when lowering QT past 300 °C. It is possible that autocatalytic nucleation ("burst martensite" [28]) accelerates the rate of transformation. The rapid martensite formation is shown in Figure 13, which displays the retained austenite content with respect to QT overlaid with the dilatation curve in the temperature regime of the martensitic transformation.

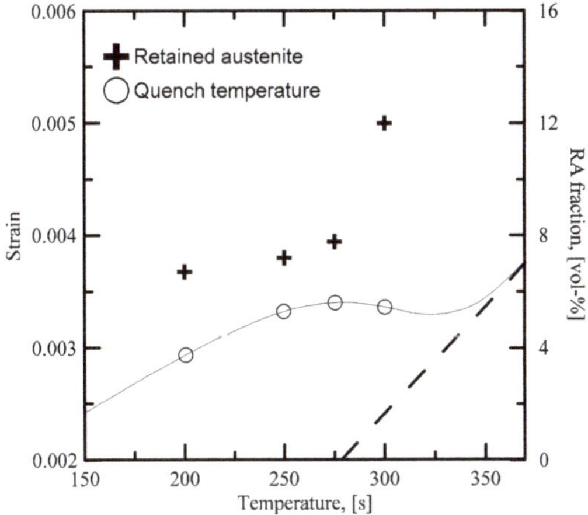

Figure 13. The dilatation curve of Steel B in the regime of martensite transformation, with the retained austenite content with respect to the initial quench temperature QT on the secondary y-axis.

Interestingly, the uniform elongation Ag does not correlate with the high retained austenite content at 300 °C. Instead, there is a significant drop in yield strength, accompanied with a significant rise in R_m, as shown by Figure 9. This behavior is likely to be caused by a combination of both high retained austenite fraction and the effect of the different austenite morphology characteristic to this QT. Figure 8b shows that the average carbon content of the austenite phase after quenching to 300 °C is lower, which should also affect mechanical stability. The presence of unstable austenite grains results in a very high degree of strain hardening at the initial stages of deformation and consequently results in a high R_m combined with a low initial yield point. The unstable austenite grains are unable to contribute to ductility during later stages of deformation, having been completely transformed at an earlier stage and resulting in a lower total elongation A, as shown by Figure 9b.

From a microstructure point of view, the expected response to the quenching and partitioning would be the partial transformation of each austenite grain into martensite, followed by the enrichment of the balance untransformed austenite with carbon. The final microstructure, shown in Figure 14a, would then be a mixture of intercritical ferrite and martensitic islands interspersed with carbon-enriched retained austenite in martensite.

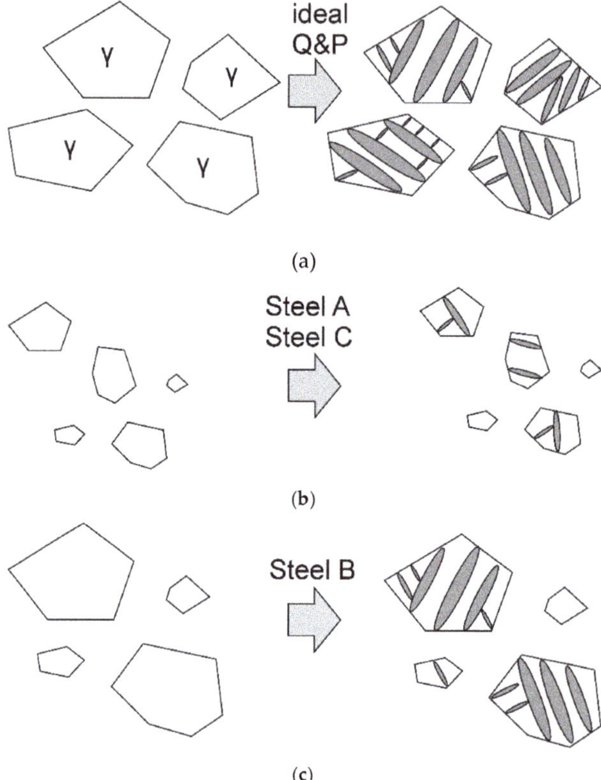

Figure 14. Austenite grains in a ferritic matrix partially transformed into martensite for (**a**) an ideal quenching and partitioning (Q&P) scenario, (**b**) the observed behavior of Steel A and Steel C and (**c**) the observed behavior of Steel B.

This type of microstructure was not observed in any of the studied grades. Instead, a complex microstructure had emerged in all cases consisting of intercritical ferrite, untransformed austenite grains and prior austenite grains in which the martensitic transformation had progressed to some degree.

This behavior can be attributed to two factors—the local chemical composition and the size of each austenite grain. During intercritical annealing, austenite will form at low-energy sites that are favorable towards nucleation [10]. In practice, this means ferrite grain boundaries and grain corners where the dissolution of cementite or other carbides has formed a carbon-rich volume suitable for nucleation. The stage of the annealing cycle at which each austenite grain nucleates will be decided by these local conditions. The growth rate of a nucleated austenite grain will, in turn, initially depend on the carbon content of the nucleus and at later stages the diffusion barrier formed by ferrite-stabilizing elements. In the case of the studied experimental steels, the primary element limiting austenite growth is aluminum, which is a strong ferrite stabilizer. As the austenite growth front advances, more and more aluminum will diffuse across the advancing front, until the aluminum content is high enough in the interfacial ferrite neighborhood and the growth slows down significantly.

The behavior during intercritical annealing can thus be characterized by a slow and uneven growth of austenite. Just prior to cooling, a microstructure has formed where exists a range of austenite grains with different sizes and chemical compositions. Such intercritical austenite microstructures are outlined schematically in Figure 14b,c.

Steel A and Steel C exhibited microstructures and mechanical properties quite similar and will be discussed together. Both steels have a similar chemical composition, with the notable differences being an elevated manganese content and a lower aluminum content in Steel A. This difference, along with the lower aluminum content, resulted in a slightly elevated austenite volume fraction after intercritical annealing in Steel A. In any case, both steels exhibit the type of intercritical austenite structure shown in Figure 14b, in which there are both very small and slightly larger austenite grains in the microstructure after annealing. When the steel in this condition is quenched to the quench temperature *QT*, primarily the larger, less stable austenite grains undergo a martensitic transformation, while the smaller, more stable grains remain unchanged. The less stable retained austenite is then stabilized with carbon during the partitioning stage. The result is a microstructure where the large, less stable austenite grains have become more refined and chemically stable due to martensitic transformation and subsequent rejection of supersaturated carbon during partitioning. Although only the larger austenite grains exhibit the expected quenched and partitioned response, the final result is a more homogenous, refined microstructure in terms of austenite stability, leading to the observed improvement in uniform elongation. Similar results have also been previously observed for steels with higher aluminum contents [3].

The quenching and partitioning response of Steel B is mostly explained by a greater fraction of intercritical austenite, along with a large distribution of austenite grain sizes and a larger average grain size overall. For Steel B, it is likely that bainite or isothermal martensite formation has occurred during partitioning, indicated by the large, irregularly shaped laths in Figure 11b. Another factor supporting bainite formation during partitioning is the high volume fraction of intercritical austenite (approximately 65 vol %). At 65 vol % austenite, assuming full partitioning of carbon, the austenite carbon concentration would be approximately 0.17 wt %. Compared to the Steels studied here with lower intercritical austenite fractions and consequentially higher carbon concentrations, Steel B is more amenable towards bainite (or, keeping in mind the high aluminum content, carbide-free bainite) formation during partitioning.

When Steel B is quenched to a sufficiently high temperature (in this case, 300 °C), the situation is similar to that observed for Steel A and Steel C: the smaller grains are left almost fully austenitic, while the larger grains transform to a greater degree (either to martensite or to carbide-free bainite). However, in this case, the smaller grains are left more unstable in the final microstructure; these unstable grains are transformed to martensite during the early stages of deformation, becoming unable to promote ductility at later stages. The message of this result is that to increase ductility, the end goal of the heat treatment should not be a perfect quenching and partitioning response, but the presence of highly stable retained austenite that will transform at a controlled stage of deformation. For Steels A and C, the quenching and partitioning treatment can be successfully used to refine and stabilize blocky-type austenite and increase the ductility of the steel, while at the same time introducing martensite into the microstructure. For Steel B, this is also possible to some degree, even though the tendency to form bainite or isothermal martensite during partitioning affects the final austenite fraction.

5. Conclusions

The quenching and partitioning response of three aluminum-alloyed experimental steels following intercritical annealing were investigated in this study. The following conclusions can be drawn from these investigations:

1. It is possible to produce quenched and partitioned dual-phase microstructures with superior mechanical properties using alloy compositions that can be processed with the current level of technology.
2. Each austenite grain undergoes martensitic transformation to a different degree at the initial quench temperature *QT* according to its mechanical and chemical stability.
3. Sufficiently stable austenite grains may remain completely untransformed.

The successful quenching and partitioning response of an intercritically annealed steel seems to largely depend on the state of the austenite prior to quenching. A sufficient carbon content and a small austenite grain size give the steels robust behavior with regard to heat treatment, when looking at Steels A and C. The martensitic transformation of these alloys can be controlled by varying *QT*.

Author Contributions: Conceptualization, T.N., O.O., P.J. and P.P.; methodology, T.N.; software, A.S.; validation, T.N.; formal analysis, T.N., M.S. and P.P.; investigation, T.N.; resources, O.O. and P.J.; data curation, T.N.; writing—original draft preparation, T.N.; writing—review and editing, T.N., P.P. and M.S.; visualization, T.N.; supervision, P.P.; project administration, T.N.; funding acquisition, P.P.

Funding: This research received no external funding.

Conflicts of Interest: The authors declare no conflict of interest.

References

1. Garcia, C.I.; Deardo, A.J. Formation of austenite in 1.5 pct Mn steels. *Metall. Trans. A* **1981**, *12*, 521–530. [CrossRef]
2. Speer, J.G.; Streicher, A.M.; Matlock, D.K.; Rizzo, F.C.; Krauss, G. Quenching and partitioning: A fundamentally new process to create high strength TRIP sheet microstructures. In Proceedings of the Symposium on the Thermodynamics, Kinetics, Characterization and Modeling of: Austenite Formation and Decomposition, Chicago, IL, USA, 9–12 November 2003; pp. 505–522.
3. Nyyssönen, T.; Peura, P.; De Moor, E.; Williamson, D.; Kuokkala, V.-T. Crystallography and mechanical properties of intercritically annealed quench and partitioned high-aluminum steel. *Mater. Charact.* **2019**, *148*, 71–80. [CrossRef]
4. Tan, X.; Ponge, D.; Lu, W.; Xu, Y.; Yang, X.; Rao, X.; Wu, D.; Raabe, D. Carbon and strain partitioning in a quenched and partitioned steel containing ferrite. *Acta Mater.* **2019**, *165*, 561–576. [CrossRef]
5. De Meyer, M.; Vanderschueren, D.; De Cooman, B.C. The Influence of the Substitution of Si by Al on the Properties of Cold Rolled C-Mn-Si TRIP Steels. *ISIJ Int.* **1999**, *39*, 813–822. [CrossRef]
6. Jacques, P.; Girault, E.; Catlin, T.; Geerlofs, N.; Kop, T.; van der Zwaag, S.; Delannay, F. Bainite transformation of low carbon Mn–Si TRIP-assisted multiphase steels: influence of silicon content on cementite precipitation and austenite retention. *Mater. Sci. Eng. A* **1999**, *273–275*, 475–479. [CrossRef]
7. Pierce, D.T.; Coughlin, D.R.; Clarke, K.D.; De Moor, E.; Poplawsky, J.; Williamson, D.L.; Mazumder, B.; Speer, J.G.; Hood, A.; Clarke, A.J. Microstructural evolution during quenching and partitioning of 0.2C-1.5Mn-1.3Si steels with Cr or Ni additions. *Acta Mater.* **2018**, *151*, 454–469. [CrossRef]
8. Nyyssönen, T.; Peura, P.; Kuokkala, V.-T. Crystallography, Morphology, and Martensite Transformation of Prior Austenite in Intercritically Annealed High-Aluminum Steel. *Metall. Mater. Trans. A* **2018**, *49*, 6426–6441. [CrossRef]
9. Datta, D.P.; Gokhale, A.M. Austenitization kinetics of pearlite and ferrite aggregates in a low carbon steel containing 0.15 wt pct C. *Metall. Trans. A* **1981**, *12*, 443–450. [CrossRef]
10. Shtansky, D.V.; Nakai, K.; Ohmori, Y. Pearlite to austenite transformation in an Fe–2.6Cr–1C alloy. *Acta Mater.* **1999**, *47*, 2619–2632. [CrossRef]
11. Nouri, A.; Saghafian, H.; Kheirandish, S. Effects of Silicon Content and Intercritical Annealing on Manganese Partitioning in Dual Phase Steels. *J. Iron Steel Res. Int.* **2010**, *17*, 44–50. [CrossRef]
12. Garcia, C.I.; Cho, K.; Redkin, K.; Deardo, A.J.; Tan, S.; Somani, M.; Karjalainen, L.P. Influence of Critical Carbide Dissolution Temperature during Intercritical Annealing on Hardenability of Austenite and Mechanical Properties of DP-980 Steels. *ISIJ Int.* **2011**, *51*, 969–974. [CrossRef]
13. Saunders, N.; Guo, U.K.Z.; Li, X.; Miodownik, A.P.; Schillé, J.-P. Using JMatPro to model materials properties and behavior. *JOM* **2003**, *55*, 60–65. [CrossRef]
14. Kirkaldy, J.; Doane, D. *Hardenability concepts with applications to steel*; Metallurgical society of AIME: New York, NY, USA, 1978.
15. Lee, J.-L.; Bhadeshia, H.K.D.H. A methodology for the prediction of time-temperature-transformation diagrams. *Mater. Sci. Eng. A* **1993**, *171*, 223–230. [CrossRef]
16. Yang, H.; Bhadeshia, H. Austenite grain size and the martensite-start temperature. *Scr. Mater.* **2009**, *60*, 493–495. [CrossRef]

17. Dong-Woo, S.; Kim, N.J.; Yi, H.L.; Chen, P.; Hou, Z.Y.; Hong, N.; Cai, H.L.; Xu, Y.B.; Wu, D.; Wang, G.D. A novel design: Partitioning achieved by quenching and tempering (Q–T & P) in an aluminium-added low-density steel. *Scr. Mater.* **2013**, *68*, 370–374.
18. Wang, X.; Liu, L.; Liu, R.D.; Huang, M.X. Benefits of Intercritical Annealing in Quenching and Partitioning Steel. *Metall. Mater. Trans. A* **2018**, *49*, 1460–1464. [CrossRef]
19. Maki, J.; Mahieu, J.; De Cooman, B.C.; Claessens, S. Galvanisability of silicon free CMnAl TRIP steels. *Mater. Sci. Technol.* **2003**, *19*, 125–131. [CrossRef]
20. Koistinen, D.P.; Marburger, R.E. A general equation prescribing the extent of the austenite-martensite transformation in pure iron-carbon alloys and plain carbon steels. *Acta Metall.* **1959**, *7*, 59–60. [CrossRef]
21. Vander Voort, G. *Metallography: Principles and Practice*; ASM International: Geauga County, OH, USA, 1984; ISBN 978-0-87170-672-0.
22. Jatczak, C.F.; Larsen, J.A.; Shin, S.W. *Retained Austenite and Its Measurements by X-Ray Diffraction; SP-453;* Society of Automotive Engineers, Inc.: Warrendale, PA, USA, 1980.
23. Cullity, B.D.; Stock, S.R. *Elements of X-Ray Diffraction*; Prentice Hall: Upper Saddle River, NJ, USA, 2001.
24. *ISO 2566-1:1984(E) Steel—Conversion of elongation values—Part 1: Carbon and low alloy steels*; International Organization for Standardization: Vernier, Switzerland, 1984; pp. 1–28.
25. Bhadeshia, H.K.D.H. Thermodynamic extrapolation and martensite-start temperature of substitutionally alloyed steels. *Met. Sci.* **1981**, *15*, 178–180. [CrossRef]
26. Bhadeshia, H.K.D.H. Driving force for martensitic transformation in steels. *Met. Sci.* **1981**, *15*, 175–177. [CrossRef]
27. Nyyssönen, T.; Isakov, M.; Peura, P.; Kuokkala, V.-T. Iterative Determination of the Orientation Relationship between Austenite and Martensite from a Large Amount of Grain Pair Misorientations. *Metall. Mater. Trans. A* **2016**, *47*, 2587–2590. [CrossRef]
28. Bhadeshia, H.K.D.H. An aspect of the nucleation of burst martensite. *J. Mater. Sci.* **1982**, *17*, 383–386. [CrossRef]

Article

Dynamic Phase Transformation Behavior of a Nb-microalloyed Steel during Roughing Passes at Temperatures above the Ae_3

Samuel F. Rodrigues [1,2,3,*], Fulvio Siciliano [4], Clodualdo Aranas Jr. [5], Eden S. Silva [1], Gedeon S. Reis [1], Mohammad Jahazi [2] and John J. Jonas [3]

1. Department of Materials Engineering, Federal Institute of Maranhao, Sao Luis 65075-441, Maranhao, Brazil; eden.silva@ifma.edu.br (E.S.S.); gedeonreis@ifma.edu.br (G.S.R.)
2. Département de Génie Mécanique, École de Technologie Supérieure, Montreal, QC H3C 1K3, Canada; mohammad.jahazi@etsmtl.ca
3. Materials Engineering, McGill University, Montreal, QC H3A 0C5, Canada; john.jonas@mcgill.ca
4. Dynamic Systems Inc. 323 NY 355, Poestenkill, NY 12140, USA; fulvio@gleeble.com
5. Mechanical Engineering, University of New Brunswick, Fredericton, NB E3B 5A3, Canada; clod.aranas@unb.ca

* Correspondence: samuel.filgueiras@ifma.edu.br or samuel.rodrigues@mail.mcgill.ca; Tel.: +55-98-98517-9142

Received: 16 February 2019; Accepted: 12 March 2019; Published: 15 March 2019

Abstract: A five-pass torsion simulation of the roughing passes applied during hot plate rolling was performed in the single-phase austenite region of a Nb-microalloyed steel under continuous cooling conditions. The deformation temperatures were approximately half-way between the Ae_3 and the delta ferrite formation temperature (i.e., 250 °C above the Ae_3) in which the free energy difference of austenite and ferrite is at maximum. The microstructures in-between passes were analyzed to characterize and quantify the occurrence of deformation-induced dynamic phase transformation. It was observed that about 7% of austenite transforms into ferrite right after the final pass. The results are consistent with the calculated critical strains and driving forces which indicate that dynamic transformation (DT) can take place at any temperature above the Ae_3. This mechanism occurs even with the presence of high Nb in the material, which is known to retard and hinder the occurrence of DT by means of pinning and solute drag effects. The calculated cooling rate during quenching and the time–temperature–transformation curves of the present material further verified the existence of dynamically transformed ferrite.

Keywords: dynamic transformation; Nb-microalloyed steel; roughing passes

---

1. Introduction

The thermomechanical processing of steels is carried out primarily within the austenite phase field. Previous work has shown that hot rolling produces partial phase transformation of austenite into ferrite in the roll bite inside the single austenite phase field [1,2]. This has been referred to as dynamic transformation (DT) and was first investigated by Yada and co-workers in the 1980's [3,4]. In their study, fine grains of ferrite were produced when three plain carbon were strained during compression testing above the Ae_3 temperature.

In order to provide real-time evidence for the occurrence of DT, Yada and co-workers returned to its study in early 2000s. They used the in-situ X-ray diffraction technique coupled to a torsion machine and deformed three Fe-C alloys above the Ae_3 temperature [5]. They captured the diffraction patterns associated with α-ferrite during deformation. Chen and Chen [6] performed experiments in 2003 by using a laser dilatometry technique and observed the reverse transformation of dynamically transformed ferrite into austenite at temperatures above the Ae_3. Liu et al. [7] performed tests in a Gleeble thermomechanical simulator in 2007 on a low-carbon steel and obtained similar metallographic results. In 2008, Sun et al. also employed a laser dilatometer in order to follow the reverse DT both below and above the Ae_3 [8]. The former authors deformed a 0.17% C plain carbon steel using a Gleeble thermomechanical simulator above the Ae_3. They confirmed the existence of both the forward and reverse transformation at temperatures up to 115 °C above the Ae_3.

In 2010, Basabe and Jonas [9] conducted torsion tests on a 0.036% Nb microalloyed steel in order to study the effects of strain, strain rate, and temperature on DT. They concluded that the reverse transformation was retarded by the addition of niobium in comparison with a plain C steel. This phenomenon was a result of dislocation pinning and solute drag of the niobium carbonitride precipitates and Nb in solution, respectively.

In 2013, Ghosh et al. [10], showed that DT ferrite can be formed as high as 130 °C above the Ae_3. In their work, they allowed for the inhomogeneous distribution of dislocations, leading to driving forces for ferrite formation as high as 197 J/mol when their materials were submitted to large strains. In 2015, Aranas et al. [1,11,12], presented a new approach involving thermodynamic features to explain the occurrence of DT as much as 500 °C above the Ae_3. The driving force for DT was redefined to consist only of the softening that takes place during transformation due to the applied stress. The free energy barrier against the driving force for DT consists of the Gibbs free energy difference between the phases as well as the lattice dilatation work and shear accommodation work. According to this model, DT takes place when the driving force overcomes the total barrier preventing its formation.

More recently, Rodrigues et al. [13–15], investigated the phenomenon of DT under various industrial plate rolling simulation conditions. They found that the application of roughing passes under isothermal conditions can lead to the presence of around 8% of transformed ferrite after deformation. The occurrence of DT above the Ae_3 temperature during thermomechanical processing is known to generate lower rolling loads and mean flow stresses (MFS) [16]. Moreover, the volume flow rate (as the bar passes through a rolling mill) increases due to formation of less dense ferrite. This type of transformation involves carbon partitioning, which can generate undesirable volume fractions of martensite. Thus, an accurate account of phases during high temperature deformation leads to better mechanical properties of the material.

Nevertheless, the phenomenon of DT has not been studied under the cooling stage of roughing rolling simulation, which represents more realistic industrial conditions. Thus, the present study represents an advance over the previous investigations in that the DT behavior of a Nb-microalloyed steel is investigated under continuous cooling condition of 2 °C/s, comparable to industrial roughing rolling schedules. The results obtained are described and discussed in following sections.

2. Materials and Methods

A Nb-microalloyed steel was investigated in the present work. This material was provided by EVRAZ North America in the form of hot-rolled plates. The complete chemical composition of the material (in wt%) is displayed in Table 1. The orthoequilibrium and paraequilibrium Ae_3 temperatures were identified by employing the FSstel database of the FactSage thermodynamic software 7.1 [17]. For the present analysis, the orthoequilibrium Ae_3 temperature will be considered, where both the substitutional and interstitial atoms are assumed to participate during phase transformation. The hot-rolled plates were machined into torsion samples with diameters and gauge lengths of 10 mm and 20 mm, respectively. The cylindrical axis of all the samples were parallel to the rolling direction of the plate. The torsion experiments were carried out using a Gleeble 3800 thermomechanical simulator (Dynamic Systems Inc. Poestenkill, NY, USA) with a Hot Torsion Mobile Conversion Unit (MCU). The samples were heated by flowing alternated current to the desired reheat temperature. This ensured that each deformation occurred at the proper temperature simulated by that rotation step. Thermal gradients in the sample were controlled by making the specimen solid in the torsion span and hollow on both sides to minimize gradients in the sample that would otherwise develop due to non-uniform current densities in the shoulders versus the reduced center section of the sample. A thermocouple was welded to the sample to accurately track the deformation temperatures at every pass.

Table 1. Chemical composition (mass%) and equilibrium transformation temperatures (°C).

C	Mn	Si	Cr	Nb	N	Orthoequilibrium Ae_3	Paraequilibrium Ae_3
0.047	1.56	0.25	0.21	0.092	0.008	845 °C	810 °C

The thermomechanical schedule of the present work is shown in Figure 1. The torsion samples were heated to 1200 °C at a rate of 1 °C/s and were isothermally held for 5 min to attain a single-phase austenite microstructure. The samples were then cooled to 1100 °C at a rate of 1 °C/s. The first deformation was applied after 60 s at 1100 °C, followed by controlled cooling at a rate of approximately 2 °C/s. The succeeding deformations (from 2nd pass up to the 5th pass) were applied during continuous cooling conditions with an interpass time of 10 s, mimicking the actual plate rolling process. The deformation temperatures were 1080 °C (2nd), 1060 °C (3rd), 1040 °C (4th), and 1020 °C (5th). Note that the samples were strained to 0.3 during each pass applied at a strain rate of 1 s^{-1}. A strain higher than the critical stains for the onset of DT [9–15] was selected so as to allow for DT to take place. Water spray quenching after R1, R3, and R5 were performed at minimum rate of 500 °C/s using the Gleeble high flow quenching system. All experiments were performed under argon atmosphere to minimize the oxidation and decarburization, which can affect the results. Additionally, the experiments were repeated three times to validate the results. In general, less than 3% difference in the level flow curves was observed.

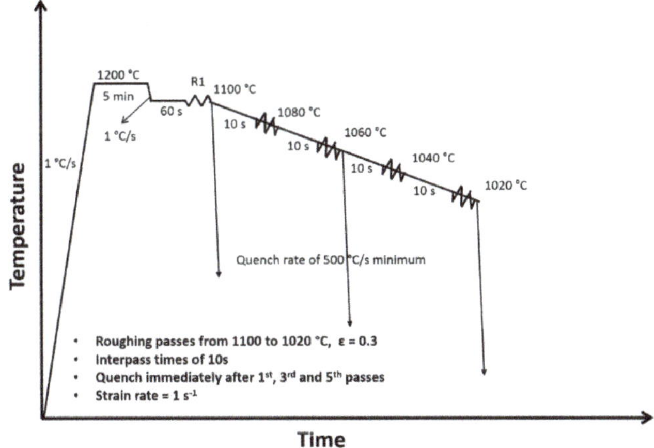

Figure 1. Thermomechanical schedule employed in the Gleeble torsion simulations of roughing passes. The deformation temperatures were 1100, 1080, 1060, 1040, and 1020 °C, strain of 0.3 was employed in each pass and interpass time of 10 s.

The deformed samples were water quenched before the 1st pass, and after 1st, 3rd, and 5th pass to accurately track the evolution of microstructure during multi-pass rolling simulation. The deformed and quenched torsion samples were sectioned longitudinally to reveal the changes in grain shape that accompany straining for microscopy analysis. These analyses were carried out at about 150 μm below the surface of the samples so as to avoid the oxidized outer layer. A conductive hot phenolic resin was used to mount the samples. These were polished using SiC grits 400, 600, 800, and 1200 lubricated with water. The final polishing was carried out using 3 and 1 μm diamond suspensions. The polished surfaces were etched with 2% nital (to reveal the microstructure) followed by 10% aqueous metabisulfite ($Na_2S_2O_5$) solution (to provide contrast between ferrite and martensite).

3. Results

3.1. Flow Curves and Mean Flow Stress

The torque-twist curves obtained were converted into stress-strain curves using Fields and Backofen formulation [18]. The stress–strain curves associated with the five-pass torsion simulation are displayed in Figure 2a. The peak stresses of the 1st (1100 °C) 2nd (1080 °C), 3rd (1060 °C), 4th (1040 °C), and 5th (1020 °C) passes are 83, 93, 98, 105, and 112 MPa, respectively. Although the increasing trend of peak stresses during cooling appears to be typical behavior of a material during cooling, note that the rate of increase from 1st pass going to the 2nd pass is about 0.5 MPa/°C. This rate is higher than the ones in succeeding passes due to strain accumulation. For example, from 2nd to 5th pass, the average rate of increase from pass to pass is approximately 0.3 MPa/°C. This is 40% lower than the rate of increase from the initial two passes. It appears that the low rate of peak stress increase indicates that dynamic softening is taking place in the material. To further analyze these observations from the stress–strain curves, the mean flow stresses (MFS) were calculated by measuring the area of the stress–strain curves, normalized by the amount of strain. The dependence of MFS on temperature and on the total applied strain is shown in Figure 2b. Note that the slope of the linear relation between the 1st and 2nd passes are considerably higher than the slope of the 3rd, 4th, and 5th passes. This is an indication of softening by combination of recrystallization and phase transformation occurring in the material during deformation.

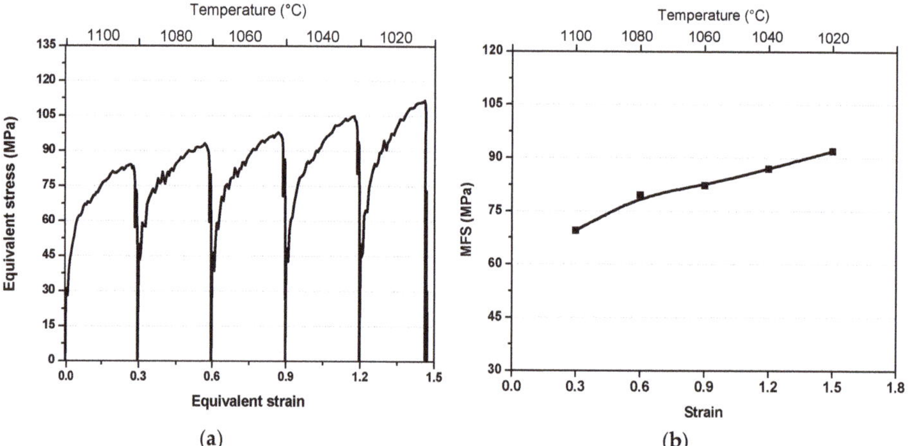

Figure 2. (**a**) Roughing stress–strain curves determined according to the schedule of Figure 1 using pass strains of 0.3 applied at 1 s^{-1}: (**b**) mean flow stresses (MFS) curve derived from the stress–strain curves of Figure 2a.

3.2. Microstrutural Results

The microstructures before the 1st pass (Figure 3a), after the 1st pass (Figure 3b), after the 3rd pass (Figure 3c), and after the 5th (Figure 3d) are displayed in Figure 3. The martensite phase appears dark and characterized by needle-shape structures inside the grains, while ferrite is the lighter structure and commonly identified by its polygonal structure. Here the grain sizes before the initial deformation are quite large, which was measured to be around 54 ± 15 μm (see Figure 3a). These grain sizes are slightly smaller than the typical sizes of austenite phase in industrial plate rolling before applying the roughing passes. The phase consists of entirely martensitic structure (prior austenite phase at elevated temperature). Note that after the five-pass simulation (see Figure 3d), the grain sizes decreased to less than 10 μm, which suggests the occurrence of either static recrystallization (SRX), metadynamic recrystallization (MDRX), dynamic transformation (DT), or a combination of these softening mechanisms. Although a long interpass time of 10 s was employed, it is important to note that the presence of Nb can significantly delay SRX in-between passes. A more interesting observation is the presence of light structures (ferrite), which forms dynamically. It is well-known that ferrite is softer than austenite due to its higher stacking fault energy [19]. Therefore, the presence of dynamically transformed ferrite (in combination with DRX) can generate significant softening of the material.

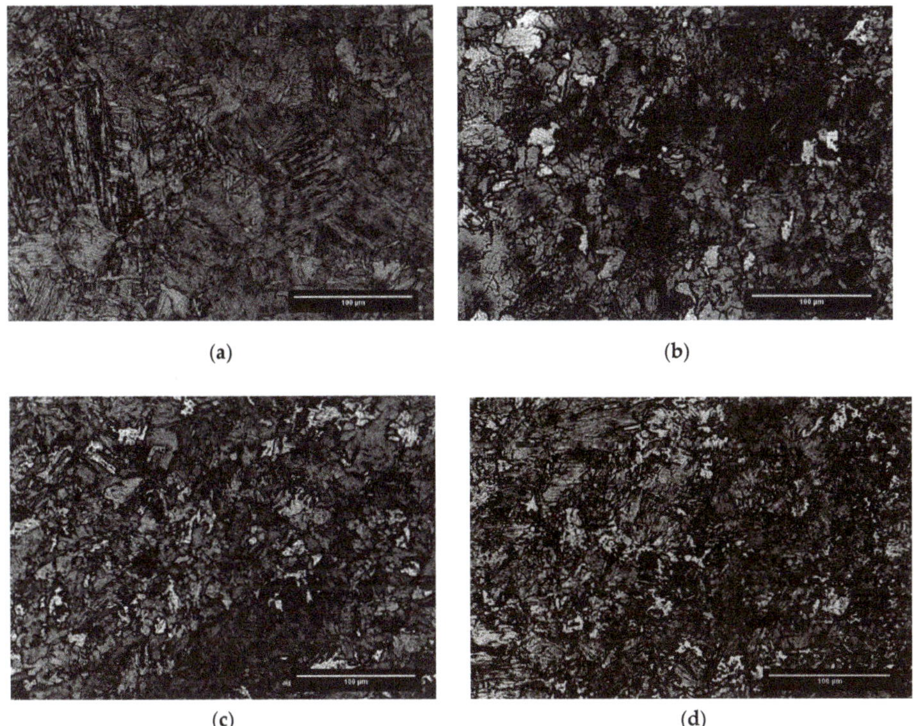

Figure 3. Optical microscopy images of steel subjected to the roughing simulation. The samples were quenched immediately: (**a**) Before the first pass; (**b**) after the first pass; (**c**) after the third pass and (**d**) after the fifth pass. Light regions are ferrite while the dark regions are martensite (prior austenite).

3.3. Volume Fraction of Transformed Ferrite

The volume fraction of ferrite was measured using the ImageJ software [20] to quantify the contribution of DT ferrite on the unusual behavior of the stress–strain curves, as shown above. The results are plotted and presented in Figure 4. The volume fraction of ferrite was measured based on the cumulative strain. Even though the reverse transformation of ferrite back into austenite can take place during the pass intervals, the amount of ferrite continously increases with applied strain. Note that the volume fraction of ferrite after the 2nd and 4th passes were interpolated. These values demonstrate that the rate of peak stress increase during cooling can be significantly affected by the occurrence of DT once the volume fraction of ferrite reached above 5%. This observation is consistent with the results of an earlier work of the present authors [13–15].

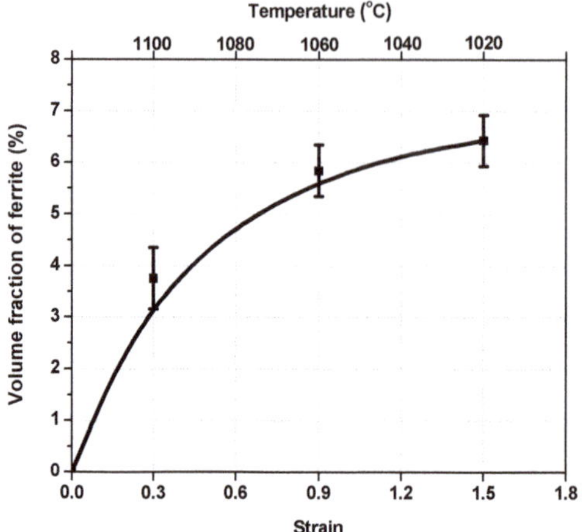

Figure 4. Dependence of the cumulative volume fraction of ferrite formed on the cumulative strain and temperature.

4. Discussion

4.1. Cooling Rate and Continuous Cooling Transformation Curves

Although the microstructures after quenching can be considered as a direct evidence of dynamic transformation of austenite to ferrite above the Ae_3 temperature, the cooling rate during quenching is a critical data that should be obtained to verify the validity of the observations. This is due to the fact that ferrite can also form during cooling of austenite. In the present experiments, the quenching rate was measured to be above 1200 °C/s. For example, water quenching from 1020 °C to 200 °C (well below the Ae_3) took 0.7 s (see Figure 5). These measurements were taken using a welded thermocouple attached to the surface of the sample and it remained connected even after deformation. Although the cooling rate of the cylindrical sample may vary from the surface going to its center, the microstructures shown in the previous section were obtained from near the surface of the samples, which can accurately represent the microstructures with cooling rates of more than 1200 °C/s.

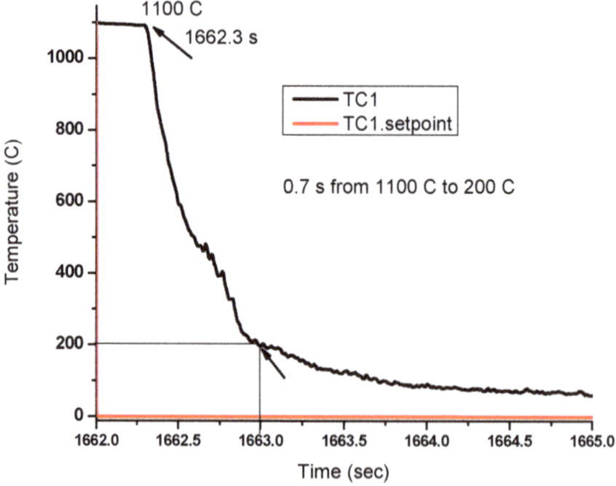

Figure 5. Setpoint (red line) and measured temperature (black line) by the thermocouple on the surface of the samples. It is observed that took less than 1 s for the temperature to drop well below the Ae_3.

The continuous cooling transformation (CCT) curves of the present material with austenite grain sizes of 10 µm and 26 µm are displayed in Figure 6a,b, respectively. These were calculated using the Fe Alloys module of JMatPro software 10.2 (Guildford, UK). Based on these plots, it can be seen that it requires approximately 4 s to obtain 1% ferrite. This required cooling time of austenite is applicable to both coarse and fine grain austenite structure. In the present work, the cooling time is always below 1 s. This means that the ferrite phase observed and measured from the previous section was solely due to the applied deformation at elevated temperature.

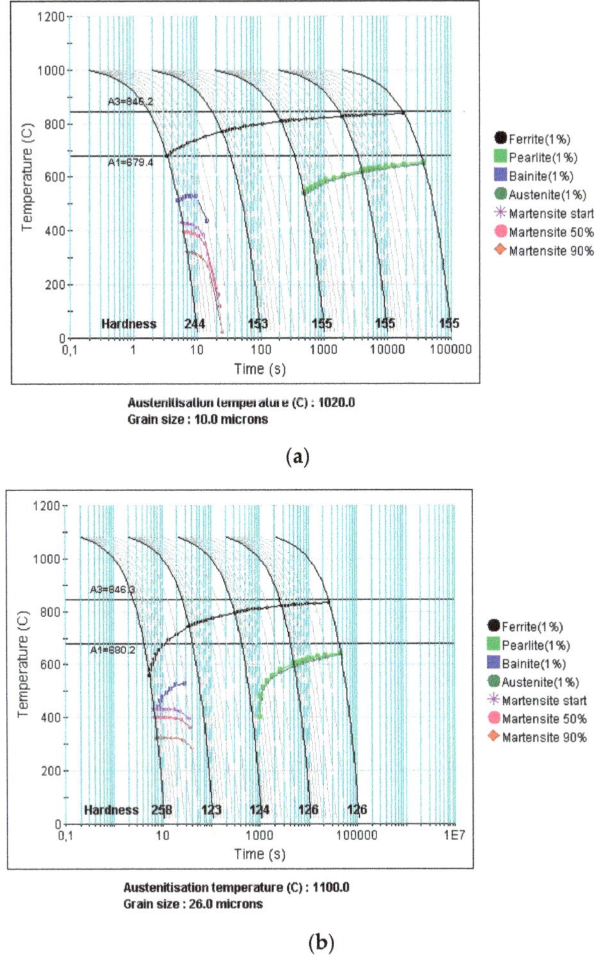

Figure 6. Calculated continuous cooling transformation (CCT) of the material quenched from: (**a**) 1020 °C and grain size of 10 µm and; (**b**) 1100 °C and grain size of 26 µm.

4.2. Critical Stresses and Strains

The stress–strain curves associated with 5-pass rolling simulation of Figure 2a are fitted with a 9th order polynomial using a Matlab software R2018b (The MathWorks, Natick, MA, USA) followed by application of the double differentiation method [18]. In this process, the critical stresses and strains were determined by initially plotting the θ versus σ curves where θ is strain hardening rate calculated by $\delta\sigma/\delta\varepsilon$ at a fixed strain rate (first derivative). The critical stresses are characterized by inflection points of the θ-σ plot. This can be described by the equation:

$$\frac{\delta}{\delta\sigma}\left(\frac{\delta\theta}{\delta\sigma}\right) = 0, \tag{1}$$

The minima in the plot of $\delta\theta/\delta\sigma$ versus σ (see Figure 7) are related to the softening mechanisms during deformation. The lower critical stresses of each curve are associated with dynamic transformation while the higher critical stresses are for dynamic recrystallization [20]. The critical strains can also be identified from critical stresses of Figure 7.

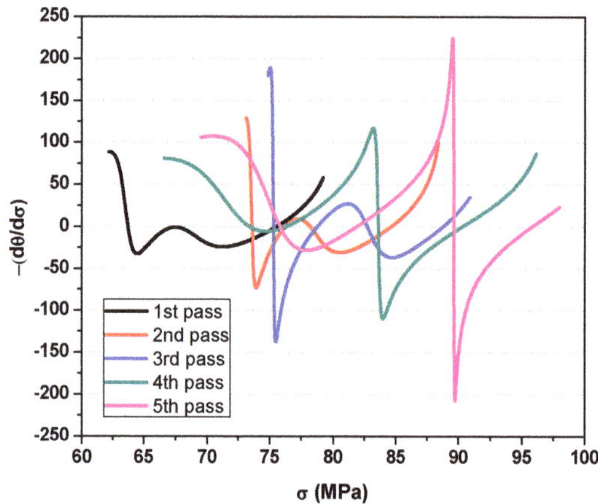

Figure 7. Plots of $-(\mathrm{d}\theta/\mathrm{d}\sigma)$ versus σ used to identify the minima related to the initiation of dynamic transformation (DT) and dynamic recrystallization (DRX).

The dependence of critical strains on pass number is displayed in Figure 8. The critical strains for DT are estimated to be around 0.06 while the critical strain for DRX is 0.12. These values are consistent with previous investigations of the present authors in the present material [21]. Moreover, the critical strains for DRX falls within the range of values shown in the literature [22] using the same method employed in the current work [19].

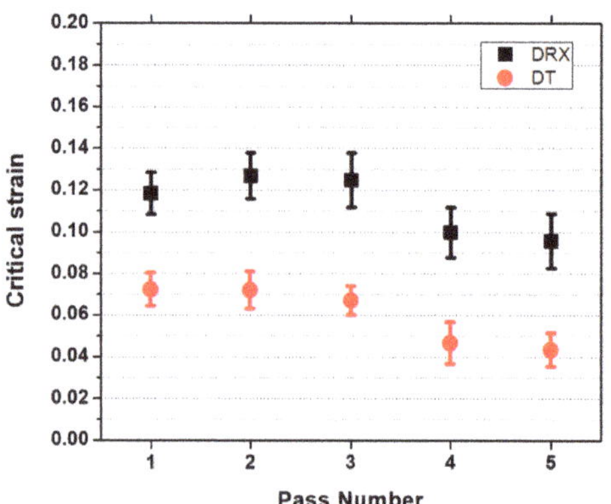

Figure 8. Critical strains for dynamic transformation (DT) and dynamic recrystallization (DRX) as a function of roughing pass number.

The critical strains for both DT and DRX display a slight decrease with increasing pass number. Note that for DRX, the critical strains should increase during cooling from one pass to another. The discrepancy in the trend of critical strains can be attributed to retained work hardening from previous pass. This is quite noticeable in the flow curves displayed in Figure 2. Since it is expected that NbC precipitates can form and pin down the dislocations, there is less recovery in-between passes. This leads to higher retained work hardening, which provides additional energy on top of the applied stress. A lower retained work hardening is expected if the material is low-alloyed steel or if the interpass time is increased [16]. In the present work, the attention will be focused on DT and the energies associated with this phenomenon.

4.3. Total Energy Obstacles and Driving Force

The occurrence of dynamic transformation can be explained in terms of transformation softening model [21]. In this model, the driving force to transformation (*DF*) is the flow stress difference between the critical stress of the parent phase (σ_C) and the yield stress of the product phase (σ_{YS}) defined by Equation (2) below:

$$DF = \sigma_C - \sigma_{YS}, \quad (2)$$

The critical stresses from austenite phase were obtained from the previous section while the yield stress of ferrite was estimated in a previous work of the present authors [23]. The dependence of the calculated driving force with temperature is shown in Figure 9 (see solid circles). The work per unit volume are converted into thermodynamic quantities using the conversion factor 1 MPa = 7.2 J/mol [10].

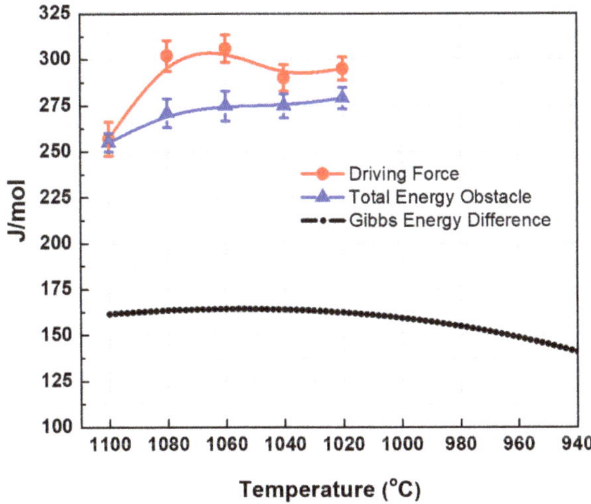

Figure 9. Calculated driving force (red line) over the total energy obstacle with the temperature.

The ***total energy obstacles*** consist of the work of dilatation (W_D) and shear accommodation (W_{SA}) associated with the phase transformation of austenite to ferrite, and the free energy difference between the phases ($\Delta G\gamma\text{-}\alpha$). This is defined by Equations (3)–(5) below:

$$\textit{Total Energy Obstacles} = W_{SA} + W_D + \Delta G\gamma\text{-}\alpha, \quad (3)$$

$$W_{SA} = \sigma_C \times 0.36 \times m, \quad (4)$$

$$W_D = \sigma_C \times 0.03 \times m, \quad (5)$$

The work of dilatation and shear accommodation are dependent on the critical stresses, which were obtained from the previous section. Since the dilatation and shear accommodation strains are difficult to measure experimentally, the present work assumed values of 0.03 and 0.36, respectively [24]. These values are based on the theoretical required deformation strains to transform austenite into ferrite. The Schmid factor (m) is based on the transformation habit plane and the shear direction when austenite transforms into ferrite, which are $(0.506, 0.452, 0.735)$ and $(-0.867, 0.414, 0.277)$, respectively. The present calculation employs the highest Schmid factor of 0.5 since the most oriented grains (with respect to the applied stress) are expected to transform first. Note that the critical strains are normally associated with the transformation of grains that are well oriented with the direction of applied stress [21,23,25].

The free energy difference between the austenite and ferrite was calculated using the FSstel database of the FactSage thermodynamic software. The sum of the components of the energy obstacles was calculated and are presented in Figure 9 (see solid triangle). Since the driving force is greater than the total energy obstacles for the present temperature range of 1020 °C to 1100 °C, it is expected that transformation of austenite to ferrite can take place. Although the calculated values may not represent the exact driving force and total energy obstacles due to the assumptions specified above, the difference between the curves look reasonable. This is supported by the microstructures which show that austenite transforms into ferrite. The kinetics of transformation may possibly be hindered by niobium which can delay the progression of transformation due to pinning and/or solute drag effects [16]. For this reason, the present material only obtained less than 10% ferrite at total applied torsional strain of 1.5. This may also be the reason for lower difference between the calculated driving forces and total energy obstacle.

It is also important to note that the free energy difference curve for the present material shows a peak, see Figure 10. This behavior indicates that austenite can easily transform if the temperatures are either close to the Ae_3 or near the delta-ferrite formation temperature. The present temperature range covers the peak free energy difference; thus, these temperatures requires higher amount of driving force to initiate the occurrence of dynamic phase transformation. Based on the present results at the selected temperature range, it is expected that austenite can be transformed into ferrite at any temperature above the Ae_3 for the present material [1].

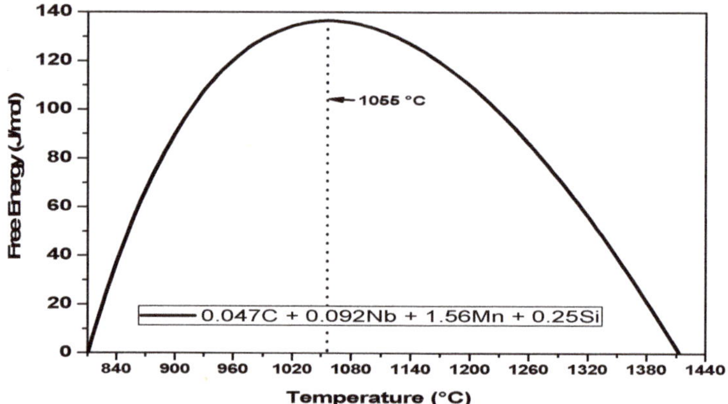

Figure 10. $\Delta G_{(\alpha-\gamma)}$ vs ΔT for the present Nb-microalloyed steel showing the Gibbs energy obstacle opposing dynamic transformation.

5. Conclusions

1. Dynamic transformation of austenite to ferrite can take place during roughing passes of the plate rolling process and its volume fractions formed and retained in the simulations increase as the pass

number increases. The present work showed that this can take place on a Nb-microalloyed steel subjected to roughing passes at temperature range 1020 to 1100 °C.

2. The calculated CCT diagrams confirmed that the measured cooling rate of 1200 °C/s is enough to prevent the formation of ferrite by cooling. Therefore, the volume fraction of ferrite measured in present work is only attributed to the applied deformation.

3. The thermodynamic calculations show that the driving force for DT is higher than the total barrier in the present investigation. Assumptions were made on the values of dilatation and shear accommodation strains.

4. The critical strains to dynamic transformation were shown to be in the range 0.04 to 0.07. This value decreases from pass to pass due to retained work hardening from the previous pass. The highest critical strain pertains to the first pass (where there is no prior deformation) during multi-pass high-temperature deformation experiments.

Author Contributions: All the authors contributed to this research work: Conceptualization, S.F.R., F.S., C.A.J., M.J. and J.J.J.; Formal analysis, S.F.R., C.A.J. and J.J.J.; Data curation, S.F.R., F.S., G.S.R. and E.S.S.; Methodology, S.F.R., F.S. and J.J.J.; Software, S.F.R., C.A.J. and E.S.S.; Validation, S.F.R., M.Z. and J.J.J.; Investigation, S.F.R., C.A.J., G.S.R. and J.J.J.; Resources, S.F.R., F.S., G.S.R., and E.S.S.; Writing—original draft preparation, S.F.R., C.A.J., F.S. and J.J.J.; Writing—review and editing, C.A.J., F.S. and J.J.J.; Visualization, C.A.J., F.S. and E.S.S.; Supervision, S.F.R., M.Z., J.J.J. and F.S.; Project administration, S.F. and J.J.J.; Funding acquisition, S.F.R., F.S. and C.A.J.

Acknowledgments: The authors acknowledge with gratitude funding received from the Brazilian National Council for Scientific and Technological Development (CNPq), the Industrial Research Chair in Forming Technologies of High Strength Materials of the École de Technologie Supérieure of Montréal, the New Brunswick Innovation Foundation (NBIF) and the Natural Sciences and Engineering Research Council of Canada. They also thank Jon Jackson and Laurie Collins from the EVRAZ North America Research and Development Centre (Regina) for providing the steel investigated in this research.

Conflicts of Interest: The authors declare no conflict of interest.

References

1. Grewal, R.; Aranas, C.; Chadha, K.; Shahriari, D.; Jahazi, M.; Jonas, J.J. Formation of Widmastatten ferrite at very high temperatures in the austenite phase field. *Acta Mater.* **2016**, *109*, 23–31. [CrossRef]
2. Aranas, C.; Jung, I.H.; Yue, S.; Rodrigues, S.F.; Jonas, J.J. A metastable phase diagram for the dynamic transformation of austenite at temperatures above the Ae_3. *Int. J. Mater. Res.* **2016**, *107*, 881–886. [CrossRef]
3. Matsumura, Y.; Yada, H. Evolution deformation of ultrafine- grained ferrite in hot successive deformation. *Trans. ISIJ.* **1987**, *27*, 492–498. [CrossRef]
4. Yada, H.; Matsumura, Y.; Senuma, T. International conference on physical metallurgy of thermomechanical processing of steels and other metals. Thermec-88. Vol. 2. In Proceedings of the 1st Conference Physical Metallurgy of Thermomechanical Processing of Steels and other Metals, Tokyo, Japan, 6–10 June 1988; Tamura, I., Ed.; 1988; pp. 200–207.
5. Yada, H.; Li, C.M.; Yamagata, H. Dynamic γ—A transformation during hot deformation in Iron-Nickel-Carbon alloys. *ISIJ Int.* **2000**, *40*, 200–206. [CrossRef]
6. Chen, Y.; Chen, Q. Dilatometric investigation on isothermal transformation after hot deformation. *J. Iron Steel Res. Int.* **2003**, *10*, 46–48.
7. Liu, Z.; Li, D.; Lu, S.; Qiao, G. Thermal stability of high temperature deformation induced ferrite in a low carbon steel. *ISIJ Int.* **2007**, *47*, 289–293. [CrossRef]
8. Sun, X.; Luo, H.; Dong, H.; Liu, Q.; Weng, Y. Microstructural evolution and kinetics for post-dynamic transformation in a plain low carbon steel. *ISIJ Int.* **2008**, *48*, 994–1000. [CrossRef]
9. Basabe, V.V.; Jonas, J.J. The ferrite transformation in hot deformed 0.036% Nb austenite at temperature above the Ae_3. *ISIJ Int.* **2010**, *50*, 1185–1192. [CrossRef]
10. Ghosh, C.; Basabe, V.V.; Jonas, J.J.; Kim, Y.; Jung, I.; Yue, S. The dynamic transformation of deformed austenite at temperatures above the Ae_3. *Acta Mater.* **2013**, *61*, 2348–2362. [CrossRef]
11. Aranas, C.; Jonas, J.J. Effect of Mn and Si on dynamic transformation of austenite above the Ae_3 temperature. *Acta Mater.* **2015**, *82*, 1–10. [CrossRef]

12. Aranas, C.; Nguyen-Minh, T.; Grewal, R.; Jonas, J.J. Flow softening-based formation of Widmanstatten ferrite in a 0.06%C steel deformed above the Ae_3. *ISIJ Int.* **2015**, *55*, 300–307. [CrossRef]
13. Rodrigues, S.F.; Aranas, C.; Sun, B.; Siciliano, F.; Yue, S.; Jonas, J.J. Effect of grain size and residual strain on the dynamic transformation of austenite under plate rolling conditions. *Steel Res. Int.* **2018**, *89*, 1–7. [CrossRef]
14. Rodrigues, S.F.; Aranas, C.; Wang, T.; Jonas, J.J. Dynamic transformation of an X70 steel under plate rolling conditions. *ISIJ Int.* **2017**, *57*, 162–169. [CrossRef]
15. Rodrigues, S.F.; Aranas, C.; Jonas, J.J. Retransformation behavior of dynamically transformed ferrite during the simulated plate rolling of a low C and an X70 Nb steel. *ISIJ Int.* **2017**, *57*, 929–936. [CrossRef]
16. Aranas, C.; Wang, T.; Jonas, J.J. Effect of interpass time on the dynamic transformation of a plain C-Mn and Nb-microalloyed steel. *ISIJ Int.* **2015**, *55*, 647–654. [CrossRef]
17. Bale, C.W.; Belisle, E.; Chartrand, P.; Decterov, S.A.; Eriksson, G.; Hack, K.; Jung, I.H.; Kang, Y.Y.; Melancon, J.; Pelton, A.D.; et al. FactSage thermomechanical software and database-recent developments. *Calphad* **2009**, *33*, 295–311. [CrossRef]
18. Fields, D.S.; Backofen, W.A. Determination of strain hardening characteristics by torsion testing. *Proc. Am. Soc. Test. Mater.* **1957**, *57*, 1259–1272.
19. Poliak, E.I.; Jonas, J.J. A one-parameter approach to determining the critical conditions for the initiation of dynamic recrystallization. *Acta Mater.* **1996**, *44*, 127–136. [CrossRef]
20. Schneider, C.A.; Rasband, W.S.; Eliceiri, K.W. NIH Image to ImageJ: 25 years of image analysis. *Nat. Meth.* **2012**, *9*, 671–675. [CrossRef]
21. Ghosh, C.; Aranas, C.; Jonas, J.J. Dynamic transformation of deformed austenite at temperatures above the Ae. *Prog. Mater. Sci.* **2016**, *82*, 151–233. [CrossRef]
22. Rodrigues, S.F.; Aranas, C.; Jonas, J.J. Dynamic Transformation during the Simulated Plate Rolling of a 0.09% Nb Steel. *ISIJ Int.* **2017**, *57*, 1102–1111. [CrossRef]
23. Poliak, E.I.; Jonas, J.J. Critical strain for dynamic recrystallization in variable strain rate hot deformation. *ISIJ Int.* **2003**, *43*, 692–700. [CrossRef]
24. Aranas, C.; Rodrigues, S.F.; Fall, A.; Jahazi, M.; Jonas, J.J. Determination of the critical stress associated with dynamic phase transformation in steels by means of free energy method. *Metals* **2018**, *8*, 360. [CrossRef]
25. Jonas, J.J.; Ghosh, C. Role of mechanical activation in the dynamic transformation of austenite. *Acta Mater.* **2013**, *61*, 6125–6131. [CrossRef]

Article

Transformation-Induced Plasticity in Super Duplex Stainless Steel F55-UNS S32760

Andrea Francesco Ciuffini [1,*], Silvia Barella [1], Cosmo Di Cecca [1], Andrea Di Schino [2], Andrea Gruttadauria [1], Giuseppe Napoli [2] and Carlo Mapelli [1]

1 Dipartimento di Meccanica, Politecnico di Milano, via La Masa 34, 20156 Milano, Italy; silvia.barella@polimi.it (S.B.); cosmo.dicecca@polimi.it (C.D.C.); andrea.gruttadauria@polimi.it (A.G.); carlo.mapelli@polimi.it (C.M.)

2 Dipartimento di Ingegneria, Università degli Studi di Perugia, via G. Duranti, 06125 Perugia, Italy; andrea.dischino@unipg.it (A.D.S.); giuseppe.napoli@studenti.unipg.it (G.N.)

* Correspondence: andreafrancesco.ciuffini@polimi.it; Tel.: +39-340-125-3162

Received: 17 January 2019; Accepted: 3 February 2019; Published: 6 February 2019

Abstract: Due to their unique combination of properties, Super Duplex Stainless Steels (SDSSs) are materials of choice in many industries. Their applications and markets are growing continuously, and without any doubt, there is a great potential for further volume increase. In recent years, intensive research has been performed on lean SDSSs improving mechanical properties exploiting the lack of nickel to generate metastable γ-austenite, resulting in transformation-induced plasticity (TRIP) effect. In the present work, a commercial F55-UNS S32760 SDSS have been studied coupling its microstructural features, especially secondary austenitic precipitates, and tensile properties, after different thermal treatments. First, the investigated specimens have been undergone to a thermal treatment solution, and then, to an annealing treatment with different holding times, in order to simulate the common hot-forming industrial practice. The results of microstructural investigations and mechanical testing highlight the occurrence of TRIP processes. This feature has been related to the Magee effect, concerning the secondary austenitic precipitates nucleated via martensitic-shear transformation.

Keywords: steel; austenite; mechanical characterization; martensitic transformations; phase transformation

---

1. Introduction

Duplex stainless steels (DSSs), and consequently, Super Duplex stainless steels (SDSSs) consist of austenite and ferrite phases. The resultant microstructure exhibits good combinations of strength, ductility, and corrosion resistance, since it takes advantages of the single-phase counterparts [1], with the main difference from more common austenitic stainless steels [2,3]. The steels not only inherit the mechanical properties of the completely ferritic or completely austenitic alloys, but they also exceed them. A factor of economic importance is the low content of expensive nickel, usually 4–7% compared with 10% or more in austenitic grades, as a result of which the life cycle cost of the DSSs is the lowest in many applications [4–7]. In DSSs, the two structure components b.c.c α-ferrite and f.c.c. γ-austenite lies as crystals of the same size statistically distributed next to each other [8]. However, adjustment of the two-phase microstructure of duplex stainless steels is complicated because a balanced phase ratio does not only depend on alloy components [9]. Indeed, the decomposition of ferrite to austenite can occur over a wide temperature range. This phenomenon can be understood on the basis that the duplex structure is quenched from a higher temperature, at which the equilibrium fraction of α-ferrite is higher. There appear to be three mechanisms by which austenite can precipitate within α-ferrite grains: By the eutectoid reaction, as Widmannstätten precipitates, and via a martensitic shear process [9–11]. Martensitic transformation in solids provides an unusual mechanical behavior ranging from the superelastic behavior typical of shape-memory alloys to non-thermoelastic behavior where the transformation induced plasticity (TRIP) phenomenon allows the development of steels with a good compromise between ductility and toughness [12–15]. This typical property of TRIP-aided steels results from the strong couplings between plasticity by dislocation motion and martensitic phase transformation through the internal stresses generated by both inelastic processes. The TRIP mechanism is based on deformation-stimulated displacive transformation of a metastable former phase. In such materials, the overall behavior depends deeply on the so-called chemical energy, which leads the martensitic shear transformation. The motivation for that is twofold. First, the former phase should be sufficiently unstable such that a transformation-induced plasticity effect is initiated upon loading. Second, the former phase should be sufficiently stable that the TRIP effect occurs over a wide strain regime, specifically at high strains, where strain-hardening reserves are usually more desirable than at low strains. In order to obtain this phenomenon in DSS steels, different studies have been recently performed; designing new alloys compositions [16–18]. The aim of this work is to achieve the occurrence of the TRIP effect in SDSSs commercial alloys, exploiting the martensitic shear transformation of the γ-austenite within α-ferritic grains, just by tuning the proper thermal treatment.

2. Materials and Methods

This study has been performed on a commercial F55-UNS S32760 super duplex stainless steel, which had a chemical composition, as designed by standards and measured via optical emission spectroscopy (OES), is reported in Table 1. Samples have been drawn from a bulk ingot and thermally treated. First, a solution thermal treatment (STT) at 1573 K (1300 °C) for 145 s/mm has been executed, in order to erase the previous thermal and stress history of the specimens and to provide the super-saturation of the α-ferritic matrix with γ-formers elements, such as Ni, Mn, Cu, and N. Then, an annealing thermal treatment (ATT) has been performed at 1353 K (1080 °C) for different holding times: 36, 72, 215, 355, 710, and 1135 s/mm (Figure 1). The purpose of this heat treatment is the supply of the energy needed to trigger the γ-austenite precipitation within the α-matrix. This temperature range grants to avoid precipitation of embrittling phases within α-ferritic grains, such as σ (Fe-Cr-Mo), χ ($Fe_{36}Cr_{12}Mo_{10}$), and nitrides (CrN and Cr_2N). Both thermal treatments have been followed by water quench [19]. Afterwards, specimens have been machined into Round Tension Test Specimen shapes and tensile tests have been performed; both these operations have been executed, following the ASTM E8/E8M standard. The specimens have been treated for the microstructural examination using Beraha's tint etching (5 mL H_2O, 1 mL HCl, 0.06 g $K_2S_2O_5$, 0.06 g NH_4FHF). Subsequently, the samples have been analyzed via stereoscopy, optical microscopy, and electron microscopy. The volume fraction of γ-austenite within the material has been calculated through automatic image analysis of 10 micrographs measuring 1 mm^2 randomly taken on the samples, following ASTM E1245 standard. Electron microscopy has been used in order to obtain morphological information via Secondary Electron (SEM/SE) imaging, chemical composition data through Energy-Dispersive X-ray Spectroscopy (SEM/EDS), and crystallographic data by Electron Backscatter Diffraction (SEM/EBSD) analysis. The beam spot has 1 µm radius. The chemical composition data have been obtained, averaging five measures. The Electron Backscatter Diffraction (SEM/EBSD) analysis has been calculated through the software INCA provided by Oxford Instruments (INCA Oxford Instrument, Oxford, UK). The crystallographic data have been used to highlight the influence of grain boundary distribution on the onset of secondary recrystallization. The presence of special Coincidence Site Lattice (CSL) boundaries between primary and secondary grains the development of recrystallization.

Table 1. UNS S32760 super duplex stainless steel chemical composition (expressed in wt.%), as designed by standards and measured via optical emission spectroscopy (OES).

F55-UNS S32766	Wt. %									
	C	Mn	Si	Cr	Ni	Mo	N	Cu	W	Fe
Designation	<0.03	<1.00	<1.00	24.0–26.0	6.0–8.0	3.0–4.0	0.2–0.3	0.5–1.0	0.5–1.0	Bal.
OES measure	0.027	0.63	0.51	24.37	6.69	5.45	0.22	0.72	0.85	Bal.

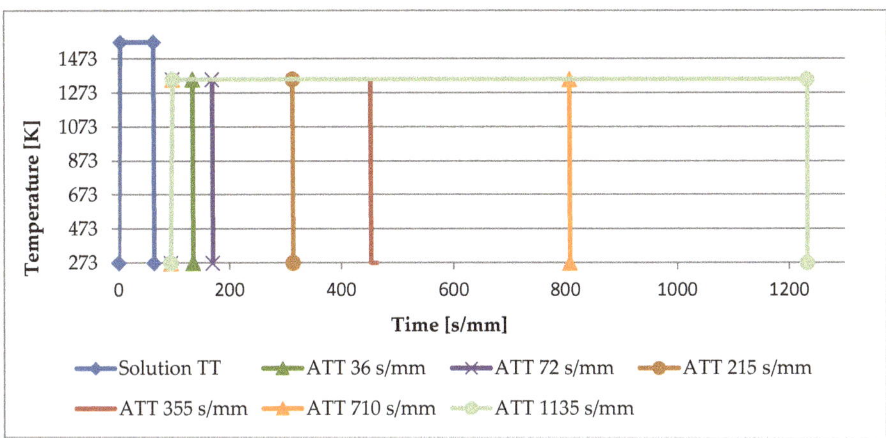

Figure 1. Diagram representing the thermal treatments parameters.

3. Results

Optical microscopy analysis of the F55-UNS S32760 super duplex stainless steel samples have been performed after the execution of the heat treatments. The resulting microstructures of the different specimens have been reported in Figure 2. Micrographs display different phase fraction ratios in accordance with the different holding time of the heat treatments (Table 2).

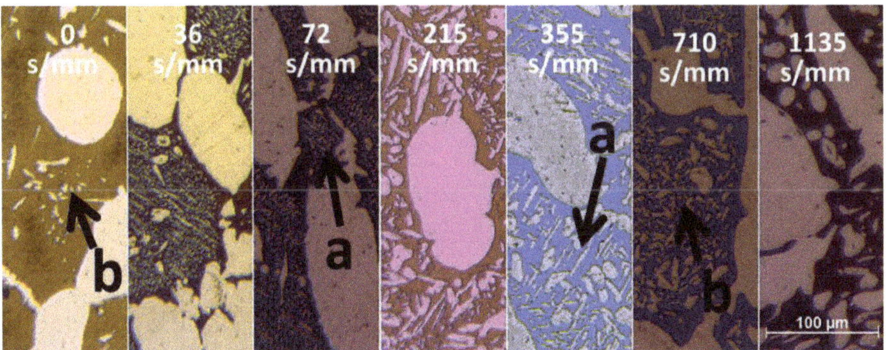

Figure 2. γ-Austenite content variation during isothermal annealing treatment assessed via metallographic method and sample micrographs of each stage of UNS S32760 evolution. The precipitation of γ-austenitic nuclei, as Widmannstätten precipitates (**a**) and via a martensitic shear process (**b**) within the α-ferritic matrix, is displayed in Reference [7].

Table 2. γ volume fraction of the γ-austenitic grains during the isothermal annealing at 1353 K (1080 °C).

Annealing Soaking Times [s/mm]	0	36	72	215	355	710	1135
γ Volume Fraction	0.36	0.49	0.54	0.40	0.54	0.41	0.41
Error	0.03	0.02	0.05	0.01	0.01	0.04	0.02

The scanning electron microscope through Energy-Dispersive X-ray Spectroscopy (SEM/EDS) can investigate the chemical features of the phases belonging to the material. In particular, the way in which alloying elements distributes inside ferrite rather than inside austenite and can be appreciated [20,21]. The most significant results of the chemical composition analysis are hence briefly reported in the following (Table 3), considering that the concentration of each element is expressed as a percentage by weight.

Table 3. SEM/EDX chemical analysis of the α-ferritic matrix. The profile of Ni concentration, with the ongoing of the annealing thermal treatment, testifies the supersaturation of α-ferritic matrix with γ-formers elements, such as Ni, Mn, and Cu.

Specimen	Area	Wt.%							
		Fe	Cr	Ni	Mo	W	Cu	Mn	Si
STT 1300 °C	α-ferrite	60.85 ± 0.31	24.92 ± 0.46	6.23 ± 0.28	5.11 ± 0.33	0.80 ± 0.11	0.90 ± 0.07	0.67 ± 0.32	0.52 ± 0.19
ATT 36 s/mm	α-ferrite	61.68 ± 0.42	23.45 ± 0.34	4.98 ± 0.32	7.02 ± 0.28	0.80 ± 0.12	0.88 ± 0.9	0.61 ± 0.19	0.58 ± 0.17
ATT 1135 s/mm	α-ferrite	63.37 ± 0.35	23.24 ± 0.41	4.37 ± 0.27	6.49 ± 0.22	1.08 ± 0.8	0.54 ± 0.15	0.51 ± 0.14	0.40 ± 0.10

The tensile tests results, provided in Table 4, show the interesting mechanical features typical of SDSSs, which not only match the mechanical properties of the completely ferritic or completely austenitic alloys, but also overcoming them. A wide strain regime, specifically at high strains ($\varepsilon > 0.15$), where strain hardening reserves are usually more desirable than at low strains, is displayed by all the annealing treated samples, resulting in good plastic behavior [22]. These specimens all result in a good compromise between ductility, toughness, and elastic regime properties. Thus, the best combination of both strength and plasticity is displayed by the samples annealed treated for 36 s/mm.

Table 4. UNS S32760 tensile test results: the specimen annealed thermal treated at 1353 K (1080 °C) for 36 s/mm holding time show the best combination of elastic and plastic properties.

Annealing Time [s/mm]	γ%	σ_y [MPa]	σ_{Max} [MPa]	ε_R	E [GPa]	Adsorbed Energy [J/mm^3]
0	35.7	653	828	0.225	255	176
36	49.3	603	849	0.334	211	260
72	54.0	572	839	0.302	211	230
215	39.7	533	826	0.315	112	237
355	53.8	554	827	0.282	181	212
710	41.4	560	803	0.293	190	215
1135	41.3	547	824	0.300	245	226

Further, a post-mortem morphological analysis of the tensile specimens has been performed. A visual inspection and stereoscopic analysis, as reported in Figure 3a, can highlight that all the samples show a cup-cone fracture and surface orange peel features. These results are in a good agreement with the highly ductile behavior observed during the tensile tests. Fully ductile fractures, showing very high dimples densities, are observed in all the specimens via SEM/SE morphological imaging, as reported in Figure 3b. Again, these results confirm the tensile tests data. Via optical micrographic analysis of the fracture surface sections, different features can be observed. As shown in Figure 3c, the crack path should be highlighted; the growing cracks appear to preferentially contour the γ-austenitic grains, or eventually, to cross them in a straightforward way. As a consequence, it can be pointed out that a lower cohesive strength of the α/γ interfaces, with respect to the γ/γ internal interfaces or twinning interfaces within γ grains, and even more, with respect to the continuous lattice of the grains interior. Further, as reported in Figure 3d, along the fracture surface, a greater density of small γ-austenitic nuclei within the α-matrix with respect to the un-deformed areas of the specimens has been observed.

Finally, dislocations slip bands, generated by the great-imposed strain, became visible even through optical microscopy, underlining the local direction of the stress field and the response of the single grains. As expected, γ-austenitic grains show a much higher density of dislocations slip bands, testifying to the intense strain experienced preferentially by γ-grains, as displayed in Figure 3e [23].

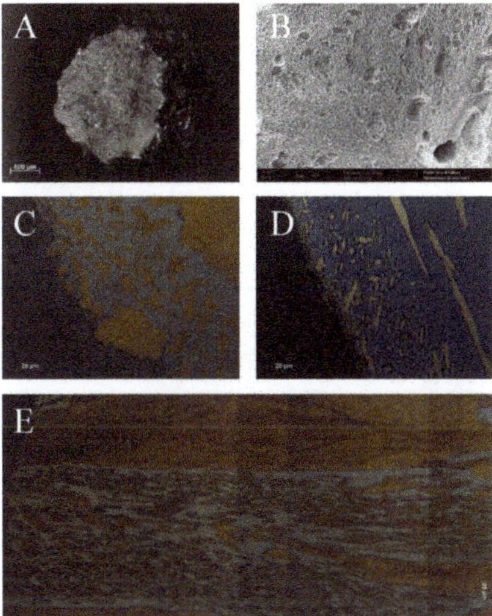

Figure 3. Post-mortem analysis of the tensile samples: stereoscopy morphological imaging of the sample annealing treated for 355 s/mm (**A**), SEM/SE morphological imaging of the specimen annealing treated for 710 s/mm (**B**), optical micrography of the fracture surface section of the specimen annealing treated for 1135 s/mm (**C**), optical micrography of the fracture surface section of the sample annealing treated just solution heat treated (**D**), and optical micrography of the fracture surface section of the specimen annealing treated for 36 s/mm (**E**).

Finally, the occurrence of a transformation induced phenomenon can be definitively confirmed by the assessment of the newly nucleation of the martensitic shear γ-nuclei along dislocations slip bands aligned to the dislocations structure via SEM/EBSD analysis, occurring after high strains have been experienced by the material. Taking into account the infinity of possible orientations of two grains relative to each other, some special orientations may be found. When some lattice points of the two lattices exactly coincide, a kind of superstructure, called coincidence site lattice (CSL), develop. The coincidence site lattice (CSL) model can be used as a standard for the characterization of grain boundary structure for poly-crystals. The frequency of occurrence of a particular type of CSL grain boundary is inherently associated with the newly recrystallized grains form a preferred crystallographic orientation relative to the deformed grains. The CSL frequency occurrence has been determined for the martensitic shear γ-nuclei aligned to the dislocations structure shown in detail in Figure 7 and for other martensitic shear γ-precipitates present within un-deformed α-ferritic grains, as reported in Figure 4. In detail, CSL boundaries $\Sigma13$ for f.c.c. lattices are associated to fast moving boundaries, testifying to the occurrence of nucleation and growth phenomena [24–26].

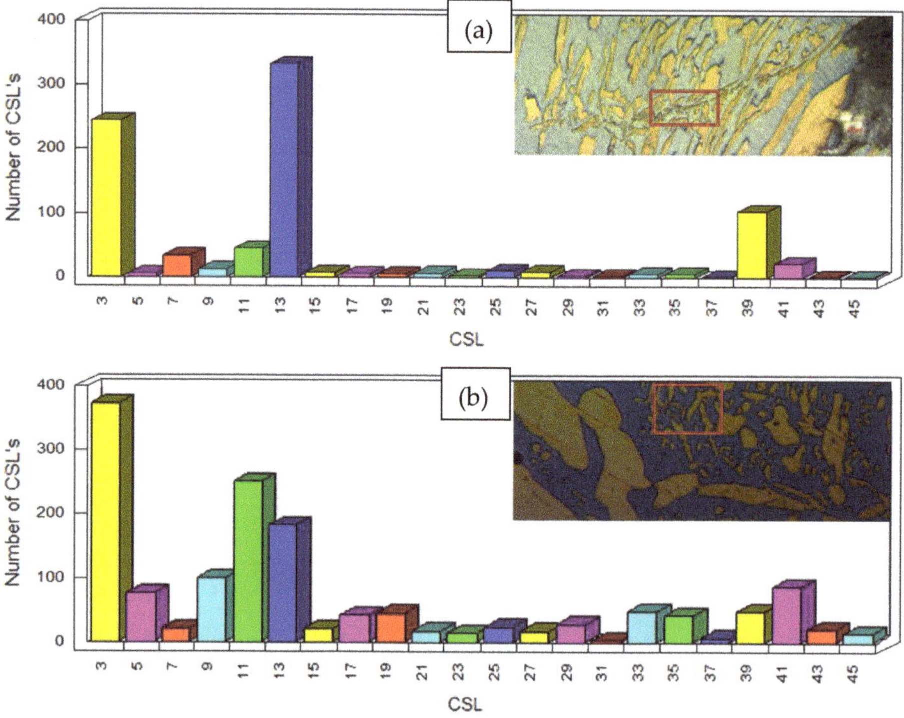

Figure 4. Comparison between CSL frequency occurrences of the martensitic shear γ-nuclei aligned to the dislocations structure shown in Figure 6 (**a**) and for other martensitic shear γ-precipitates present within un-deformed α-ferritic grains (**b**). The increase in CSL boundaries Σ13, testifying to the occurrence of nucleation and growth phenomena, and thus, of transformation induced plasticity phenomena.

4. Discussion

From the results of the experimental campaign, it is possible to express different considerations pointing out the occurrence of a TRIP phenomenon during the deformation of a commercial F55-UNS S32760 SDSS. In detail, micrographs, collected through optical microscopy, display different microstructures in accordance to the different holding time of the heat treatments. After the solution thermal treatments, the samples show a strong misproportion in the phases volume fraction from the ideal 1:1. The ferritic phase has a strong increase, as reported in Table 2. Further, specimens are featured by γ-austenite grains, embedded in α-ferrite matrix, supersaturated in γ-stabilizers elements, such as Ni, Mn, Cu, and N (Table 3). A consequence of this super-saturation is the segregation of γ-stabilizers elements, resulting in γ-austenite thick plates located at α-grain boundaries, which interconnect the γ-austenitic grains. Another result of the super-saturation is the presence of the early stages of the α-intragranular γ-nuclei nucleated during the cooling path; both Widmanstätten grain boundaries saw-teeth structures and Martensitic Shear products nuclei can be detected (Figure 2) [8,27]. The annealing thermal treatment produces the precipitation of γ-austenitic nuclei as Widmannstätten precipitates and via a martensitic-shear process within the α-ferritic matrix. Further, the γ-precipitates display different sizes, which result in becoming larger with the ongoing of the annealing thermal treatment, according to the well-known Ostwald ripening model [7,10,28,29]. The super-saturation of the α-matrix with γ-formers elements has been proven via SEM/EDX chemical analysis (Table 3). The data show a very high nickel concentration within the α-matrix after the solution thermal treatment.

Further, after the 36 s/mm long annealing heat treatment, the nickel concentration still results much higher with respect to the 1135 s/mm long annealing heat treatment, which can be considered as an equilibrium value due to such long holding time at a constant annealing temperature. Such a strong difference in nickel concentration testifies to the residual presence of the chemical free-energy, which leads the martensitic shear transformation from α-ferrite to γ-austenite [30]. This condition would allow further nucleation and growth phenomena of the γ-austenitic precipitates, generated by martensitic shear transformation. Since this kind of phase transformation can also be activated at room temperature by the presence of a stress or strain field within the former lattice. Thus, the α-ferritic matrix could act as a former phase, via a deformation-stimulated displacive martensitic shear transformation, of a transformation induced plasticity (TRIP) phenomenon [18,21,31].

Focusing the attention on the tensile tests results reported in Table 4, the best combination of both strength and plasticity is displayed by the samples annealed treated for 36 s/mm. In these specimens, as shown before (Figure 2), the precipitation phenomena of the γ nuclei within the α-ferritic matrix via martensitic shear transformation have already occurred. However, since the Oswald ripening processes have just started during the annealing thermal treatment, the γ-former elements are still enriching the α-matrix and they are still present within the α-matrix at high concentrations (Table 3). This condition allows the material to undergo further phase transformations, suggesting, coupled with the mechanical response in the plastic field of the other annealed treated samples, the occurrence of a transformation induced plasticity effect. Further, combining the mechanical data with the main microstructural feature, the austenitic phase volume fraction allows for the observation that there is not a good correspondence between them. Moreover, no specific trends or correspondences can be highlighted, as shown in Figure 5. Indeed, the different microstructural phase ratio deeply influences the mechanical properties, since the two constituent phases own different micromechanical features, but their influence acts in a complex way [32].

Different hardening mechanisms are known to act in DSSs. In keeping with the additivity law, ferrite hardens the austenitic matrix. Additionally, DSSs usually have a fine grain microstructure, which additionally contributed to the hardening effect, according to the Hall-Petch relationship. In the present case, the γ-precipitation within the α-matrix in both its forms leads to a grain refinement. Moreover, SDSSs additionally harden due to the increased percentage of alloying elements, such as chromium and molybdenum in solid solution [33,34].

While these hardening phenomena have been taken into account, again, a good correspondence with the tensile tests results cannot be pointed out. Thus, the high mechanical response during the tensile tests has to be investigated under a different point of view, in order to explain the obtained results.

Regarding the data of the morphological observation of the specimens undergone to the tensile tests, it is possible to observe other aspects featuring the behavior of this material. As shown in Figure 3c, the crack path preferentially contours the γ-austenitic grains. As a consequence, a lower bonding strength of the α/γ interfaces can be seen, with respect to other sites (γ/γ internal interfaces, twinning, and the lattice of the grains interior). Since the extremely ductile behavior displayed by the studied steels, this lower resistance of α/γ interfaces has to be linked to two different mechanisms. The presence of tiny particles, usually carbides or nitrides, at the grain boundaries generates the dimples that will cause the final rupture; however, no particles have been detected, and then, this mechanism would not be the main one involved (Figure 3b). The second possible process is intrinsically related to the duplex structure nature: Dislocations in γ-austenite pile up against the α/γ phase boundaries and create a local stress concentration, which generates both dislocations in α-ferrite or steps at the surface grains near the α/γ phase boundaries, which coarsen into the dimples [35]. Moreover, as reported in Figure 4d, the density and the dimensions of the small γ-austenitic nuclei within the α-matrix drastically increase along the fracture surface with respect to the un-deformed areas of the tensile specimens. These areas are subjected to very intense strains. A much higher density of dislocations slip bands have been generated, especially within γ-grains, testifying the intense strain experienced preferentially by γ-grains, as displayed in Figure 4e, underlining the local direction of the stress field and the response of the single grains. Coupling these last two considerations, the definition of the Magee effect can be obtained, which is one of the two possible causes of a transformation induced plasticity phenomenon. This event consists in a stress-assisted nucleation or growth of crystallographic variants, which are favorably aligned with the orientation of the applied stress, and it may be predominant in particular for the diffusion-less martensitic shear phase transformations in steels [32,36].

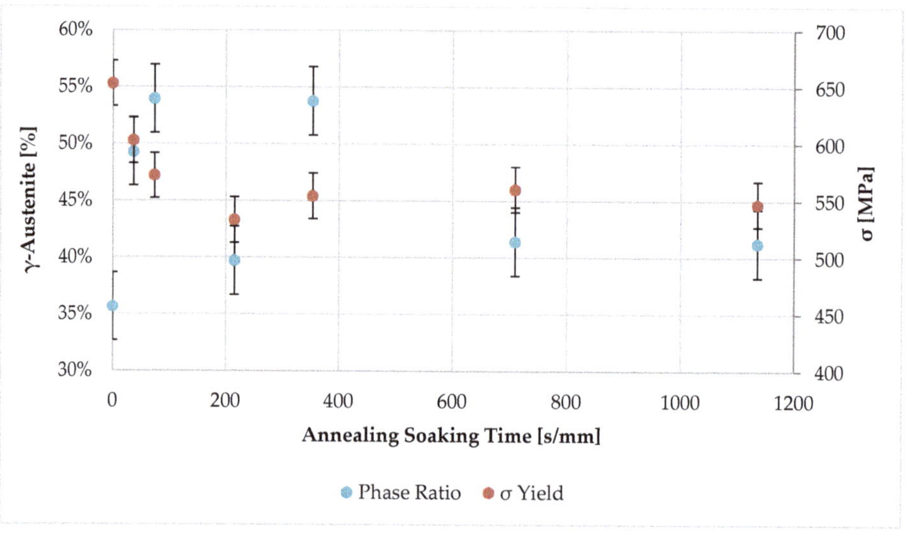

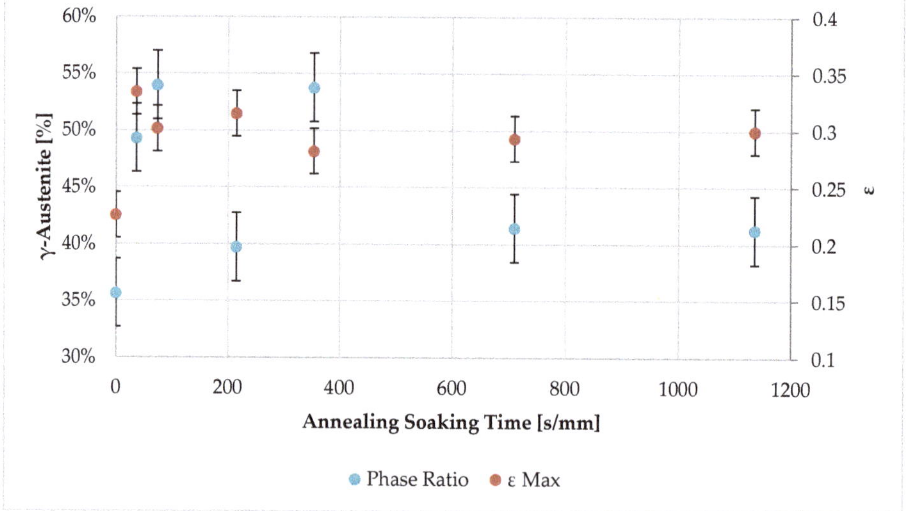

Figure 5. UNS S32760 tensile test results combined with the evolution of the phase ratio with the ongoing of the annealing thermal treatment. No specific trend appears; moreover, no correspondence can be underlined.

The phase fraction ratio of the un-deformed areas of the samples and of the highly deformed area next to the fracture surface has been evaluated via metallographical analysis, as reported in Figure 6.

A robust increase in the γ-austenite content has been measured in all the samples with a lower phase fraction ratio than the ideal equilibrium relationship of 50:50 ferrite:austenite microstructure. Moreover, in all the samples has been recorded a increment of the γ-austenite phase ratio associated to the diffusion-less growth martensitic shear precipitates. These increases match with a strain-assisted nucleation and growth phenomenon described by the Magee effect, testifying to the occurrence of a transformation induced plasticity event. Since martensitic phase transformation occurs without diffusion through a cooperative shear movement of atoms, it is recognized that the applied, as well as internal stresses, assist the transformation.

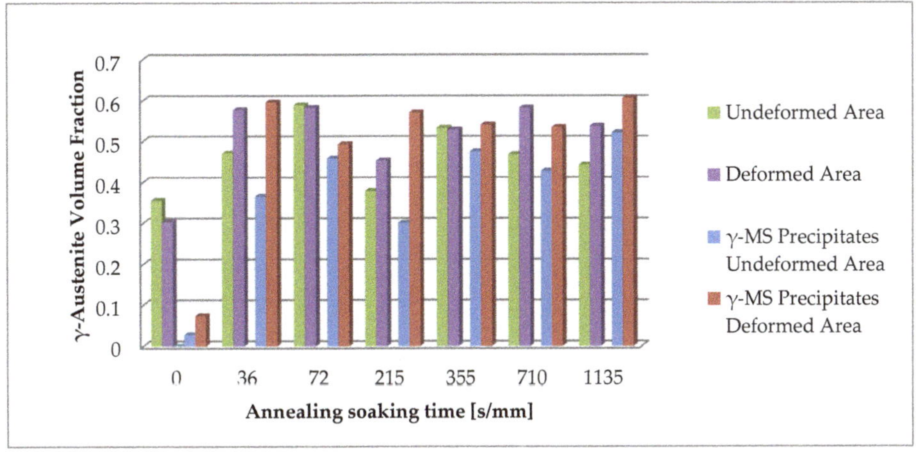

Figure 6. Comparison the γ-austenite phase ratio between the un-deformed and the deformed areas of the frature surface sections of the tensile samples. The increase in γ-austenite volume fraction between the two examinated ares, in particular regarding the Martensitic Shear precipitates, testifies to the occurrence of a strain-aided phase transformation phenomenon.

The presence of newly nucleated martensitic shear γ-nuclei, as can be stated by the much lower dimensions with respect to the other martensitic shear precipitates present nearby, along dislocations slip bands and the alignment of these particles to the dislocations structure testify to the occurrence of the Magee effect characteristic of TRIP effect phenomena, as reported in Figure 7.

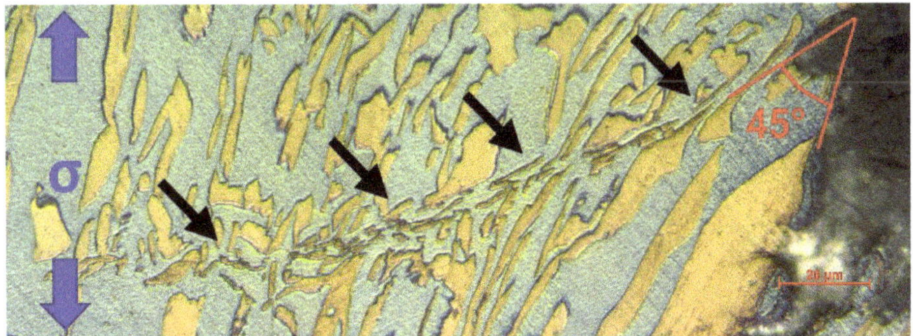

Figure 7. Newly nucleated martensitic shear γ-nuclei along dislocations slip bands and the alignment of these particles to the dislocations structure on the fracture surface section of specimen annealing treated for 355 s/mm at 1353 K (1080 °C), due to the Magee effect.

Since this phenomenon is related to the ability to nucleate martensitic shear γ-nuclei, it is consequently linked to the α-matrix super-saturation in γ-former elements. Thus, the greatest ductility shown by the sample annealed for 36 min/inch at 1353 K (1080 °C) has to be attributed to its ability to exploit the TRIP effect due to the highest α-ferritic matrix supersaturation. Moreover, as could be expected by the relationship between the TRIP effect and the α-matrix super-saturation, there are fewer occurrences of these processes in the samples corresponding to the maxima in the phase fraction ratio evolution during the annealing thermal treatment at 1353 K (1080 °C) can be explained.

Further, the particular microstructural features, shown in Figure 7, evidences the strain-induced character of the occurring martensitic shear transformations, since the lattice defects generated by the strain act as nucleation and growth sites for the diffusion-less phase transformation. It can be confirmed also by the fact that the strain-induced martensitic transformation owns preferential shear directions, relative to the applied stress in just two variants accompanying the larger shear strain. Since this transformation results to be strain-induced, it can be remarked its occurrence at high strains regimes, where strain hardening reserves are usually more desirable [37,38]. Finally, the presence of newly nucleation of the martensitic shear γ-nuclei along dislocations slip bands can be definitively proven by the SEM/EBSD analysis (Figure 4). Since the frequency of occurrence of CSL boundaries $\Sigma 13$ for f.c.c. lattices is associated to the fast moving boundaries, testifying to the occurrence of nucleation and growth phenomena. Thus, the strong increase in CSL boundaries $\Sigma 13$ for f.c.c. γ-austenitic martensitic shear precipitates aligned to the dislocations structure shown in Figure 7, proves the occurrence of transformation induced plasticity phenomena [26,39].

5. Conclusions

The microstructure and deformation mechanisms of the commercial Super Duplex stainless steel F55-UNS S32760 were analyzed and discussed, after thermal heat treatments and tensile tests. The best combination of mechanical properties have been displayed after a solution thermal treatment at 1573 K (1300 °C) for 145 s/mm and an annealing thermal treatment performed at 1353 K (1080 °C) for 36 s/mm holding time. These high results have been related to the occurrence of a transformation induced plasticity (TRIP) phenomena. This feature is generated by the Magee effect during the deformation. In detail, it results as a strain-induced martensitic shear transformation, triggered by the high density of lattice defects (as slip bands). Since this TRIP phenomenon results need to be strain-induced, it would occur at high strains regimes, where it is usually more desirable. Further, these effects are observed for the first time in commercial SDSSs, involve the martensitic shear precipitation transformation $\alpha \rightarrow \gamma$, and occur in a favorable phase ratio range for the corrosion resistance properties. All these aspects and easy thermal processing could grant the industrial exploitation of this phenomenon in the near future.

Author Contributions: A.F.C. conceived and designed the experiments, performed the experiments, analyzed the data and wrote the paper; S.B. conceived and designed the experiments and review the paper; C.D.C. performed the experiments and analyzed the data; A.D.S. contributed reagents/materials/analysis tools and review the paper; A.G. conceived and designed the experiments and review the paper; G.N. performed the experiments and analyzed the data; C.M. contributed reagents/materials/analysis tools and review the paper.

Funding: This research did not receive any specific grant from funding agencies in the public, commercial, or not-for-profit sectors.

Conflicts of Interest: The authors declare no conflict of interest.

References

1. Corradi, M.; Di Schino, A.; Borri, A.; Rufini, R. A review of the use of stainless steel for masonry repair and reinforcement. *Constr. Build. Mater.* **2018**, *18*, 335–346. [CrossRef]
2. Rufini, R.; Di Pietro, O.; Di Schino, A. Predictive simulation of plastic processing of welded stainless steel pipes. *Metals* **2018**, *8*, 519. [CrossRef]

3. Di Schino, A.; Longobardo, M.; Porcu, G.; Turconi, G.L.; Scoppio, L. Metallurgical design and development of C125 grade for mild sour application. In Proceedings of the NACE-International CORROSION 2006, San Diego, CA, USA, 12–16 March 2006; NACE-06125. pp. 1–14.
4. Gunn, R.N. *Duplex Stainless Steels, Microstructure, Properties and Applications*; Woodhead Publishing: Cambridge, UK, 1994.
5. Alvarez-armas, I. Duplex Stainless Steels: Brief History and Some Recent Alloys. *Recent Patents Mech. Eng.* **2008**, *1*, 51–57. [CrossRef]
6. Nilsson, J.O. Super duplex stainless steels. *Mater. Sci. Technol.* **1992**, *8*, 685–700. [CrossRef]
7. Ciuffini, A.F.; Barella, S.; Di Cecca, C.; Gruttadauria, A.; Mapelli, C.; Mombelli, D. Isothermal Austenite–Ferrite Phase Transformations and Microstructural Evolution during Annealing in Super Duplex Stainless Steels. *Metals* **2017**, *7*, 368. [CrossRef]
8. Knyazeva, M.; Pohl, M. Duplex Steels: Part I: Genesis, Formation, Structure. *Metallogr. Microstruct. Anal.* **2013**, *2*, 113–121. [CrossRef]
9. Charles, J.; Chemelle, P.M. The history of duplex developments, nowadays DSS properties and duplex market future trends. In Proceedings of the 3rd Duplex World Conference, Beaune, France, 13–15 October 2010.
10. Southwick, P.D.; Honeycombe, R.W.K. Decomposition of ferrite to austenite in 26%Cr-5%Ni stainless steel. *Metal Sci.* **1980**, *14*, 253–261. [CrossRef]
11. Ohmori, Y.; Nakai, K.; Ohtsubo, H.; Isshiki, Y. Mechanism of Widmanstatten austenite formation in a d/y duplex phase stainless steel. *ISIJ Int.* **1995**, *35*, 969–975. [CrossRef]
12. Tomida, T.; Maehara, Y.; Ohmori, Y. Martensitic transformation from δ-ferrite during the Melt-Quenching process of δ-γ duplex stainless steel. *Mater. Trans. JIM* **1989**, *5*, 326–336. [CrossRef]
13. Safdel, A.; Zarei-Hanzaki, A.; Shamsolhodaei, A.; Krooß, P.; Niendorf, T. Room temperature superelastic responses of NiTi alloy treated by two distinct thermomechanical processing schemes. *Mater. Sci. Eng. A* **2016**, *684*, 303–311. [CrossRef]
14. Sato, A.; Chishima, E.; Soma, K.; Mori, T. Shape memory effect in $\gamma \leftrightarrow \epsilon$ transformation in Fe-30Mn-1Si alloy single crystals. *Acta Metal.* **1982**, *32*, 1177–1183. [CrossRef]
15. Otsuka, K.; Ren, X.; Murakami, Y.; Kawano, K.; Ishii, T.; Ohba, T. Composition dependence of the rubber-like behavior in $\zeta 2'$-martensite of AuCd alloys. *Mater. Sci. Eng. A* **1999**, *273–275*, 558–563. [CrossRef]
16. Dan, W.J.; Li, S.H.; Zhang, W.G.; Lin, Z.Q. The effect of strain-induced martensitic transformation on mechanical properties of TRIP steel. *Mater. Des.* **2008**, *29*, 604–612. [CrossRef]
17. Sohn, S.S.; Choi, K.; Kwak, J.H.; Kim, N.J.; Lee, S. Novel ferrite-austenite duplex lightweight steel with 77% ductility by transformation induced plasticity and twinning induced plasticity mechanisms. *Acta Mater.* **2014**, *78*, 181–189. [CrossRef]
18. Patra, S.; Ghosh, A.; Kumar, V.; Chakrabarti, D.; Singhal, L.K. Deformation induced austenite formation in as-cast 2101 duplex stainless steel and its effect on hot-ductility. *Mater. Sci. Eng. A* **2016**, *660*, 61–70. [CrossRef]
19. Choi, J.Y.; Ji, J.H.; Hwang, S.W.; Park, K.T. TRIP aided deformation of a near-Ni-free, Mn-N bearing duplex stainless steel. *Mater. Sci. Eng. A* **2012**, *535*, 32–39. [CrossRef]
20. Knyazeva, M.; Pohl, M. Duplex Steels. Part II: Carbides and Nitrides. *Metallogr. Microstruct. Anal.* **2013**, *2*, 343–351. [CrossRef]
21. Pareige, C.; Novy, S.; Saillet, S.; Pareige, P. Study of phase transformation and mechanical properties evolution of duplex stainless steels after long term thermal ageing (>20 years). *J. Nucl. Mater.* **2011**, *411*, 90–96. [CrossRef]
22. Herrera, C.; Ponge, D.; Raabe, D. Design of a novel Mn-based 1 GPa duplex stainless TRIP steel with 60% ductility by a reduction of austenite stability. *Acta Mater.* **2011**, *59*, 4653–4664. [CrossRef]
23. Kocks, U.F.; Mecking, H. Physics and phenomenology of strain hardening: The FCC case. *Prog. Mater. Sci.* **2003**, *48*, 171–273. [CrossRef]
24. Caleyo, F.; Cruz, F.; Baudin, T.; Penelle, R. Texture and grain size dependence of grain boundary character distribution in recrystallized Fe-50% Ni. *Scr. Mater.* **1999**, *41*, 847–853. [CrossRef]
25. Huang, J.C.; Hsiao, I.C.; Wang, T.D.; Lou, B.Y. EBSD study on grain boundary characteristics in fine-grained Al alloys. *Scr. Mater.* **2000**, *43*, 213–220. [CrossRef]
26. Beck, P.A. Annealing of cold worked metals. *Adv. Phys.* **1954**, *3*, 245–324. [CrossRef]

27. Kobayashi, S.; Nakai, K.; Ohmori, Y. Isothermal decomposition of δ-ferrite in a 25Cr-7Ni-0.14N stainless steel. *Acta Mater.* **2001**, *49*, 1891–1902. [CrossRef]
28. Zhang, J.; Huang, F.; Lin, Z. Progress of nanocrystalline growth kinetics based on oriented attachment. *Nanoscale* **2010**, *2*, 18–34. [CrossRef] [PubMed]
29. Balluffi, R.W.; Allen, S.; Carter, W.C. *Kinetics of Materials*; John Wiley & Sons: Hoboken, NJ, USA, 2005.
30. Daróczi, L.; Palánki, Z.; Beke, D.L. Determination of the non-chemical free energy terms in martensitic transformations. *Mater. Sci. Eng. A* **2006**, *438–440*, 80–84.
31. Wan, J.; Ruan, H.; Shi, S. Excellent combination of strength and ductility in 15Cr-2Ni duplex stainless steel based on ultrafine-grained austenite phase. *Mater. Sci. Eng. A* **2017**, *690*, 96–103. [CrossRef]
32. Videau, J.-C.; Cailletaud, G.; Pineau, A. Experimental Study of the Transformation-Induced Plasticity in a Cr-Ni-Mo-Al-Ti Steel. *J. Phys. IV* **1996**, *6*, 465–474. [CrossRef]
33. Voronenko, B.I. Austenitic-ferritic stainless steels: A state-of-the-art review. *Met. Sci. Heat Treat.* **1998**, *39*, 428–437. [CrossRef]
34. Horvath, W.; Prantl, W.; Stroißnigg, H.; Werner, E. Micro hardness and microstructure of austenite and ferrite in nitrogen alloyed duplex steels between 20 and 500 °C. *Mater. Sci. En. A* **1998**, *256*, 227–236. [CrossRef]
35. Fréchard, S.; Martin, F.; Clément, C.; Cousty, J. AFM and EBSD combined studies of plastic deformation in a duplex stainless steel. *Mater. Sci. Eng. A* **2006**, *418*, 312–319. [CrossRef]
36. Taleb, L.; Petit, S. New investigations on transformation induced plasticity and its interaction with classical plasticity. *Int. J. Plast.* **2006**, *22*, 110–130. [CrossRef]
37. Huang, B.X.; Wang, X.D.; Rong, Y.H. A method of discrimination between stress-assisted and strain-induced martensitic transformation using atomic force microscopy. *Scr. Mater.* **2007**, *57*, 501–504. [CrossRef]
38. Chatterjee, S.; Bhadeshia, H.K.D.H. Transformation induced plasticity assisted steels: Stress or strain affected martensitic transformation? *Mater. Sci. Technol.* **2007**, *23*, 1101–1104. [CrossRef]
39. Rutter, J.W.; Aust, K.T. Migration of ‹100› tilt grain boundaries in high purity lead Wanderung von ‹100›-neidungskorngrenzen in hochreinem blei. *Acta. Metall.* **1965**, *13*, 181–186. [CrossRef]

Article

Effect of Carbon Partitioning, Carbide Precipitation, and Grain Size on Brittle Fracture of Ultra-High-Strength, Low-Carbon Steel after Welding by a Quenching and Partitioning Process

Farnoosh Forouzan [1,2,*], M. Agustina Guitar [2], Esa Vuorinen [1] and Frank Mücklich [2]

[1] Department of Engineering Sciences and Mathematics, Luleå University of Technology, SE-97187 Luleå, Sweden; esa.vuorinen@ltu.se

[2] Department of Materials Science, Functional Materials, Saarland University, D-66041 Saarbrücken, Germany; a.guitar@mx.uni-saarland.de (M.A.G.), muecke@matsci.uni-sb.de (F.M.)

* Correspondence: farnoosh.forouzan@ltu.se; Tel.: +46-920-493-217

Received: 16 August 2018; Accepted: 18 September 2018; Published: 23 September 2018

Abstract: To improve the weld zone properties of Advanced High Strength Steel (AHSS), quenching and partitioning (Q&P) has been used immediately after laser welding of a low-carbon steel. However, the mechanical properties can be affected for several reasons: (i) The carbon content and amount of retained austenite, bainite, and fresh martensite; (ii) Precipitate size and distribution; (iii) Grain size. In this work, carbon movements during the partitioning stage and prediction of Ti (C, N), and MoC precipitation at different partitioning temperatures have been simulated by using Thermocalc, Dictra, and TC-PRISMA. Verification and comparison of the experimental results were performed by optical microscopy, X-ray diffraction (XRD), Scanning Electron Microscop (SEM), and Scanning Transmission Electron Microscopy (STEM), and Energy Dispersive Spectroscopy (EDS) and Electron Backscatter Scanning Diffraction (EBSD) analysis were used to investigate the effect of martensitic/bainitic packet size. Results show that the increase in the number density of small precipitates in the sample partitioned at 640 °C compensates for the increase in crystallographic packets size. The strength and ductility values are kept at a high level, but the impact toughness will decrease considerably.

Keywords: low-carbon AHSS; Q&P; toughness; modelling; precipitation; martensite packet

1. Introduction

The automotive industry focuses on increasing the use of advanced high-strength steels (AHSS) in order to satisfy the current demand for decreasing the fuel consumption by reduced weight and increasing vehicle safety by using these steels in different energy-absorbing components [1]. Usually, these AHSS are produced by thermomechanical processes, which control the microstructure and grain size as well as the precipitation hardening of micro-alloyed steels.

During welding of AHSS, the weld area will be completely changed and the excellent properties (i.e., tensile strength, toughness) will be lost. This means that the welded area could be the best area for crack propagation [2]. Therefore, pre-and/or post-welding treatment is necessary to improve the properties of this zone. Laser welding is a popular method in the industry because it is fast, creates narrow and deep welds, and can be used for different materials and shapes. So, in this work, a quenching and partitioning [3] method has been applied for post-welding treatment in order to control the microstructure [4]. The final structure will contain tempered martensite (which increases the yield strength) with retained austenite (which improves the ductility), and, depending on the Q&P conditions and chemical composition of the steel, some bainite and fresh martensite can also be formed [3–5].

However, carbide/nitride precipitation during the partitioning step cannot be avoided, even in low-carbon [6] or in high-carbon, high-silicon steels [7]. Therefore, one of the critical aspects is to monitor the behavior of precipitates during the process. Although precipitates and/or particles are designed to strengthen the material by a precipitation hardening mechanism, they could act as cleavage initiation sites and deteriorate the toughness [8–10]. Fairchild et al. [11] showed that a strong inclusion-matrix bond is why TiN inclusions are potent cleavage initiators in steels even with modest Ti contents. Another study by Di Schino et al. [12] on the effect of Nb microalloying on the heat-affected zone (HAZ) showed that a small difference in Nb content is able to influence the size of the bainitic packet, which results in both toughness and hardness.

Several works have been done on the mechanism of Q&P up to now [4,13–17], while only a few papers have investigated the fracture mechanisms and toughness of Q&P steels [18,19]. Fracture causes could mainly be related to: (i) Kinetics of carbon partitioning and stabilization of retained austenite (RA) in the structure, because it has been shown that increasing volume fraction of RA due to the TRIP effect delays the crack propagation [20]; (ii) Kinetics, size, number density and shape of secondary precipitates; and (iii) Microstructural refinement can be very effective for improving the toughness. Wang et al. [21] found that, when the cleavage crack encounters another packet of martensite, it may be arrested and then largely changes its propagation direction.

Previous studies by the authors [4,22] on the effect of Q&P after laser welding of a low-carbon steel showed that samples partitioned at a higher temperature had the best tensile properties but very low Charpy V impact toughness results. In the present work, in order to investigate the origins of such contradictory results, the abovementioned main issues are studied. Carbon diffusion from α' to retained austenite was modeled by a diffusion-controlled transformation tool (DICTRA) [23] during quenching and partitioning. The type, size, and distribution of the precipitates were evaluated using a scanning transmission electron microscope (STEM) in order to verify the precipitation prediction modeling by TC–PRISMA. Finally, the effect of the packet size of martensite/bainite laths on crack propagation has been investigated by electron backscatter diffraction (EBSD).

2. Materials and Methods

Laser welding was conducted on 5.5 mm thick sheets of 'Domex 960' advanced high strength steel from SSAB (Stockholm, Sweden). This steel is thermomechanically processed (TMP) and was received in TMP condition with a yield strength of 960 MPa and elongation (A5) of approximately 8%. The chemical composition of the steel is given in Table 1, but for Thermocalc simulations, a simplified composition of Fe–0.08 wt % C–1.78 wt % Mn–0.5 wt % Mo–0.187 wt % Ti–0.004 wt % N is assumed. All calculations were carried out with the thermodynamic database TCFE8 [24] and mobility database of MOBFE3 [25].

Table 1. The chemical composition of the steel used in the experiments (wt %).

C	Si	Mn	P	S	Al	Ti	Mo	Cr	Ni	Cu	V	N	Fe
0.082	0.23	1.79	0.008	0.001	0.038	0.184	0.503	0.064	0.296	0.016	0.012	0.004	Balance

For the welding, process the keyhole penetration mode, with ytterbium Fiber Laser (YLR Laser–15000, IPG Photonics, Oxford, MA, USA), 5 kW power, travel speed of 1.1 m/min, and Argon shielding gas with a flow rate of 20 L/min, was used. The welded specimens were subjected to post heat treatment immediately after the laser welding. For that, an induction heater was placed above the specimen. The temperature was monitored during the whole procedure of welding and Q&P, by using welded thermocouples located 1.5 mm from the fusion zone (FZ). For the same quenching temperature (QT) of 355 °C (approx. equal to 60% initial martensite), three partitioning temperatures (PT) of 440 °C (20 °C above M_s), 540 °C (half between M_s and B_s), and 640 °C (B_s) and three partitioning times (Pt) were tested.

Tensile testing was performed on cross weld specimens; the tensile axis was kept perpendicular to the rolling direction of the steel sheet, while the weld zone is located exactly in the middle of the dog-bone-shaped A50 samples. Charpy V notch samples were prepared according to the standard EN ISO 6892–1:2009 and testing was conducted on transverse specimens with a notch in the center of FZ. Results showed that all the samples that were partitioned at 640 °C had very low impact toughness.

In order to investigate the reasons, the samples that showed the best and the worst toughness were selected for comparison, see Table 2.

Table 2. Mechanical properties of samples Q&P treated immediately after laser welding. (Energy = Absorbed energy during Charpy V test at 20 °C.)

Sample Code	QT (°C)	PT (°C)	Pt (s)	A_{80} (%)	A_5 (%)	$Rp_{0.2}$ (MPa)	R_m (MPa)	Energy (J/cm^2)
S(540, 5)	355	540	5	4	6	1005	1025	109
S(640, 50)	355	640	50	6	8	1034	1051	4

The precipitate characterization was determined from C–replicas taken from weld area (FZ and HAZ) with a FE–SEM Helios Nanolab 600 (FEI, Hillsboro, OR, USA) working with an acceleration voltage of 30 kV and a 1.4 nA beam current. The precipitate sizes were determined by image analysis (I–A) of STEM micrographs using the software A4i© (Aquinto AG, Berlin, Germany). The carbide chemistry was identified in transmission using energy-dispersive X-rays (EDX) (Jeol Ltd., Tokyo, Japan) at 30 kV accelerating voltage. Electron Backscatter Diffraction (EBSD) was carried out on HAZ using an accelerating voltage of 20 kV, a beam current of 11 nA, and a step size of 150 nm and 100 nm for Q&P and base metal (BM) samples, respectively. The raw data were filtered by confidence index (CI) of CI > 0.09 after a CI cleanup and grain dilation (grain tolerance angle 5° and minimum grain size 2). The grains were defined as at least two adjacent points with similar orientation within a range of 15° misorientation.

3. Results and Discussion

3.1. Fracture Results

Figure 1 shows dimple and cleavage mix-mode fracture in all fractographies, but the average radius of dimples in S(540, 5) (Figure 1a) is smaller than the other samples including the reference weld without any heat treatment (Figure 1c), and S(640, 50) (Figure 1b).

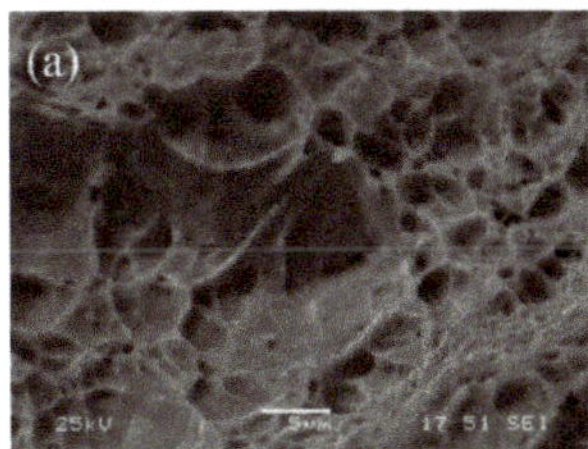

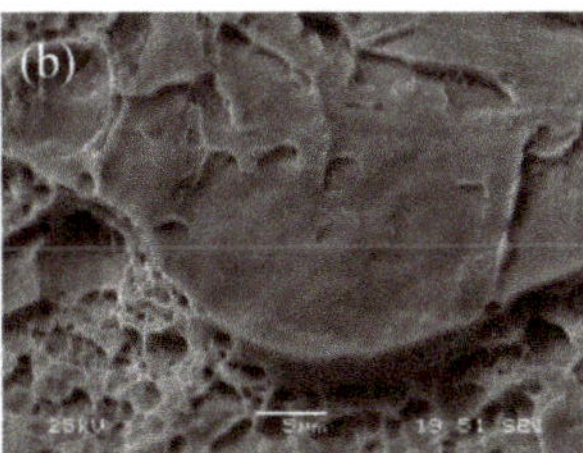

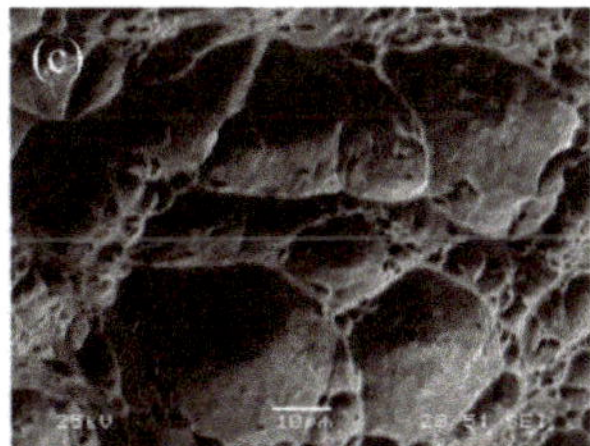

Figure 1. Fracture surface after tensile test of following samples; (**a**) S(540, 5), shows ductile fracture of HAZ (**b**) S(640, 50), shows brittle fracture of HAZ and (**c**) reference weld.

3.2. Thermodynamic Modeling

In order to understand the critical temperatures of phase transformation and formation of stable precipitates in an equilibrium condition of Domex 960, the amount of phases at different temperatures was calculated with Thermocalc [23]. Although the temperatures and amount of phases could be far from the real conditions during Q&P, this calculation could give a good overview of the most stable microconstituents. Figure 2 shows that Ti(C, N) has the highest tendency to form the first carbonitrides in this system. It will start when there is still some liquid in the system and the precipitation will continue until it reaches room temperature. After that, two other carbides (M_7C_3) and cementite will nucleate around 700 °C, but they are not stable and will disappear very soon around 600 °C, while their amount is also very small (e.g., 0.001 mol). The next important precipitate in this system in the critical temperature range of 440 °C to 640 °C, is MoC, as expected from its enthalpy for carbide formation in comparison with other alloying elements in this steel [26]. The kinetics of nucleation and growth of these elements modeled by the TC–Prisma [23] module are illustrated in Figure 3.

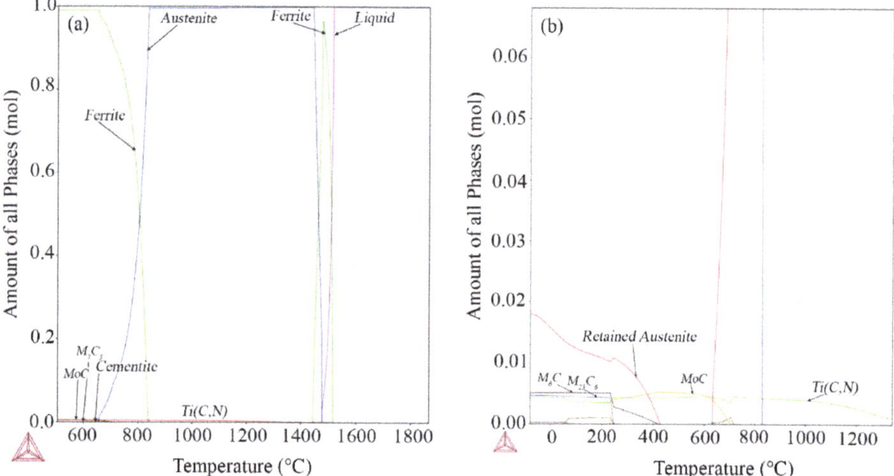

Figure 2. Amount of all phases calculated with Thermocalc: (**a**) full range of phases; (**b**) focus on carbides.

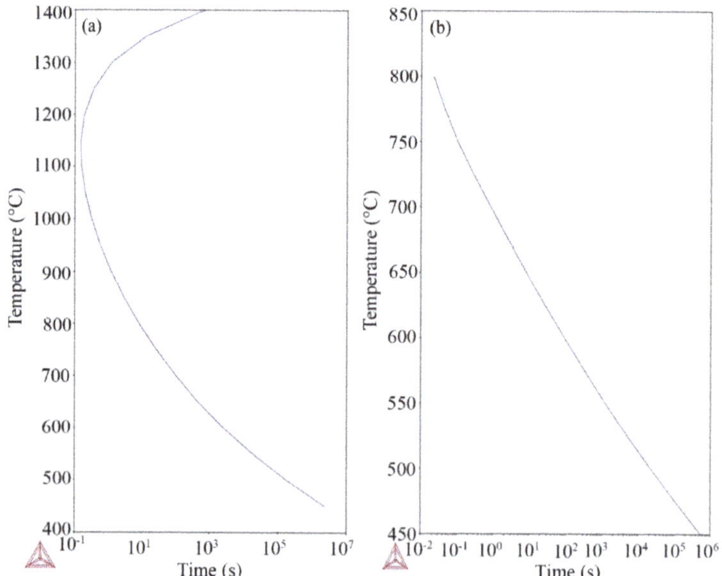

Figure 3. Time–temperature–precipitation (TTP) diagrams of (**a**) Ti(C, N); (**b**) MoC.

According to Figure 3a, Ti(C, N) nucleates at very high temperature, with a maximum rate at ca. 1150 °C, at which it takes only 0.2 s. At lower temperatures up to 800 °C, the nucleation takes under 10 s. The time for nucleation grows exponentially with decreasing temperature and nucleation takes a longer time than the longest partitioning time used in this work (maximum 50 s). On the other hand, Figure 3b shows that, although nucleation of this carbide (MoC) is very fast, 600 °C is a critical temperature for this Q&P, because below this temperature nucleation takes more than 100 s, which means that nucleation of MoC cannot occur during this heat treatment. In summary, Ti(C, N) precipitation could be an issue during welding and previous Ti(C, N) particles from casting could still remain in the structure, but MoC precipitation occurs during the partitioning stage at 640 °C.

3.3. Carbon Partitioning

Understanding the carbon movement during the partitioning stage at different temperatures with regard to the first quenched martensite and retained austenite grain boundaries has an important role in the prediction of the phase transformations. Comparison between the diffusion coefficient of carbon in austenite (D_c^{γ}) and ferrite (D_c^{α}) in Table 3 shows that for the temperature range of $\gamma + \alpha$ (440–640 °C in this case) the equilibrium 'D' is more than 100 times higher in ferrite than austenite. This means that carbon can partition out of martensite rapidly but will then pile up behind the α'/γ grain boundary.

Table 3. Carbon diffusion coefficient in austenite (D_c^{γ}) and ferrite (D_c^{α}) and at different temperatures.

Carbon Diffusion Coefficient	440 °C	540 °C	640 °C
D_c^{γ}	3.30×10^{16}	7.12×10^{14}	2.82×10^{-21}
D_c^{α}	8.54×10^{-13}	4.49×10^{-17}	1.22×10^{-18}

In other words, this velocity is critical for the determination of the area of retained austenite around tempered martensite plates, especially for low-carbon steels, in order to design the structural and mechanical properties of the material, since three different phenomena could occur during the partitioning stage: (i) The amount of austenite's carbon enrichment to stabilize it after the final quench; (ii) Nucleation and growth of third phases (e.g., bainitic ferrite); and (iii) Carbide precipitation. Nishikawa et al. [27] modeled the influence of the bainite reaction on the kinetics of carbon redistribution during the Q&P process. Simulations indicate that the kinetics of carbon partitioning from martensite to austenite is controlled by carbon diffusion in austenite and is affected only to a small extent by the decomposition of austenite into bainitic ferrite.

Based on microscopy pictures of the samples, a model with 3 μm space for austenite until the next lath of martensite and a 2 μm space for ferrite, which represents martensite in this simulation, in a rectangular linear model with 50 nodes for calculations that are more frequent close to the interface, is assumed (see Figure 4).

Figure 5 shows the simulation results of carbon partitioning at 640 °C and 540 °C. Regarding the fact that Mn drops the chemical potential of carbon in austenite, it is expected that regions with high Mn concentration attract more carbon [13]. Therefore, adding 1.78 wt % Mn to the system makes carbon atoms pile up at the border of γ/α and causes a constant increase of the carbon content of austenite until 0.15 wt % after 50 s at 640 °C. If the partitioning process stops after 2 or 5 s, a small distance of approximately 0.5 μm could be enriched with up to 0.3 wt % C (Figure 5a). As can be seen in Figure 5b, diffusion at 540 °C is much slower and makes the boundary full of carbon up to 1.6 wt %.

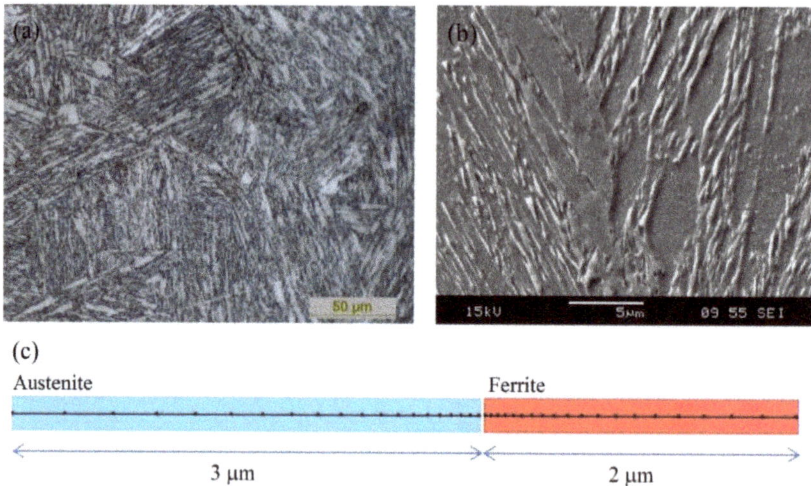

Figure 4. (**a**) Optical microscopy (OM) and (**b**) SEM pictures of samples quenched to 350 °C; (**c**) rectangular model with linear mesh assumed for 3 μm austenite and 2 μm martensite.

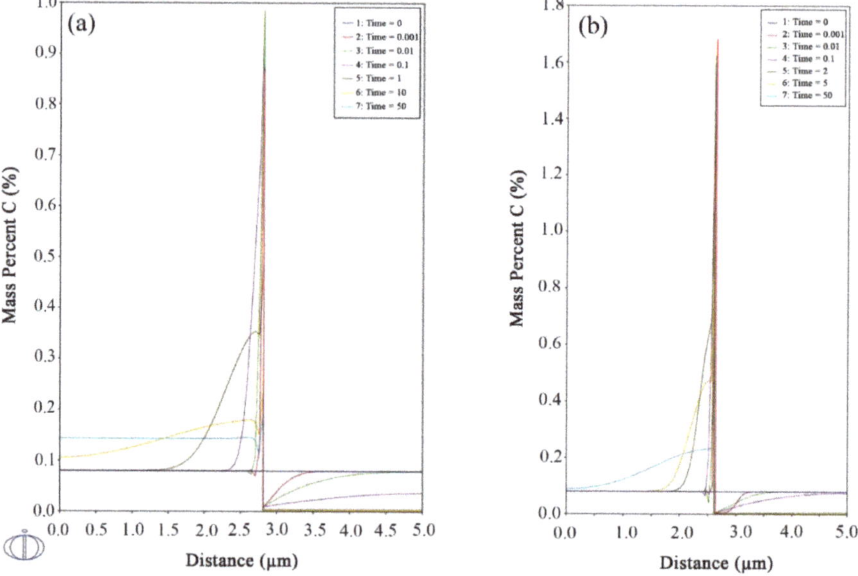

Figure 5. Carbon content of (**a**) Fe–0.08 wt % C–1.78 wt % Mn system at 640 °C; (**b**) Fe–0.08 wt % C–1.78 wt % Mn system at 540 °C.

In order to find the reason for the very low impact toughness results of samples partitioned at 640 °C while they have very good tensile properties, simulations were focused on S(640, 50) and S(540, 5) for comparison. The effect of temperature is shown in Figure 6a. Figure 6b shows the carbon movement inside austenite close to the martensite interface for these two samples, and the existence of retained austenite after the final quench after partitioning can be predicted.

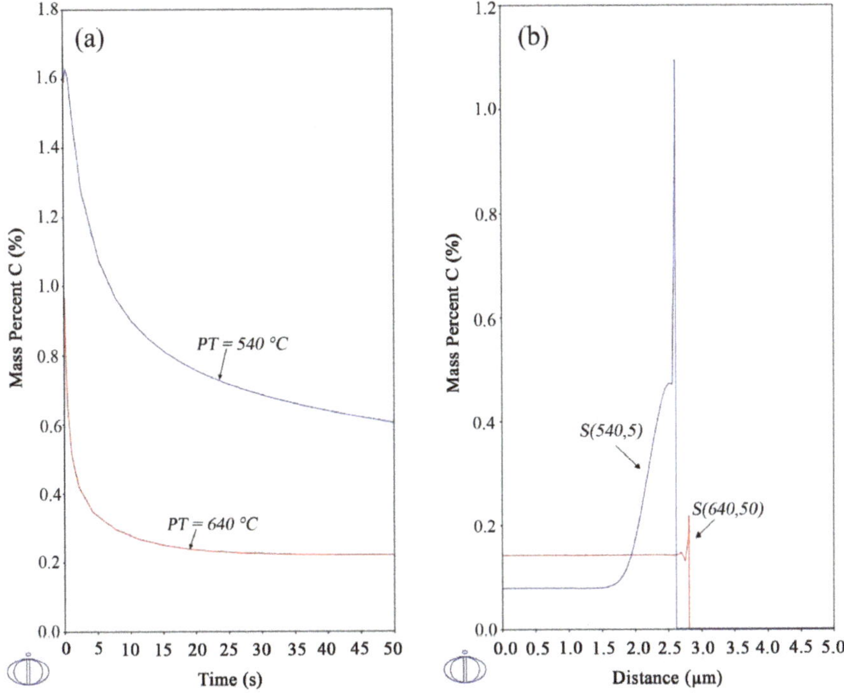

Figure 6. Comparison of samples partitioned at 540 °C, 5 s (blue) and 640 °C, 50 s (red), calculated using Dictra. (**a**) The composition of the interface as a function of the time; (**b**) carbon content vs. location γ/α interface is at 3 μm distance).

Minimum estimated amount of C to stabilize austenite, based on different equations [28–33] for this steel is 0.9 to 1.2 wt % C at ambient temperature. So, comparing with Figure 6b implies that there is no chance for stabilizing the retained austenite in S(640, 50), but there is some possibility in samples treated at 540 °C or less. This is also confirmed by XRD measurements of these two samples S(540, 5) and S(640, 50).

Wu et al. [34] investigated the effect of austenite on fracture resistance of Q&P and showed that the energy absorption by transformation from austenite to martensite postponed the crack propagation and enhanced the fracture resistance of Q&P steels. Even a small amount of retained austenite at room temperature could have a significant effect on energy absorption during crack propagation.

3.4. Kinetics of Precipitation

In Figure 3 it was shown that precipitation of Ti(C, N) can start from the liquid. So, it is difficult to control the size and distribution of the Ti(C, N) particles in the structure. However, SEM pictures of the Q&P-treated sample shows small precipitates as well (Figure 7). EDS study of the particles confirmed the Ti content of these large and sharp-edged particles.

Beside these very large precipitates form during casting, there are some other particles that can nucleate, grow, or coarsen during welding and Q&P treatment. For example, in Figure 7 different particle types can be seen in the HAZ. For example, in region (b) a Ti(C, N) particle nucleated and coarsened during welding and Q&P; region (c) shows accumulation of nucleated carbides at the grain boundary and (d) shows transition carbides that are created during tempering of martensite laths in the partitioning stage.

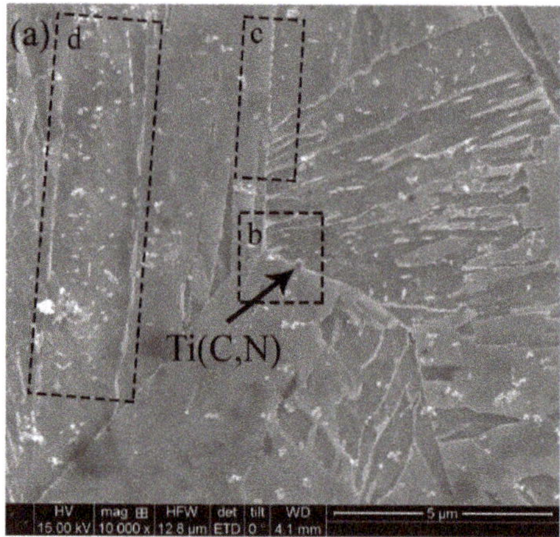

Figure 7. SEM pictures of S(640, 50) displays secondary precipitates at different size and shapes.

As reported by Gustafson [35], there are two sizes of TiC particles, one with sparsely distributed particles of micrometer size and a second with densely distributed particles with sizes of a few tens or hundreds of nanometers. In this work, precipitates with the smaller size were studied, under the assumption that the large primary particles are so sparsely distributed that they will not affect the coarsening of the secondary ones, since the coarsening of the large particles is expected to follow a much slower process and has no important influence on the mechanical properties.

Since the precipitates (especially the small ones) from casting will melt in FZ during welding, the most critical area will be HAZ. In order to model the nucleation and growth of precipitation using TC–Prisma, isothermal heating at 1350 °C for 5 s is considered. Figure 8, shows that the equivalent average diameter of Ti(C, N) particles is around 150 nm.

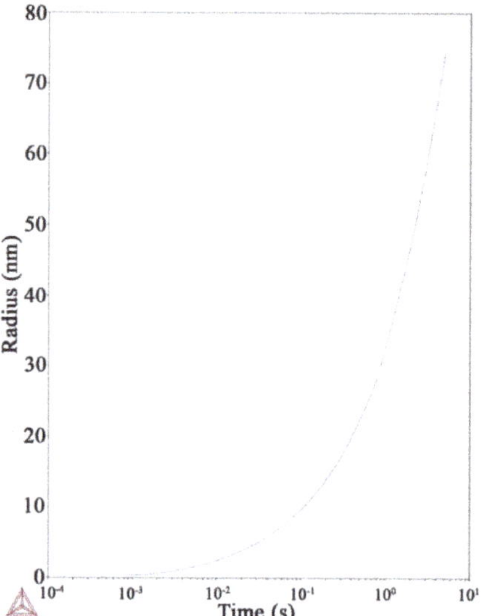

Figure 8. Prediction of precipitate size in HAZ considering isothermal heating at 1350 °C for 5 s, resulting in a particle size of around 150 nm.

Comparison of the precipitate coarsening rates in corresponding matrix phases using the Thermocalc database shows that increasing the temperature from 540 °C to 640 °C will increase the coarsening rate of Ti(C, N) 1000 times to 1.27 × 10^{-35} m^3/s and of MoC 100 times to 3.875 × 10^{-31} m^3/s.

Quantification of the precipitate size distribution in base material (BM) and in those partitioned at 540 °C and 640 °C was carried out by STEM from carbon replicas of the samples.

Figure 9 shows the particle size distribution for both partitioned samples (at 540 °C and 640 °C) and for the BM. The average particle size was calculated to be 0.263 ± 0.108 μm, 0.207 ± 0.089 μm, and 0.07 ± 0.11 μm for S(540, 5), S(640, 50), and BM, respectively. The precipitates quantification in the BM became more complicated since it presents particles over a large range of sizes. SEM analysis of precipitates show a few large particles (between 0.2 and 0.65 μm) at relatively low magnification (20,000×). However, at higher magnifications (e.g., 80,000×), a large quantity of small particles (<0.2 μm) can be distinguished. For this reason, the analysis of the precipitates in this sample was performed at two different magnifications.

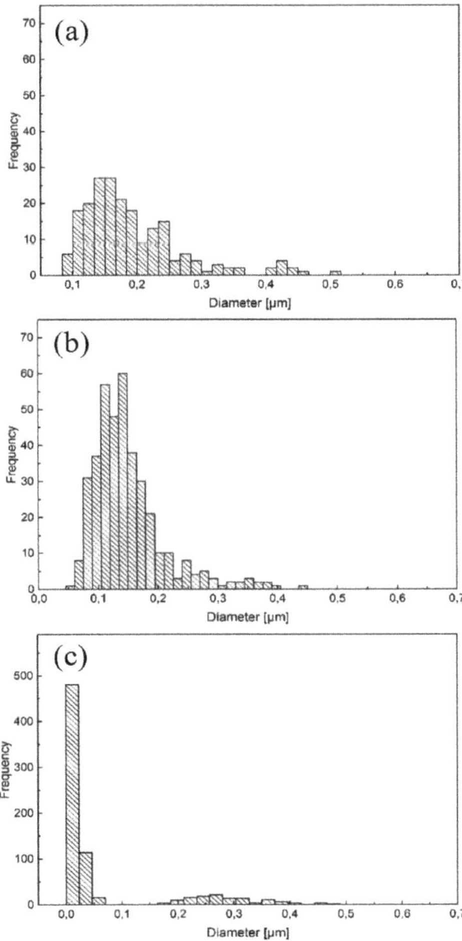

Figure 9. Measured size distribution of precipitates vs. number of particles for post-welding heat-treated samples (**a**) at QT = 355 °C and PT = 540 °C for 5 s; (**b**) at QT = 355 °C and PT = 640 °C for 50 s; (**c**) base metal.

Figure 10 shows STEM images from C replicas corresponding to the samples. An apparently larger particle density can be observed in the sample partitioned at 640 °C, when comparing the same area of both samples (Figure 10a,b). However, the particle density has not been systematically analyzed in this case. Two hundred six and 388 particles were considered for S(540, 5) and S(640, 50), respectively, and this was compared with the base metal (see Figure 9).

Results indicate that the approximate average size of particles in treated samples are 0.263 ± 0.108 μm for the sample partitioned at 540 °C and 0.207 ± 0.089 μm for the one partitioned at 640 °C, while the number density of particles after partitioning at 640 °C is around 1.5 times higher than for 540 °C. This can be seen when comparing Figure 10a,b.

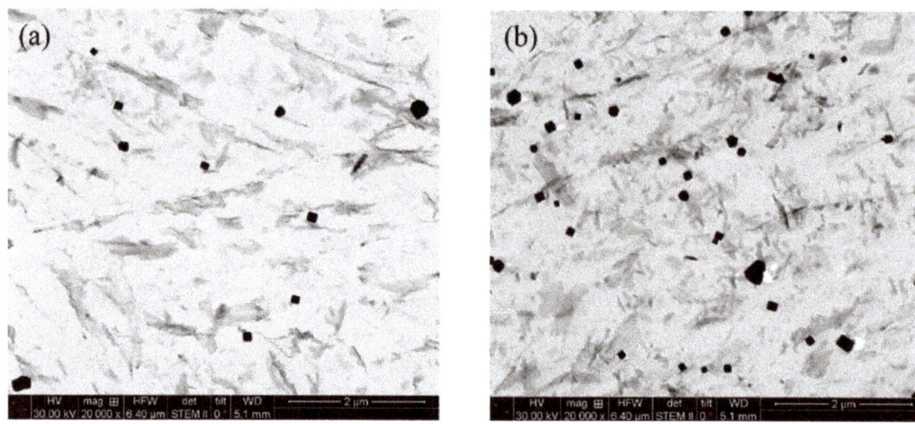

Figure 10. Comparison between distribution, number density, and shape of the particles in carbon replicas made from the sample: (**a**) S(540, 5) and (**b**) S(640, 50).

From EDS measurements, it was determined that the particles in samples partitioned at 540 °C and 640 °C are principally MoC (round shaped particles) and Ti(C, N) (rectangular shaped particles).

3.5. Grain Size Effect

As mentioned before, the structure is made of tempered and fresh martensite lathes from the first and second quench and bainite from the partitioning stage. So, the toughness of such structure is increased by a high density of the high angle boundaries created, because this kind of boundary acts as an obstacle for cleavage propagation, forcing the cleavage crack to change its microscopic plane of propagation in order to accommodate the new local crystallography [22,36,37]. In addition, the presence of coarse martensite laths leads to early strain localization, especially when they are in the vicinity of untempered martensite islands [38]. However, the complexity of the lath martensitic/bainitic structure makes the grain size measurement very difficult. For this reason, the region of martensite/bainite lathes with a determined crystallographic orientation is defined as a block [39]. Clusters of blocks form a packet when they share the same $(111)_\gamma$, to which the corresponding $(001)_\alpha$ is almost parallel [40]. Since the blocks and packages present high angle grain boundaries (>11°), the toughness of the material might be influenced by their size [41]. In this study, the effective grain size of the martensite/bainite was determined from EBSD measurements, considering grain boundary misorientation ≥15°. Figure 11 shows the inverse pole figure (IPF) superimposed on the image quality map (IQ) for the treated samples. Results shows that partitioning at a lower temperature (540 °C) leads to a refinement of the effective grain size (block packages), as can be noticed by the change in the color orientation. Kawata et al. [42] showed that the bainite block is coarsened by ausforming, especially for a higher temperature, by formation of preferred variants in a packet. In contrast, the same ausforming treatments refine the lath martensite block. In fact, packet size will influence the ductile–brittle transition temperature (DBTT). As mentioned in Equation (1), DBTT (*T*) is inversely proportional to the root square of the distance between high-angle grain boundaries (*d*), where K is a constant [36]:

$$T = T_0 - \mathrm{K}d^{-2}. \quad (1)$$

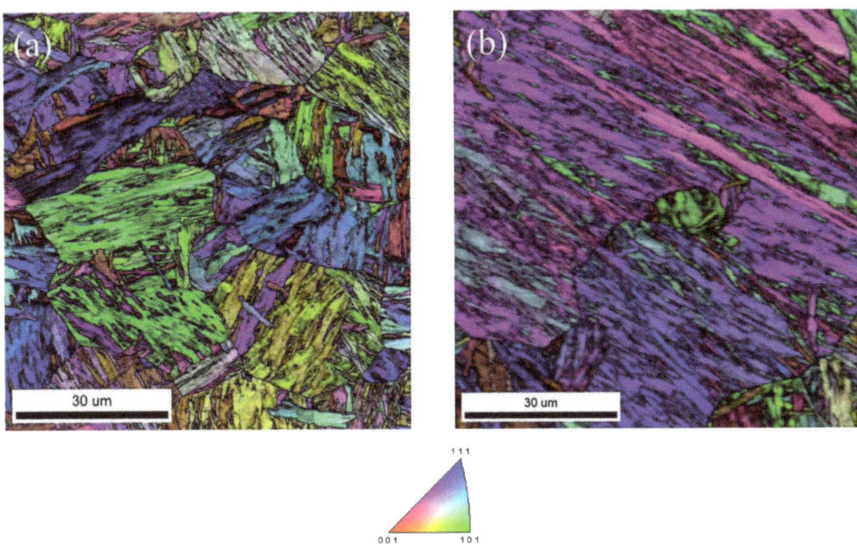

Figure 11. IPF + IQ picture from EBSD analysis of samples: (**a**) S(540, 5); (**b**) S(640, 50).

4. Conclusions

The aim of this work was to investigate the reason for brittle fracture of the samples from a low-carbon, low-Si AHSS, which were post-weld heat-treated by the Q&P method around B_s temperature (640 °C). Their strength and ductility were the best in comparison with other samples partitioned at lower temperatures, i.e., 540 °C and 440 °C. Results shows that since $D_c^{\alpha'}$ is much higher than D_c^{γ} carbon diffuses out from martensite and piles up behind the border of α'/γ. The partitioning temperature controls the rate of carbon diffusion, but this amount at the α'/γ border cannot reach the level necessary to stabilize austenite at 640 °C, while it can occur for a small area next to the border at 540 °C. Subsequently, this RA can contribute to eliminating the brittle fracture due to the TRIP effect and increasing the impact toughness.

STEM results of carbon replicas revealed that there are many more very small precipitates (<0.1 μm) in the sample partitioned at 640 °C; this can increase the strength of the material via precipitation strengthening mechanism. On the other hand, EBSD results showed much larger crystallographic packets in this sample, which can result in a lowering of the material strength. The increase of strength by precipitation strengthening and the decrease of the strength due to larger packet size will counteract each other so that the tensile test results of the samples partitioned at 640 °C are still good.

Author Contributions: Concept and design of experiments: F.F. and E.V.; Experiments, simulations, analysis of data and writing the original draft: F.F.; Experiments and revising the paper: M.A.G.; Supervision: E.V. and F.M.

Funding: This research was funded by Erasmus+: Erasmus Mundus Joint Doctorate (EMJD)-Advanced Materials Engineering-DOCMASE, grant number "2011-0020".

Acknowledgments: The support of the EUSMAT (European School of Materials) via the Ph.D. program 'DOCMASE' and Flavio Soldera is gratefully acknowledged by the authors.

Conflicts of Interest: The authors declare no conflict of interest.

References

1. Opbroek, E. *UltraLight Steel: A Global Consortium Changes the Future of Automotive Steel*; Xlibris: Bloomington, IN, USA, 2013; ISBN 1479773441.
2. Shome, M.; Tumuluru, M. *Welding and Joining of Advanced High Strength Steels (AHSS)*, 1st ed.; Woodhead Publishing: Cambridge, UK, 2015; ISBN 9780857098580.
3. Speer, J.; Matlock, D.K.; De Cooman, B.C.; Schroth, J.G. Carbon partitioning into austenite after martensite transformation. *Acta Mater.* **2003**, *51*, 2611–2622. [CrossRef]
4. Forouzan, F.; Vuorinen, E.; Mücklich, F. Post weld–treatment of laser welded AHSS by application of quenching and partitioning technique. *Mater. Sci. Eng. A* **2017**, *698*, 174–182. [CrossRef]
5. Clarke, A.J.; Speer, J.G.; Miller, M.K.; Hackenberg, R.E.; Edmonds, D.V.; Matlock, D.K.; Rizzo, F.C.; Clarke, K.D.; De Moor, E. Carbon partitioning to austenite from martensite or bainite during the quench and partition (Q&P) process: A critical assessment. *Acta Mater.* **2008**, *56*, 16–22. [CrossRef]
6. Santofimia, M.J.; Zhao, L.; Sietsma, J. Microstructural Evolution of a Low–Carbon Steel during Application of Quenching and Partitioning Heat Treatments after Partial Austenitization. *Metall. Mater. Trans. A* **2009**, *40*, 46–57. [CrossRef]
7. Toji, Y.; Miyamoto, G.; Raabe, D. Carbon partitioning during quenching and partitioning heat treatment accompanied by carbide precipitation. *Acta Mater.* **2015**, *86*, 137–147. [CrossRef]
8. Charleux, M.; Poole, W.J.; Militzer, M.; Deschamps, A. Precipitation behavior and its effect on strengthening of an HSLA–Nb/Ti steel. *Metall. Mater. Trans. A* **2001**, *32*, 1635–1647. [CrossRef]
9. Soto, R.; Saikaly, W.; Bano, X.; Issartel, C.; Rigaut, G.; Charai, A. Statistical and theoretical analysis of precipitates in dual–phase steels microalloyed with titanium and their effect on mechanical properties. *Acta Mater.* **1999**, *47*, 3475–3481. [CrossRef]
10. Toji, Y.; Matsuda, H.; Herbig, M.; Choi, P.; Raabe, D. Atomic–scale analysis of carbon partitioning between martensite and austenite by atom probe tomography and correlative transmission electron microscopy. *Acta Mater.* **2014**, *65*, 215–228. [CrossRef]
11. Fairchild, D.P.; Howden, D.G.; Clark, W.A.T. The mechanism of brittle fracture in a microalloyed steel: Part I. Inclusion–induced cleavage. *Metall. Mater. Trans. A* **2000**, *31*, 641–652. [CrossRef]
12. Di Schino, A.; Di Nunzio, P.E. Effect of Nb microalloying on the heat affected zone microstructure of girth welded joints. *Mater. Lett.* **2017**, *186*, 86–89. [CrossRef]
13. HajyAkbary, F.; Sietsma, J.; Miyamoto, G.; Furuhara, T.; Santofimia, M.J. Interaction of carbon partitioning, carbide precipitation and bainite formation during the Q&P process in a low C steel. *Acta Mater.* **2016**, *104*, 72–83. [CrossRef]
14. Clarke, A.J.; Speer, J.G.; Matlock, D.K.; Rizzo, F.C.; Edmonds, D.V.; Santofimia, M.J. Influence of carbon partitioning kinetics on final austenite fraction during quenching and partitioning. *Scr. Mater.* **2009**, *61*, 149–152. [CrossRef]
15. Somani, M.C.; Porter, D.A.; Karjalainen, L.P.; Misra, R.D.K. On Various Aspects of Decomposition of Austenite in a High–Silicon Steel During Quenching and Partitioning. *Metall. Mater. Trans. A* **2014**, *45*, 1247–1257. [CrossRef]
16. Santofimia, M.J.; Zhao, L.; Sietsma, J. Overview of Mechanisms Involved During the Quenching and Partitioning Process in Steels. *Metall. Mater. Trans. A* **2011**, *42*, 3620–3626. [CrossRef]
17. Thomas, G.A.; Speer, J.G. Interface migration during partitioning of Q&P Steel. *Mater. Sci. Technol.* **2014**, *30*, 998–1007. [CrossRef]
18. De Knijf, D.; Petrov, R.; Föjer, C.; Kestens, L.A. Effect of fresh martensite on the stability of retained austenite in quenching and partitioning steel. *Mater. Sci. Eng. A* **2014**, *615*, 107–115. [CrossRef]
19. De Knijf, D.; Puype, A.; Föjer, C.; Petrov, R. The influence of ultra–fast annealing prior to quenching and partitioning on the microstructure and mechanical properties. *Mater. Sci. Eng. A* **2015**, *627*, 182–190. [CrossRef]
20. De Diego-Calderón, I.; Sabirov, I.; Molina-Aldareguia, J.M.; Föjer, C.; Thiessen, R.; Petrov, R.H. Microstructural design in quenched and partitioned (Q&P) steels to improve their fracture properties. *Mater. Sci. Eng. A* **2016**, *657*, 136–146. [CrossRef]
21. Wang, C.; Wang, M.; Shi, J.; Hui, W.; Dong, H. Effect of microstructural refinement on the toughness of low carbon martensitic steel. *Scr. Mater.* **2008**, *58*, 492–495. [CrossRef]

22. Forouzan, F.; Gunasekaran, S.; Hedayati, A.; Vuorinen, E.; Mucklich, F. Microstructure analysis and mechanical properties of Low alloy High strength Quenched and Partitioned Steel. *DiVA* **2016**, *258*, 574–578. [CrossRef]
23. Andersson, J.O.; Helander, T.; Höglund, L.; Shi, P.F.; Sundman, B. Thermo-Calc & DICTRA, Computational tools for materials science. *Calphad* **2002**, *26*, 273–312.
24. Thermo-Calc Software TCFE9 Steels/Fe-alloys Database. Available online: https://www.thermocalc.com/products-services/databases/thermodynamic/ (accessed on 20 September 2018).
25. A.B. MOBFE3: TCS Steels/Fe–Alloys Mobility Database. Thermo–Calc Software. Available online: https://www.thermocalc.com/products-services/databases/mobility/ (accessed on 20 September 2018).
26. Bhadeshia, H.; Honeycombe, R. Microstructure and properties. In *Steels: Microstructure and Properties*, 3rd ed.; Butterworth–Heinemann: Cambridge, UK, 2011; ISBN 0080462928, 9780080462929.
27. Nishikawa, A.S.; Santofimia, M.J.; Sietsma, J.; Goldenstein, H. Influence of bainite reaction on the kinetics of carbon redistribution during the Quenching and Partitioning process. *Acta Mater.* **2018**, *142*, 142–151. [CrossRef]
28. Payson, P.; Savage, C.H. Martensite reactions in alloy steels. *Trans. ASM* **1944**, *33*, 261–280.
29. Grange, R.A.; Stewart, H.M. The temperature range of martensite formation. *Trans. AIME* **1946**, *167*, 467–501.
30. Van Bohemen, S. Bainite and martensite start temperature calculated with exponential carbon dependence. *Mater. Sci. Technol.* **2012**, *28*, 487–495. [CrossRef]
31. Nehrenberg, A.E. Contribution to Discussion on Grange and Stewart. *Trans. Am. Inst. Min. Met. Eng.* **1946**, *167*, 494–498.
32. Haynes, A.G.; Steven, W. The temperature of formation of martensite and bainite in low–alloy steel. *J. Iron Steel Inst.* **1956**, *183*, 349–359.
33. Andrews, K.W. Empirical formulae for the calculation of some transformation temperatures. *J. Iron Steel Inst.* **1965**, *203*, 721–727.
34. Wu, R.; Li, J.; Li, W.; Wu, X.; Jin, X.; Zhou, S.; Wang, L. Effect of metastable austenite on fracture resistance of quenched and partitioned (Q&P) sheet steels. *Mater. Sci. Eng. A* **2016**, *657*, 57–63. [CrossRef]
35. Gustafson, Å. Coarsening of TiC in austenitic stainless steel-experiments and simulations in comparison. *Mater. Sci. Eng. A* **2000**, *287*, 52–58. [CrossRef]
36. Díaz–Fuentes, M.; Iza–Mendia, A.; Gutiérrez, I. Analysis of different acicular ferrite microstructures in low–carbon steels by electron backscattered diffraction. Study of their toughness behavior. *Metall. Mater. Trans. A* **2003**, *34*, 2505–2516. [CrossRef]
37. Rodriguez–Ibabe, J. The Role of Microstructure in Toughness Behaviour of Microalloyed Steels. *Mater. Sci. Forum.* **1998**, *284–286*, 51–62. [CrossRef]
38. Wang, M.; Hell, J.; Tasan, C.C. Martensite size effects on damage in quenching and partitioning steels. *Scr. Mater.* **2017**, *138*, 1–5. [CrossRef]
39. Verbeken, K.; Barbé, L.; Raabe, D. Evaluation of the crystallographic orientation relationships between FCC and BCC phases in TRIP steels. *ISIJ Int.* **2009**, *49*, 1601–1609. [CrossRef]
40. Morito, S.; Tanaka, H.; Konishi, R.; Furuhara, T.; Maki, T. The morphology and crystallography of lath martensite in Fe–C alloys. *Acta Mater.* **2003**, *51*, 1789–1799. [CrossRef]
41. Bhadeshia, H.K.D.H. Bainite in Steels. In *Theory and Practice*, 3rd ed.; Maney Publishing: Cambridge, UK, 2015; ISBN 1909662747, 9781909662742.
42. Kawata, H.; Sakamoto, K.; Moritani, T.; Morito, S.; Furuhara, T.; Maki, T. Crystallography of ausformed upper bainite structure in Fe–9Ni–C alloys. *Mater. Sci. Eng. A* **2006**, *438*, 140–144. [CrossRef]

Article

Flow-Accelerated Corrosion of Type 316L Stainless Steel Caused by Turbulent Lead–Bismuth Eutectic Flow

Tao Wan * and Shigeru Saito

J-PARC Center, Japan Atomic Energy Agency, 2-4 Shirakata, Tokai-mura, Ibaraki 319-1195, Japan; saito.shigeru@jaea.go.jp

* Correspondence: wan.tao@jaea.go.jp; Tel.: +81-029-282-6948

Received: 4 July 2018; Accepted: 8 August 2018; Published: 9 August 2018

Abstract: Lead–bismuth eutectic (LBE), a heavy liquid metal, is an ideal candidate coolant material for Generation-IV fast reactors and accelerator-driven systems (ADSs), but LBE is also known to pose a considerable corrosive threat to its container. However, the susceptibility of the candidate container material, 316L stainless steel (SS), to flow-accelerated corrosion (FAC) under turbulent LBE flow, is not well understood. In this study, an LBE loop, referred to as JLBL-1, was used to experimentally study the behavior of 316L SS when subjected to FAC for 3000 h under non-isothermal conditions. An orificed tube specimen, consisting of a straight tube that abruptly narrows and widens at each end, was installed in the loop. The specimen temperature was 450 °C, and a temperature difference between the hottest and coldest legs of the loop was 100 °C. The oxygen concentration in the LBE was lower than 10^{-8} wt %. The Reynolds number in the test specimen was approximately 5×10^4. The effects of various hydrodynamic parameters on FAC behavior were studied with the assistance of computational fluid dynamics (CFD) analyses, and then a mass transfer study was performed by integrating a corrosion model into the CFD analyses. The results show that the local turbulence level affects the mass concentration distribution in the near-wall region, and therefore, the mass transfer coefficient across the solid/liquid interface. The corrosion depth was predicted on the basis of the mass transfer coefficient obtained in the numerical simulation and was compared with that obtained in the loop. For the abrupt narrow part, the predicted corrosion depth was comparable with the measured corrosion depth, as was the abrupt wide part after involving the wall roughness effects in the prediction; for the straight tube part, the predicted corrosion depth is about 1.3–3.5 times the average experimental corrosion depth, and the possible reason for this discrepancy was provided.

Keywords: FAC; austenitic stainless steel; LBE; turbulent flow; dissolution; modelling

---

1. Introduction

Flowing liquid has the potential to accelerate the rate at which metal corrodes, a phenomenon known as flow-accelerated corrosion (FAC). In fully developed turbulent flow in simple geometries (e.g., within a pipe) and a no-slip wall condition, the flow is halted at the solid wall, meaning that the flow will transition from turbulent in the bulk fluid to laminar very close to the wall. A so-called "hydro-dynamic boundary layer" will thus be formed, with its boundary line where the flow velocity equals 0.99 of the free stream flow velocity. The boundary layer can be divided into three typical layers depending on the correlation between the dimensionless velocity ($u+$) and the dimensionless distance normal to the wall ($y+$) with approach to the solid wall: (1) the turbulent flow layer, where $u+$ and $y+$ have a logarithmic correlation; (2) the buffer layer, where flow transition from turbulent to laminar occurs; and (3) the laminar flow layer, also known as the viscous sublayer, where $u+$ is linearly proportional to $y+$, the turbulence level damps rapidly, and the viscous effects of fluid are dominant. If a dissolution constituent of a fluid has a high Schmidt number (Sc), exceeding several hundreds, the mass diffusion boundary layer will be very thin, and will be embedded deep in the hydrodynamic viscous sublayer. The characteristics of the boundary layer are illustrated schematically in Figure 1 [1,2].

Pioneering studies have linked several hydrodynamic parameters, such as flow velocity [3], the Reynolds number [2], wall shear stress [4,5], and turbulence level [6] to the rate of FAC of metal. For turbulent flow in structures with simple geometries, the occurrence of FAC is mainly attributable to wall shear stress and the turbulence level in the near-wall region. The above two factors affect the protective oxide film on the base metal and the adjacent diffusion boundary layer [6], which are considered barriers to mass transfer, due to the particularly low mass transfer rate within them [7]. Fluctuations in turbulence can disturb the mass transfer rate from the bulk fluid to the metal surface and vice versa [5,8]; it can also disturb the formation of the protective oxide and disrupt the protective oxide layer [6]. Moreover, the shear stress generated at the wall may rupture or thin the protective oxide coating. In both cases, the mass transfer coefficient across the solid/liquid interface, and thus the corrosion rate of the metal, will be affected. In practice, the profile of wall shear stress and that of the near-wall turbulence level are coincident with each other for flow in structures with simple geometries because the shear stress on the wall is the main source of local turbulence [7]. This is shown schematically in Figure 2.

When flow encounters sudden contraction or expansion in pipes, orifices, valves, elbows, and weld heads, however, the hydrodynamic mechanisms of FAC become complex. In such situations, flow diversion and flow separations, flow reattachments, and flow recirculation occur. Extensive research has aimed at clarifying these mechanisms, but they are not yet thoroughly understood. Chang et al. [9] pointed out a positive correlation between wall shear stress and wall mass transfer. Utanohara et al. [10,11] concluded that the root mean square (RMS) of instantaneous shear stress is a suitable parameter for evaluating the FAC rate, despite their results showing a better correlation between the profiles of the FAC rate and the near-wall turbulence level. Their argument for discounting this result was that the relationship between the turbulence level and FAC had not been theoretically validated. Crawford et al. [12] indicated that the pressure drop attributed to secondary flow can significantly increase the average and oscillatory wall shear stress. By contrast, Nesic et al. [6,13] and Poulson [14] found that it is the high near-wall turbulence level, not the wall shear stress, which is responsible for localized enhancement in the wall corrosion rate. In addition, many other studies have linked the turbulence level to the rate of mass transfer and corrosion [15–18].

The majority of existing research has adopted water, sea water, or slurry as the flow medium, but a few studies have employed heavy liquid metals (HLMs), e.g., lead and lead-based alloys such as lead–bismuth eutectic (LBE). LBE is an ideal candidate coolant material for Generation-IV fast reactors and accelerator-driven systems (ADSs) because of the following physical characteristics: (1) a high boiling temperature, so that the heat generated in the reaction core can be utilized more efficiently; (2) its inertness on contact with water, air, and steam, enabling safe operation [19,20]. However, it is also well known that LBE poses a severe threat of corrosion on the material of its container [21], potentially ultimately leading to structural failure. The corrosion rate, CR, under LBE flows is a complex process involving many factors, and can be expressed as follows:

$$CR = f(M_c, T, \Delta T, C_{O_2}, R_w, V_l, \tau_w, I_t, \ldots), \tag{1}$$

where M_c is the chemical components of the test specimen, T is the temperature of the test specimen, ΔT is the temperature difference between the hot and cold legs of a loop, C_{O_2} is the oxygen concentration in the LBE, R_w is the wall roughness, V_l is the flow speed of the LBE, τ_w is the shear stress on the wall, and I_t is the turbulence level in the near-wall region.

Earlier studies investigating the corrosion of structural steels by LBE looked at the loop as a whole, focusing on the corrosion pattern and the mass transfer behavior of the corrosion products [22]. The effects of $M_C, T, \Delta T, C_O$, and V_l have attracted most attention [21,23]. There has also been a report of unexpected tube failure in the Corrosion In Dynamic lead Alloys (CORRIDA) loop at the Karlsruhe Institute of Technology [24]. A hole penetrated the tube, which had an initial wall thickness of 2.5 mm, downstream of a tube junction where the flowing LBE entered the vertical tube at an angle of 30 degrees. The study mentioning this attributed the failure to complex turbulent LBE flow. However, to the best of the authors' knowledge, the local mechanism of FAC under turbulent LBE flow has not been studied in detail, either experimentally or numerically, from the perspectives of hydromechanics and wall mass transfer behavior.

If the potential of LBE as a coolant is to be exploited, it is highly desirable to investigate FAC of structural materials under turbulent LBE flow. To that end, the present study experimentally investigates FAC of type 316L stainless steel (SS), which is a candidate structural material for future ADSs, using the JLBL-1 loop installed at the Japan Atomic Energy Agency (JAEA). An orifice-type test tube with abrupt narrowing and widening at each end of a straight section was installed in the loop. Flow paths in the test tube show change in flow direction, fully developed flow, and recirculation. Investigations of the behavior of corrosion under such conditions are of interest and importance to scientific research and engineering applications. The study goes on to investigate the correlation between the hydrodynamic parameters and the corrosion profile using computational fluid dynamics (CFD) numerical simulations performed with STAR-CD. CFD analyses combined with a corrosion model are used to characterize the wall mass transfer behavior, and comparisons are drawn between the experimental and numerical results.

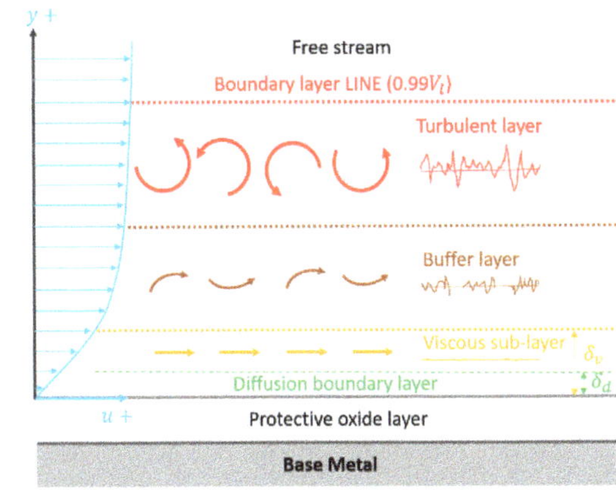

Figure 1. Schematic of the boundary layer where turbulent flow meets a no-slip wall boundary.

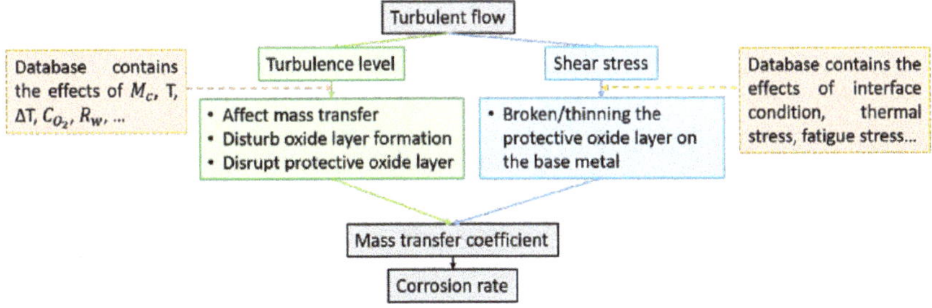

Figure 2. Schematic of the effects of turbulent flow on the corrosion rate of the base metal.

2. Physical Models

2.1. Turbulent Flow Model

The "standard" $k-\varepsilon$ low Reynolds number (LRN) eddy-viscosity turbulence model in STAR-CD was adopted to model the flow and mass transfer [25]. The mass transfer is modelled by solving the mass transport equation together with the hydrodynamic equations, where the turbulent Schmidt number, $Sc_t = \mu_t/\rho D_t$, defines the ratio of turbulent momentum transport to turbulent diffusive transport. In the present study, Sc_t was assigned a value of 0.9. The model is founded on the $k-\varepsilon$ model proposed by Launder and Spalding [26], in which modelling of fully turbulent bulk flow is achieved through the transportation equations of turbulent kinetic energy (TKE), k, and its dissipation rate, ε. Launder and Spalding model the near-wall region, where the viscous effect is dominant, by way of algebraic wall functions. However, this approach uses a velocity profile that is not applicable to recirculating flow, where there is flow separation and reversal, and does not take the detailed mass transfer characteristics in the diffusion boundary layer into account [27]. It is therefore necessary to use a turbulence model that directly solves the flow field and mass transfer deep inside the viscous sublayer to account for the changes due to the wall. The LRN model does so by solving the transport equations of k and ε everywhere, including the near-wall regions.

To extend the $k-\varepsilon$ high Reynolds number model to model low Reynolds number conditions, Lien et al. [28] have proposed damping functions that modify the transport equations of k and ε in the near-wall region. One such is the semi-viscous near-wall effect f_u, which modifies the eddy-viscosity term μ_t. An additional term for turbulence generation $P_k{}'$, which vanishes as Re_y approaches the order of 100, is added to ensure that the correct level of TKE dissipation is returned, and the destruction of TKE dissipation is modified with another damping factor f_2. These are formulated as follows in STAR-CD:

$$P_k{}' = 1.33\left[1-0.3e^{-R_t{}^2}\right]\left[P_k + 2\frac{\mu}{\mu_t}\frac{k}{y^2}\right]e^{-0.00375 Re_y{}^2}, \quad (2)$$

$$f_u = \left[1-e^{-0.0198 Re_y}\right]\left(1+\frac{5.29}{Re_y}\right), \quad (3)$$

$$f_2 = 1-0.3e^{-Re_t{}^2}, \quad (4)$$

where the turbulent Reynolds numbers are $Re_y = \rho y\sqrt{k}/\mu$ and $Re_t = \rho k^2/\mu\varepsilon$.

2.2. Corrosion Model

In an isothermal closed liquid metal loop, corrosion may come to a halt because the corrosion products achieve saturation. However, in a non-isothermal loop where the saturation concentration of the corrosion products depends on temperature, continuous dissolution will occur in the relatively high-temperature part, whereas deposition of corrosion products will occur in the relatively low-temperature part. A kinetic equilibrium will be reached in which the amount of corrosion is balanced by the amount of deposition; thus, the precipitation sustains the corrosion [29]. This is particularly pertinent where metal corrosion occurs in an HLM environment, as this is a physical or physical-chemical process, which involves species dissolution, species transport, and chemical reaction between corrosion products and impurities, rather than an electrochemical process, as is usually the case in an aqueous environment [21,22].

In LBE environment, the corrosion process of a steel can be divided into two types according to the oxygen concentration in LBE [23,29]:

(1) If the oxygen concentration in LBE is sufficiently low, as in the present experimental study, there will be no effective oxide protective layer formed on the steel surface. In this situation, the steel contacts with the LBE directly, and the main constituents of the steel are thus dissolved into the LBE directly.

(2) If the oxygen concentration is within an appropriate range, oxidation of steel will occur, and an active oxide film (Fe_3O_4-based) will eventually be formed on the steel surface. Direct dissolution of steel into LBE will be prevented due to separation of the oxide film. In this case, the iron diffuses from the base metal, and the oxygen transfers from the bulk flow to the oxide/LBE interface to engage in the oxidation–reduction chemical reaction ($3Fe + 2O_2 \Leftrightarrow Fe_3O_4$).

For long-term steady-state LBE loop operation without an oxide protective layer, the steel corrosion process in the control depends on the LBE flow speed. If the LBE flow speed is rapid enough, resulting in the mass transfer rates of constituents in liquid greater than their dissolution reaction rates at the solid/LBE interface, then the corrosion will be controlled by the dissolution rate; otherwise, it will be controlled by the mass transfer rate in fluid [21,23]. For the mass transfer-controlled corrosion, if the diffusion flux of the corrosion product in the solid is less than the mass transfer rate in the liquid, surface recession will consequently occur [22].

The selective corrosion of Cr and Ni in the steel superficial layer is common for a 316L SS contacting with LBE that contains a low oxygen concentration. Subsequently, a phase change from austenite to ferrite might occur in the selective corrosion layer, due to the depletions of Cr and Ni in this layer, and as a result, a ferritic layer might be formed on the steel surface, where the concentrations of Cr and Ni are considerably low [30,31]. Both austenite and ferrite crystal mainly consist of iron, so that the dissolution of iron atoms will liberate other element atoms of the crystal into LBE, and therefore, it is reasonable to assume that the corrosion rate of iron determines the surface recession rate [22,32]. Hereafter, the surface recession rate is referred to as corrosion rate (CR) and the surface recession depth is called corrosion depth. The concentration of Cr and Ni in the solid at the solid/LBE interface can be assumed to be zero [22]. By contrast, the concentration of Fe in the solid at the solid/LBE interface is assumed to be equal to its saturation solubility in LBE, which will be described and discussed in detail in Section 4.2.

The corrosion process is illustrated schematically in Figure 3, and described by the following expression [33]:

$$Fe_{(s)} \Leftrightarrow Fe_{(Sol)} \tag{5}$$

The dissolved iron is transferred into the diffusion boundary layer, where the transfer of iron is governed by the molecular diffusion process, and thus, the transfer rate is low. Above the diffusion boundary layer in the viscous sublayer, where the transfer rate of iron grows dramatically. The thickness of the viscous sublayer is determined by the turbulence level of LBE flow, while the thickness of the diffusion boundary layer for each constituent is determined by not only the turbulence level of LBE flow, but also the Schmidt number (Sc) of each constituent. Therefore, the thickness of diffusion boundary layers of each constituent are independently different from each other. Therefore, lack of Ni and Cr involvement in the present corrosion model has insignificant effects on the thickness of the viscous sublayer and the diffusion boundary layer of Fe.

For a fully developed flow in a simple geometry, the mass transfer coefficient of a constituent across the wall can be indicated by the Sherwood number, $Sh = aRe^bSc^c$. However, for complex geometries, there are no such empirical relationships, and CFD analysis must be used to aid its calculation.

The mass flux of iron, J_{Fe}, from the reaction location to the bulk flow can be given by

$$J_{Fe} = K_c(C_w - C_b), \tag{6}$$

where K_c is the mass transfer coefficient, C_w is the concentration of iron at the wall, and C_b is the concentration of iron in the bulk fluid. In the simulation, if the first node (the layer of node closest to the wall in the mesh structure) is located in the diffusion boundary layer, then mass transfer between the wall and the first node is controlled solely by molecular diffusion. The effects of locations of the first node are insignificant, and can be positioned extremely close to the wall or at the end of the diffusion boundary layer. According to Fick's law, the mass flux between the wall and the first node can be expressed as

$$J_{Fe} = \frac{D_m}{y_0}(C_w - C_0), \tag{7}$$

where y_0 is the distance from the wall to the first node, D_m is the molecular diffusion coefficient of iron, and C_0 is the iron concentration at the first node.

As mentioned previously, the rate of mass transfer in the diffusion boundary layer is very low, so it controls the total rate of mass transfer from the wall to the bulk flow. Therefore, the mass flux of iron is the same in Equations (7) and (6). By substituting Equation (7) into Equation (6), the mass transfer coefficient can be expressed as

$$K_c = \frac{D_m}{y_0}\frac{(C_w - C_0)}{(C_w - C_b)}. \tag{8}$$

In diffusion-controlled mass transfer, the diffusion coefficient of iron to or from the reaction site controls the rate of corrosion. The corrosion rate (in mm/h) can thus be defined as follows [34,35]:

$$CR = \frac{K_c(C_w - C_b)M_{Fe}}{\rho_{Fe}} = \frac{D_m(C_w - C_0)M_{Fe}}{y_0\rho_{Fe}} \times 60 \times 60 \times 1000, \tag{9}$$

where M_{Fe} is the molar mass of iron, and ρ_{Fe} is the density of iron. In the above equation, C_0 can be obtained in the numerical simulation by solving the mass transport equations described in the last subsection.

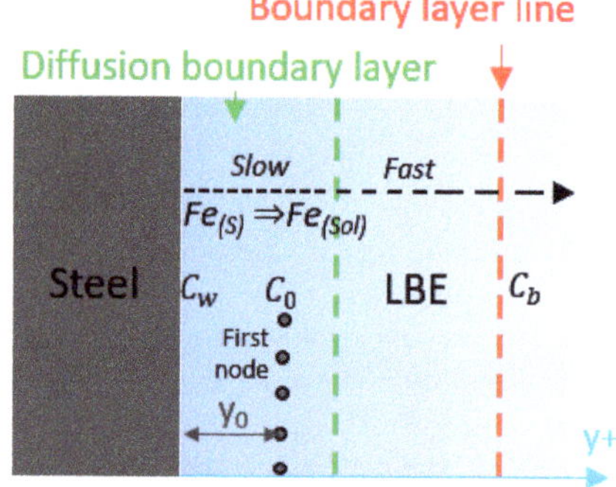

Figure 3. Schematic drawing of steel corrosion in lead–bismuth eutectic (LBE) at low oxygen concentration.

3. Experiments

3.1. JLBL-1

A corrosion test was performed with the JLBL-1 loop as specified in detail by Kikuchi et al. [36,37]. Flow in the loop is illustrated diagrammatically in Figure 4. The main circulating loop consists of an electromagnetic pump (EMP), a heater, a testing tube at high temperature (shown in Figure 5), LBE filters, a surge tank, a cooler, an electromagnetic flow meter (EMF), a surface-level meter, thermocouples, an observation window, and a drain tank. The JLBL-1 loop, including its main components, was manufactured by Sukegawa electric Co., Ltd., Takahagi, Japan. The LBE filters consist of thin, curling 430 SS foils. The EMP is linear inductive and has an annular channel. The EMF has two electrodes in contact with the LBE that detect electromotive force in the magnetic fields.

The tube and tanks that come into contact with lead–bismuth in the loop are made of 316L SS, and the tubes in the circulating loop, except for the testing tube, are 2.5 mm-thick cold-drawn products with an outer diameter of 27.2 mm. This means that the maximum velocity of the LBE will occur in the high-temperature specimen. The chemical composition of LBE is Pb-45/Bi-55 (wt %).

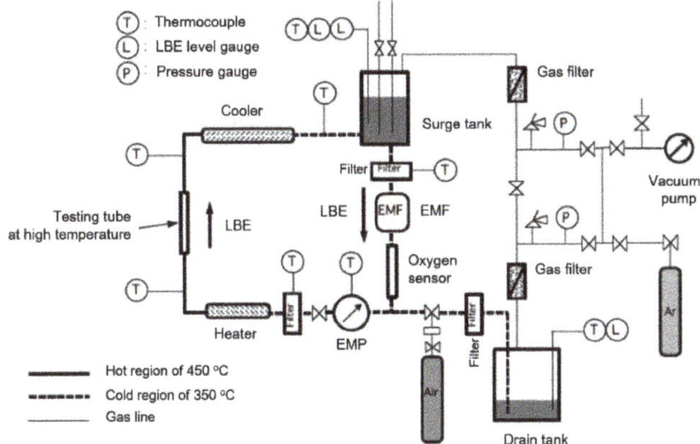

Figure 4. Flow diagram of JLBL-1.

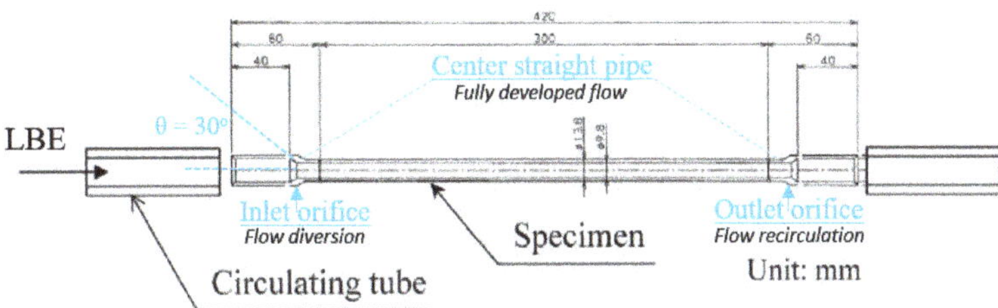

Figure 5. Geometry of specimen tubing.

3.2. Specimen Testing Tube

Figure 5 shows the testing tube. The testing tube is positioned downstream of the loop heater, as shown in Figure 4. The testing tube consists of an inlet orifice, a straight section, and an outlet orifice. The orifices are at 30° to the horizontal axis of the testing tube. The testing tube was machined from solution-annealed 316L SS plate as a tubing form of 2 mm thick, 40 cm long, and 13.8 mm outer diameter. The fabrication tolerance of the outer diameter and wall thickness is ±0.2 mm.

The chemical composition (wt %) of the type 316L austenitic SS used in this study is 16.84 Cr, 10.29 Ni, 0.73 Si, 1.02 Mn, 0.036 C, 0.030 P, and 0.005 S, balanced with Fe. The tube was solution heat-treated at 1080 °C for 1.5 h, and then cooled in water rapidly. The inner surface of the tube was polished to remove the rough layer after acid washing.

3.3. Operation Conditions

The temperature of the high- and low-temperature parts of the loop were 450 °C and 350 °C, respectively. The test duration was approximately 3000 h. The LBE flow rate was 5 L/min, corresponding to an average velocity of 0.7 m/s in the 9.8 mm diameter high-temperature testing tube. The flow rate was measured using an electromagnetic flowmeter (EMF). An oxygen sensor was placed in the loop, but did not work well, so the oxygen concentration in the LBE was neither measured nor controlled in this study. The literature [38] indicates that when the oxygen concentration is properly controlled to between 10^{-4} wt % and 10^{-7} wt %, an oxide film, with a thickness of several micrometers to tens of micrometers, will form on the surface of the stainless steel. Below that concentration range, no oxide film is formed. We can thus deduce that the oxygen concentration in the present experiment is below 10^{-8} wt %.

3.4. Post-Testing Material Characterization

After 3000 h, the specimens were cut out of the testing tube. The specimens were washed with silicon oil then cleaned with ethanol in an ultrasonic bath. They were then mounted in resin. After polishing, optical macroscopic (OM) and scanning electron microscopy (SEM) observations were carried out on the cross-sections of the specimens.

4. Numerical Simulation Conditions

4.1. Hydrodynamic Study

Steady-state CFD analysis was performed with the commercially available STAR-CD code. A half 3-dimensional (3D) model was meshed as shown schematically in Figure 6. To obtain a fully developed flow profile for the test tube, the length of each side of the specimen was extended to 250 mm in the simulations. All other dimensions are as in the specimen. The meshes were composed of tetrahedral cells, which were 0.75 mm in the straight part and 0.1875 mm in the two orifice sections. To study the hydrodynamic effects, it is necessary to understand the flow behavior deep in the viscous sublayer. This can generally be achieved by placing the first node at around $y+ = 1$. In this study, 45 layers of prism cells were meshed in the near-wall region. The distance of the first node of the prism layer from the wall is about 1.6 µm, corresponding to, approximately, $y+ = 0.5$. The expansion ratio of the prism layer is 1.1. There are approximately 7.6 million cells in the fluid field in total. The uniform flow speed at the inlet is 0.25 m/s, making the flow speed in the straight part approximately 0.7 m/s, which corresponds to that in the experiment. Good wettability is assumed for contact between LBE and the wall during the 3000 h operation, so the wall boundary is considered to be no-slip for the simulation. The $k - \varepsilon$ LRN turbulence model was adopted to obtain detailed flow characteristics in the near-wall region using the equations detailed in Section 2.1. In addition, hydrodynamic simulation was performed to study the effects of the corroded morphology and angle of the orifices.

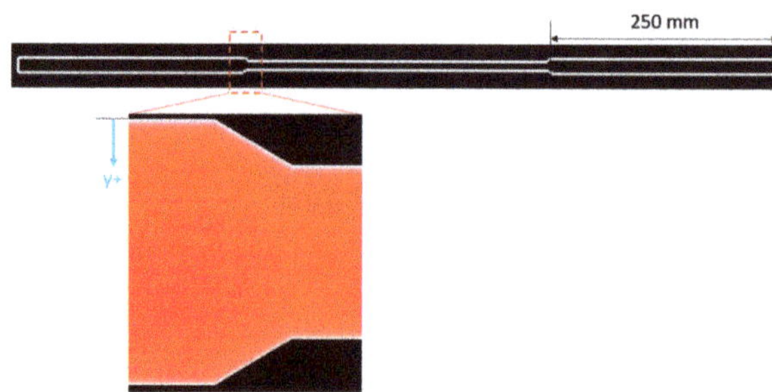

Figure 6. Example of a half 3D model used in the numerical simulation.

4.2. Mass Transfer Study

The mass transfer behavior of iron was investigated by integrating with the turbulent flow of LBE. A much finer mesh in the near-wall region is required to study the mass diffusion behavior from the wall to the bulk flow than for the hydrodynamic study, because it is necessary to place the first node in the diffusion boundary layer. In the mass transfer study, the first node of the prism layer is placed about 0.34 µm from the wall, corresponding to approximately $y+ = 0.1$, the recommended value for mass transfer studies [27]. Fifty layers of prism cells are placed beside the wall. The cell size is 0.75 mm.

As described in Section 2.2, for a long term non-isothermal closed loop, the concentration of iron in the bulk fluid cannot exceed its saturation concentration at the lowest temperature, or it will deposit on the coldest part of the wall. On the other hand, the concentration of iron increases as operation continues. Therefore, it is reasonable to assume that the iron concentration in the bulk fluid equals its saturation point at the lowest temperature [22,23]. The iron concentration at the wall in the hottest part depends on the situation. If the oxygen concentration is high enough that an Fe_3O_4-based protective film forms on the metal surface, the film can be reduced by the lead in the fluid. As a result, the equilibrium concentration of iron on the wall surface is controlled by this chemical reaction. However, if the oxygen concentration is sufficiently low that no oxide film forms, the equilibrium concentration of iron on the wall equals its saturation concentration [29]. Therefore, the concentrations of iron at the solid/liquid interface and in the bulk fluid are about 8.978×10^{-5} wt % and 9.576×10^{-6} wt %, respectively [39]. The molecular diffusion coefficient of iron in LBE is only known at certain temperatures, and has a large degree of scatter, and there is no diffusion coefficient of iron in LBE at 450 °C [38]. In this study, the iron diffusion coefficient in LBE is approximated to that of pure lead, as the diffusion coefficients of the two are considered to be similar [40,41]. Therefore, it is extrapolated from Robertson's law that the molecular diffusion coefficient of iron in LBE at 450 °C is approximately 3.16×10^{-10} m^2/s [38,41].

5. Results and Discussion

5.1. Corrosion Depth Profile of the Test Tube

Figure 7a shows an OM image of cross-section of the specimen of the central straight pipe. The corrosion depth around the pipe circumference is relatively uniform. The maximum wall thickness of the corroded specimen is about 2.1 mm, as marked, which represents that the wall thickness of the as-received tube is no less than 2.1 mm. The fabrication tolerance is ±0.2 mm, so that the wall thickness of the as-received tube ranges from 2.1 to 2.2 mm, and it deviates 0.1–0.2 mm from its design size shown in Figure 5. This also implies that the thickness of the two 30-degree orifices deviates 0.087–0.173 mm from the design size. The corrosion depth of the specimen of the central straight pipe reaches a maximum of approximately 0.2–0.3 mm, and is 0.06–0.16 mm on average. The average corrosion rate ranges from 0.02 to 0.053 μm/h. Figure 7b shows the SEM image of cross-section of the specimen of the central straight pipe. No oxide film was observed on the specimen surface. Furthermore, there is LBE penetration into the grain boundary, which illustrates a typical corrosion scenario of 316L SS in LBE as stated previously in the corrosion model (Section 2.2). That is, Ni and Cr dissolve into LBE in the superficial area of the steel, and vacancies in the steel matrix are created as a result of depletion of Ni and Cr and of lower Fe content as well, and subsequently, Pb and Bi penetrate into the steel to fill the vacancies. During the ongoing penetration, selective leaching of Ni and Cr continues [38].

Figure 7c shows the OM images of cross-sections of the two orifices at various circumferential angles. The two orifices are the sloping parts seen in Figure 5. Clearly, the two orifice parts have been severely corroded, and the corrosion profiles are not uniform along the orifice surface. The quantitative corrosion depths of the two orifices along the orifice surface at various circumferential angles are plotted in Figure 8a,b. The corrosion depth was measured from the OM images by comparing the morphologies of the corroded orifice surfaces to those of the original surfaces. Each corrosion depth profile consists of forty datapoints measured equidistantly along the original orifice surface. The deviation of the measurement is about ±0.005 mm. The measured average corrosion depth was added by a fabrication deviation of 0.087 mm, as pointed out previously. The corrosion depth profiles are similar for all circumferential angles at each orifice. The least corrosion is seen at a circumferential angle of 180 degrees. It is, at present, unclear why this is so, although it may be attributable to the deposition of corrosion products. The maximum corrosion depth is nearly 1.0 mm for both orifices. The average corrosion depth of the inlet orifice and the outlet orifice is shown in Figure 8c,d, respectively. Fabrication deviations of 0.087 mm and 0.174 mm are included in the calculation results of the average corrosion depth. The corresponding error bars are derived from the circumferential data in Figure 8a,b. The results indicate that the average depth of corrosion is larger in the outlet orifice than in the inlet orifice.

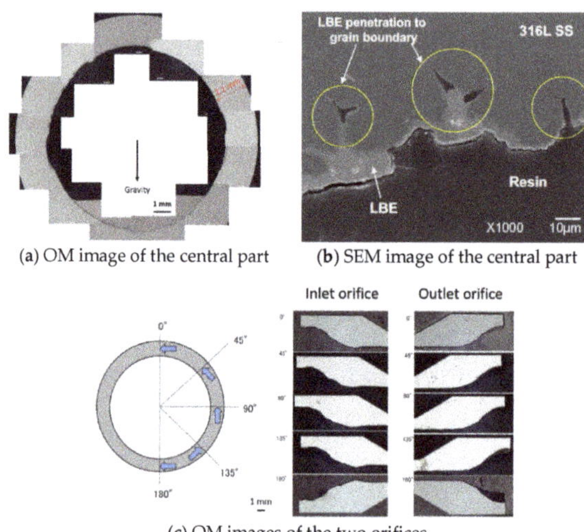

(a) OM image of the central part (b) SEM image of the central part

(c) OM images of the two orifices

Figure 7. OM and SEM images. (**a**) OM image of cross-sections of the specimen of the central straight pipe; (**b**) SEM image of cross-section of the specimen of the central straight pipe; (**c**) OM images of the two orifices at various circumferential angles.

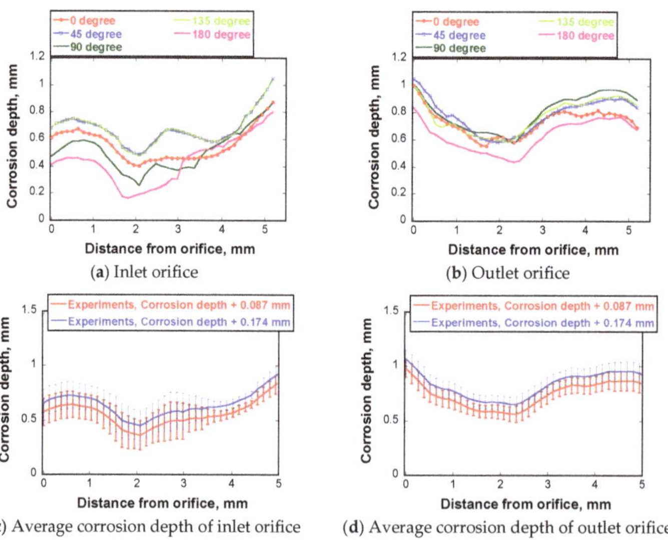

(a) Inlet orifice (b) Outlet orifice

(c) Average corrosion depth of inlet orifice (d) Average corrosion depth of outlet orifice

Figure 8. (**a**,**b**) Corrosion depth profiles of the inlet and outlet orifices at various circumferential angles as a function of the distance along the orifice surface. Fabrication deviations of 0.087 mm are added in the measured corrosion depth; (**c**,**d**) average corrosion depth of inlet orifice and outlet orifice. Fabrication deviations of 0.087 mm and 0.174 mm are added in the measured average corrosion depth, respectively.

5.2. *Effects of the TKE, Shear Stress, and Pressure*

The TKE near the wall, and the shear stress and pressure on the wall along the two orifice surfaces, are plotted in Figure 9 to investigate the effects of different hydrodynamic parameters on the corrosion depth. The average corrosion depths of the two orifices, including a fabrication deviation of 0.087 mm, are also plotted for comparison. Note that the curves obtained from the simulations have been smoothed. It can be seen that the TKE profiles are similar to those of the average corrosion depth of the two orifices, while those for shear stress and pressure are substantially different.

The maximum shear stress on the inlet and outlet orifice walls are low, approximately 40 Pa and 12 Pa, respectively. A study of copper–nickel alloys in sea water has shown that a shear stress of nearly 4650 Pa is insufficient to remove even the surface oxide film [42]. As pointed out previously, no oxygen film formed on the base metal in this study because of the low oxygen concentration in LBE. Furthermore, the critical shear stress capable of mechanically damaging the base metal is much larger than that able to peel off the oxide film. Therefore, it is considered that shear stress is not the main factor leading to FAC herein. Although the pressure fluctuation on the wall in a disturbed flow imposes shear stress on the wall and perhaps causes damage [43], the results obtained under the LBE flow conditions considered here cannot support this assumption because the complex orifice corrosion depth profiles cannot be explained well by monotonic pressure variation along the orifice wall. Rather, it is the variation in the turbulence level near the wall that best explains the corrosion profiles of the orifices.

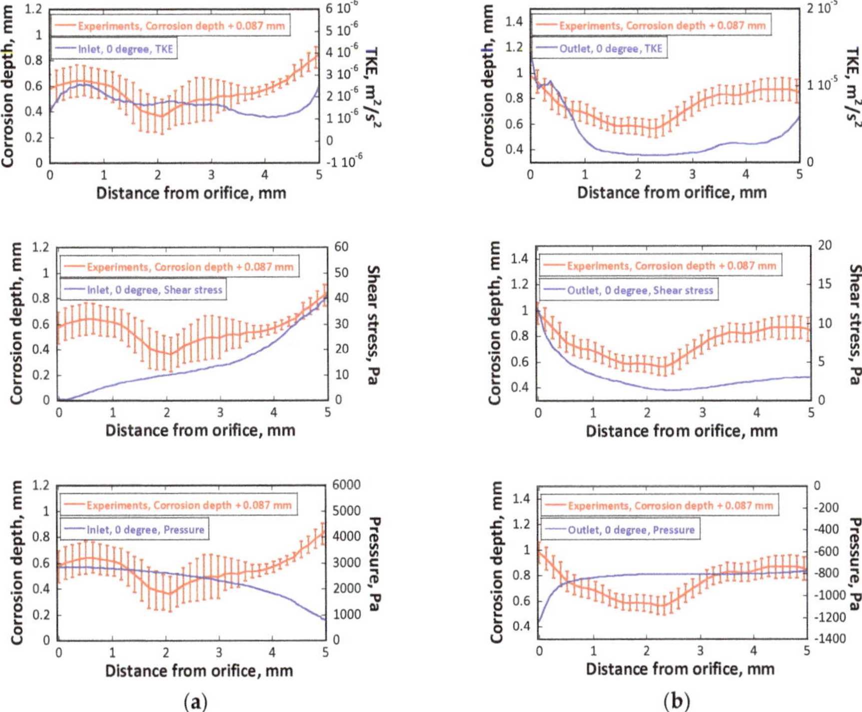

Figure 9. Comparison of the average corrosion depth with the turbulent kinetic energy (TKE) near the wall and the shear stress and pressure on the wall for the two orifices. (**a**) Inlet orifice; (**b**) Outlet orifice.

5.3. Other Effects

5.3.1. Cavitation

It is well known that the passage of fluid through an orifice leads to a sudden large pressure drop downstream, potentially decreasing the pressure below the saturated vapor pressure. As a result, the liquid may rupture and form cavitation bubbles. The subsequent collapses of the cavitation bubbles release microjets and/or shock waves at the wall, causing cavitation damage [44] that can give the wall a rough morphology. The cavitation number, which is used to judge the inception of cavitation in a liquid, is calculated by

$$C_a = \frac{P_l - P_v}{\frac{1}{2}\rho {V_l}^2}, \tag{10}$$

where C_a is the cavitation number, P_l is the local pressure, P_v is the vapor pressure of the liquid, ρ is the density of the liquid, and V_l is the characteristic flow speed of the liquid.

Under the present flow conditions, a pressure drop to about −3.58 kPa is generated downstream of the inlet orifice. The vapor pressure of LBE at 450 °C is 3.49×10^{-4} Pa [38], resulting in a cavitation number of approximately 0.19. It has been reported that the inception cavitation number is about 0.7 for PbBi-68 liquid flow with a Reynolds number in the range 5.8–7.4 $\times 10^4$ [45]. As the Reynolds number in this study is near this range, a similar inception cavitation number can be applied, indicating that cavitation damage is unlikely to have occurred. Thus, the rough profiles of the orifices can be attributed solely to corrosion damage.

5.3.2. Gravity

LBE is a high-density liquid metal, so it is necessary to consider the effects of gravity on flow. First, the direction of gravity was set to be perpendicular to the test tube. Figure 10 shows the effects of gravity on various hydrodynamic parameters. The values of 0 and 180 degrees represent the anti-gravity direction and the gravity direction, respectively. The profiles of the hydrodynamic parameters are almost the same at the two circumferential angles, and only tiny differences in the amplitude can be observed.

Secondly, gravity was set to be in the direction of the test tube, representing no gravity effect on the flow. Figure 11 shows a comparison of the TKE along the straight central part of the test tube with and without gravity effects. It is hard to distinguish any difference in the TKE profiles. The above results illustrate that the hydrodynamic parameters are insensitive to gravity. This may be because the Reynolds number is high so the strong effect of the inertial force on the flow makes the effects of gravity negligible.

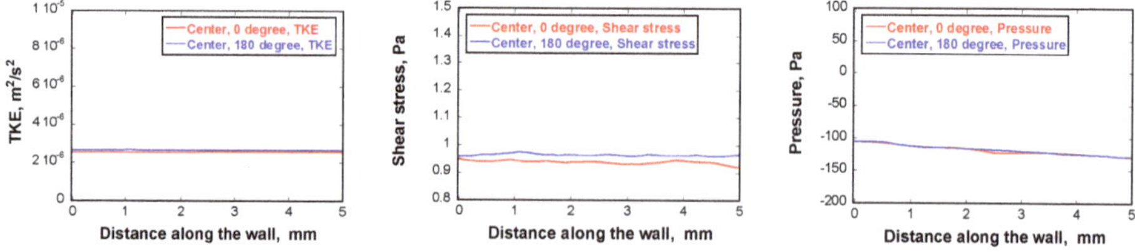

Figure 10. Effects of gravity on various hydrodynamic parameters.

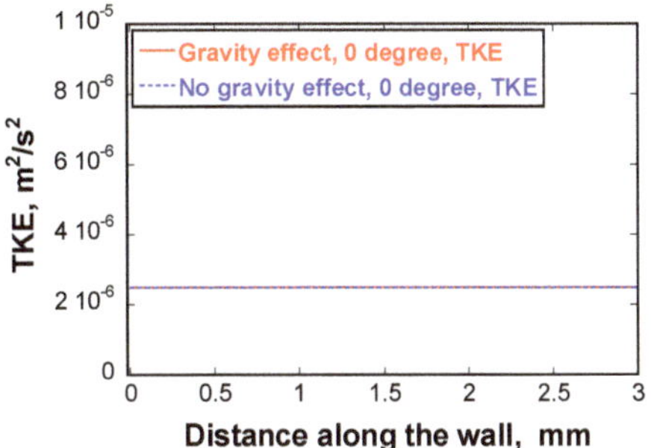

Figure 11. Comparison of TKE along the straight central part of the test tube with and without a gravity effect.

5.3.3. Corroded Morphology

Accumulation of corrosion damage on the inner surface of the wall changes its morphology. A corroded surface could be expected to disturb the flow pattern near the wall, and therefore, the TKE in the near-wall region. To investigate such effects, the experimental corroded surfaces of the two orifices were merged into the simulation models. Figure 12 shows a schematic of the corroded surface profiles used in the simulation, which are the same as the corroded depth profiles at 90 degrees in the experiments. To connect the two orifice parts and the straight central part smoothly, the straight part of the specimen was also assumed to be corroded to a uniform depth of 0.65 mm, which represents an increase in the inner diameter of the straight part of the specimen by 1.3 mm.

Figure 13 plots the TKE for a smooth versus a corroded orifice with distance along the orifice surface. For the inlet orifice, the TKE profile shows much more variability and reaches a considerably higher maximum value in the corroded profile than along the original smooth surface. The peaks in TKE usually appear downstream of sudden protuberances on the corroded surface. This is easy to understand because such convexity enhances the local turbulence level [46]. Interestingly, for the outlet orifice, the magnitude of TKE is relatively weaker for the corroded morphology than with a smooth surface. Furthermore, differences in TKE in the central section are insignificant. This can be attributed to two factors: (1) the corroded outlet orifice surface has fewer protuberances than the corroded inlet orifice surface (Figure 12), so the turbulence level is not locally enhanced; (2) the connections between the straight pipe and the corroded outlet orifice are much smoother than the original ones, which seems to ease the recirculation near the outlet orifice, and therefore, the TKE amplitude.

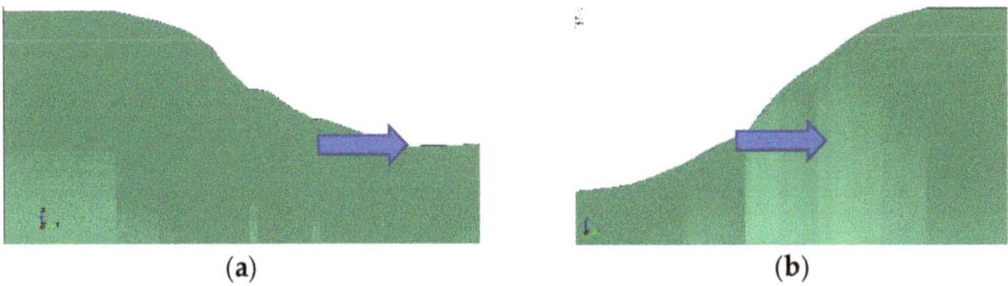

Figure 12. Schematic of the corroded surface profiles adopted in the simulations. (**a**) Inlet orifice; (**b**) Outlet orifice.

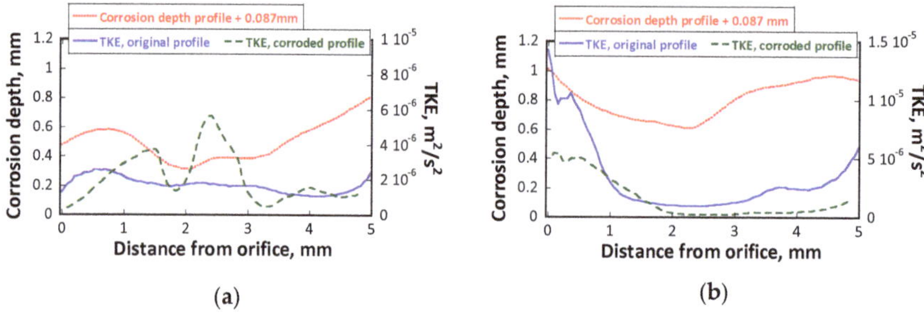

Figure 13. Effects of the corroded surface on TKE in the near-wall region at the two orifices. The utilized corrosion depth profile is plotted together for comparison. (**a**) Inlet orifice; (**b**) Outlet orifice.

5.3.4. Orifice Angle

The angle of the orifices was altered, from 30 degrees to 60 degrees and 90 degrees, to investigate the effect of orifice angle on flow behavior. This is shown schematically for the inlet orifices in Figure 14, and its effect on TKE is shown in Figure 15. For the outlet orifice, there is no significant difference in the amplitude of TKE. Shan et al. [47] pointed out that the flow field in the recirculation region is insensitive to the ratio between the orifice and pipe diameters. Therefore, it is deduced that the TKE of flow in the recirculation region close to the outlet orifice wall is insensitive to the orifice angle as well. However, for the inlet orifice, the turbulence level is very dependent on the orifice angle. The value for TKE can grow up to several magnitudes higher if the orifice angle changes from 30 to 90 degrees. As TKE has been inferred to significantly influence corrosion depth, extensive corrosion can be anticipated in the 90-degree case. These results provide guidance for system design with flowing LBE—structures with sudden geometry changes should be avoided as much as possible.

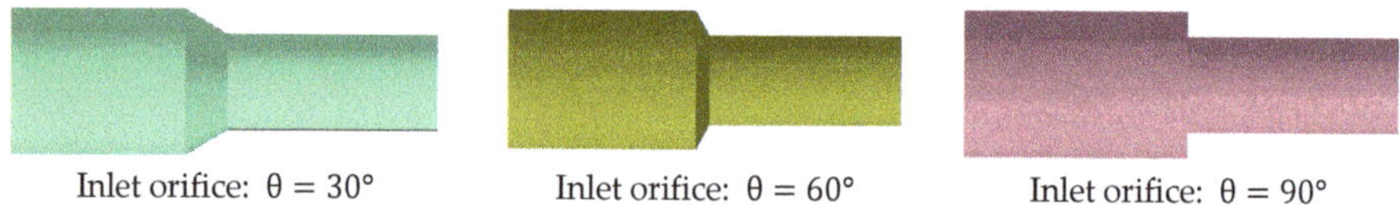

Figure 14. Schematics of inlet orifices with different angles. Note that only the outlet orifices are not shown.

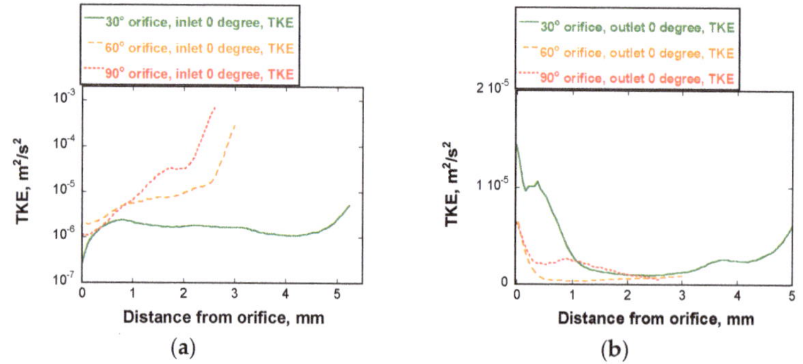

Figure 15. Effects of orifice angle on TKE in the near-wall region for the two orifices. (**a**) Inlet orifice; (**b**) Outlet orifice.

5.4. Mass Transfer

5.4.1. Effective Viscosity and Effective Diffusivity

Figure 16 shows the effective viscosity ($\mu_{eff} = \mu + \mu_t$) and effective diffusivity ($D_{eff} = D_m + D_t$) of iron upstream of the reattachment point. As expected, the effective viscosity remains constant below $y+$ = 2, which corresponds to the viscous sublayer in the near-wall region. Beyond the viscous sublayer, the turbulent viscosity is much larger than the molecular viscosity, so the turbulent-induced momentum transfer is dominant in species transfer. With approach to the wall, the turbulence level damps rapidly in the viscous sublayer and the turbulent viscosity decreases sharply to a negligible value, indicating that turbulent momentum transfer will, likewise, become insignificant. In the viscous sublayer, however, the effective diffusivity is much larger than the molecular diffusivity; in other words, even weak turbulence can strongly affect species transfer through diffusion [27]. Only when the distance from the wall reaches approximately $y+$ = 0.4 can the turbulent diffusivity be ignored and the molecular diffusivity be considered dominant in species diffusion. This is in the thin diffusion boundary layer hidden in the viscous sublayer. The relationship between the thickness of the diffusion boundary layer, δ_d, and that of the viscous sublayer, δ_v, is close to that recommended by Nesic et al. [7]:

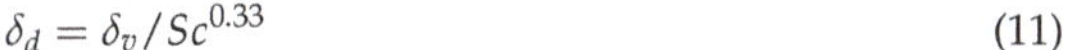

$$\delta_d = \delta_v / Sc^{0.33} \tag{11}$$

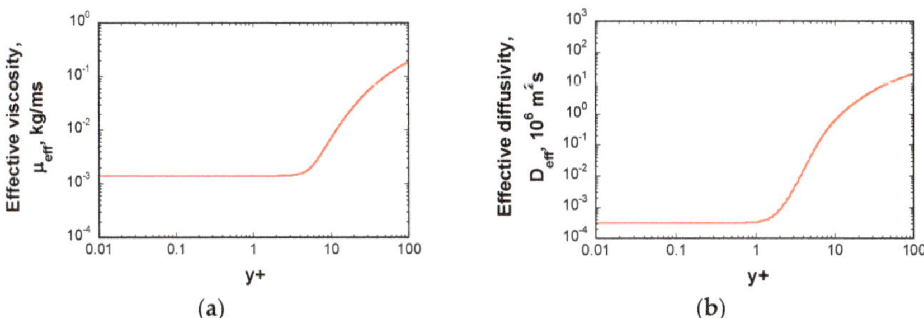

(a) (b)

Figure 16. Effective viscosity and effective diffusivity of iron upstream of the reattachment point. (**a**) Effective viscosity; (**b**) Effective diffusivity.

5.4.2. Mass Concentration Difference Close to the Wall

Figure 17 shows the difference in the mass concentration of iron at different locations. This is defined as

$$\Delta C = C_w - C_0. \tag{12}$$

In the straight central section of the pipe, the ΔC profile is relatively uniform. In comparison, the profiles of ΔC at the inlet and outlet orifices are strongly influenced by the turbulent flow. Furthermore, ΔC is lower at the outlet orifice than at the inlet orifice. This contrasts with the corrosion depth, which, as shown in Figure 8, is larger at the outlet orifice than the inlet orifice. This may be because the mass transfer coefficient is closely related to the magnitude of roughness [46,48]. Corrosion can be expected to render the smooth wall rough, and perhaps there are differences in the amplitudes of the roughness at the two orifices, resulting in different mass transfer coefficients, and thus, corrosion depths. Prediction of the corrosion depth on the basis of ΔC will be considered in the next subsection.

To clarify the correlation between turbulent flow and the mass concentration difference, the profiles of ΔC and TKE are plotted together for comparison in Figure 18. Clearly, the profile of ΔC is highly coincident with that of TKE near the wall. Therefore, the mass concentration difference can be inferred to be significantly affected by the turbulence level. As shown by Equation (9), the corrosion rate is proportional to ΔC, so it can be deduced that the turbulence level can impose direct effects on the corrosion rate of metal.

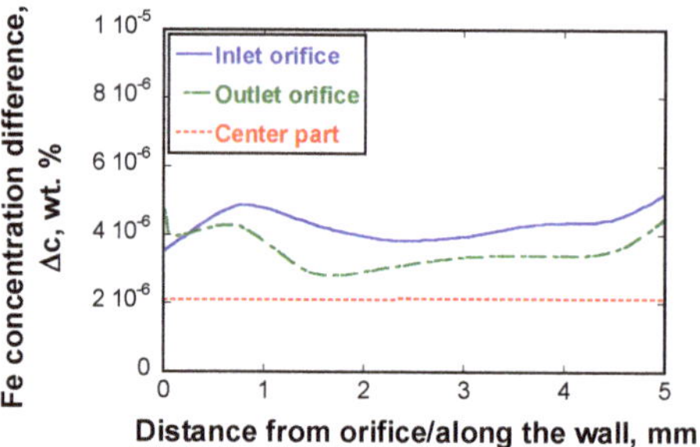

Figure 17. Mass concentration difference of iron at various locations.

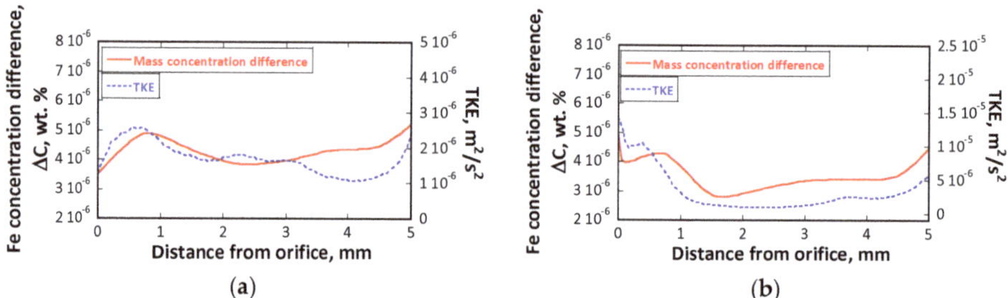

Figure 18. Comparison of mass concentration difference and TKE near the wall. (**a**) Inlet orifice; (**b**) Outlet orifice.

5.4.3. Comparison of Predicted Corrosion Depth and Measured Corrosion Depth

Figure 19 shows a comparison of the predicted corrosion depths calculated with Equation (9) by using the mass concentration differences and the corrosion depths measured from the OM images for the two orifices. The fabrication deviations of 0.087 mm and 0.174 mm are added in the experimental average corrosion depth. The magnitude of the predicted corrosion depth at the inlet orifice is slightly less than the measured corrosion depth with a deviation of 0.087 mm. Moreover, the predicted damage depth at the outlet orifice is much lower than the experimental value. One possible explanation for this is that, as pointed out previously, the effects of wall roughness should be involved, and the different wall roughness values of the inlet and outlet orifices result in different wall mass transfer coefficients; however, exact data on wall roughness is lacking in this study. As an alternative, the calculated corrosion depths of the inlet and outlet orifice were multiplied by a uniform scaler of 1.8 to take wall roughness effects into consideration [7,48]. After multiplying the scaling factor, the magnitudes of the predicted and the measured corrosion depths of the outlet orifice show good coincidence; however, the magnitude of the predicted corrosion depth is slightly higher than the measured corrosion depth, including a deviation of 0.174 mm, which indicates that a scaler less than 1.8 would be applied to the inlet orifice and illustrates that the wall roughness of the inlet orifice is different from that of the outlet orifice. In spite of the arresting coincidence, minor discrepancies can still be observed between the predicted and measured corrosion depth profiles. These may arise for two main reasons: (1) the numerical results are obtained from a model with a smooth surface, while in the experiments, the morphology of the orifice surfaces will have been reshaped gradually by the accumulation of corrosion, and accordingly the turbulence level, and thus, the mass transfer profile will have changed; (2) the wall roughness along the orifice surfaces may be non-uniform. Nevertheless, the results indicate that the turbulence level in the near-wall region has a strong influence on the corrosion depth profile. Furthermore, the corrosion depth can be predicted successfully by calculating the mass transfer coefficient by integrating the corrosion model into the turbulent flow model.

Researchers in Italy have studied the corrosion of cylindrical AISI 316L SS samples using the LECOR loop. The experiment was performed for 1500 h under the same temperature difference and similar oxygen concentration conditions as are used in this study [49]. They reported an average corrosion rate of 1.9×10^{-3} μm/h, which is one magnitude lower than that for the straight central part of the current study's experiment where flow has fully developed. This difference may result from the combined effects of three factors: (1) the absolute temperature is higher in the present study, so the solubility and diffusion coefficient of iron in LBE are higher; (2) the Reynolds number in the LECOR loop is about one-fifth of that in the present study, making the diffusion boundary layer is much thicker, and therefore, mass transfer resistance is much stronger. It has been pointed out that Sh has a nearly linear correlation with Re for roughened wall surfaces [48,50]; (3) the dimensionless wall roughness in the LECOR loop is much lower than that of the present study, and it is also known that the mass transfer coefficient depends on the dimensionless wall roughness [48]. The predicted depth of corrosion in the straight section of the pipe after 3000 h of operation is about 0.21 mm, which is about 1.3–3.5 times the average experimental corrosion depth (0.06–0.16 mm); this ratio is similar to that reported in another study [51]. Overestimation of the corrosion rate in the simulation may result from [51,52]: (1) the diffusion coefficient of iron in liquid LBE at 450 °C not being known sufficiently precisely; (2) the assumption that the liquid metal is saturated at the solid/liquid interface.

Furthermore, the absolute corrosion depth of the specimen is extremely high in the present experiment, which would not be acceptable for real engineering applications. For example, one of the future ADS concepts proposed by JAEA is an LBE spallation target that requires a significant contraction of the LBE flow path at the beam window [53]. The Reynolds number at the beam window is much higher than that of the present study, and the beam window would be required to survive extreme service conditions [54]. Moreover, considering the poor wettability of LBE to 316L SS, the corrosion depth could be larger because slip at the solid/liquid boundary enhances the turbulence level in the near-wall region [55]. In addition, it is known that the structural steel will withstand strong irradiation during its service period, and it has been also reported that the cold-worked steel experiences different corrosion behaviors than the solution-annealed steel in LBE under static and flow LBE conditions, which can be attributed to the alteration of the dominant corrosion process [56,57]. It has also been pointed out that the effects of steel microstructure on FAC depend on the relative thickness between the selective corrosion layer in the solid and the diffusion boundary layer in the fluid. If the former is thicker, the microstructure effects can be visible; otherwise, it can be ignored [30]. Nevertheless, the effects of microstructure on FAC under LBE flowing condition requires thorough and systematical investigation in the future.

Those above-combined factors mean that it is extremely important to devise ways to increase the mass transfer resistance between the base metal and the LBE. Several studies have indicated that the formation of a protective oxide layer can greatly decrease the corrosion rate of 316L SS in static LBE [58,59]. Controlling the oxygen concentration in LBE to a specific range is critical to form an effective self-healing oxide film, and meanwhile avoid the deposition of uncalled-for oxide products (e.g., PbO) on the cold legs, potentially causing blockage [39].

For future ADSs that adopt LBE as the coolant material, even once an effective protective oxide film has formed on the structural material, it may be broken or thinned by turbulence pounding [6] and various stresses. These include not only the shear stress resulting from turbulent flow, but the thermal stress caused by pulsed proton beam injections and fatigue stress from thermal shocks. Such stresses are generally very large, and their effects on oxide film should be investigated in detail. The effects of absolute temperature and temperature difference on corrosion behavior should also be investigated.

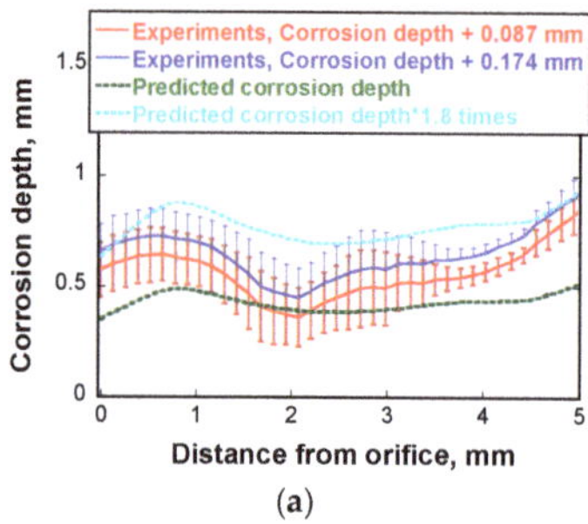

(a)

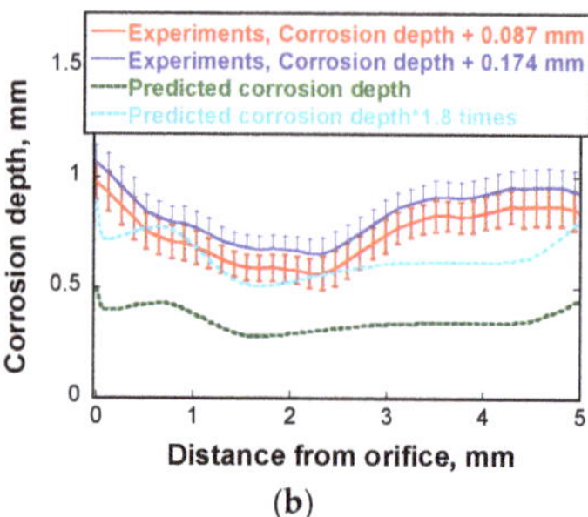

(b)

Figure 19. Comparison of the predicted corrosion depths calculated from Equation (9) using the mass concentration difference and the corrosion depths measured from the OM images at the two orifices. Note that the fabrication deviations of 0.087 mm and 0.174 mm are added in the experimental corrosion depth, respectively, and the predicted corrosion depths of the inlet orifice and outlet orifice were multiplied by a scaler of 1.8 to take the wall roughness into consideration. (**a**) Inlet orifice; (**b**) Outlet orifice.

6. Conclusions

FAC behavior of 316L SS under LBE turbulent flow was studied by using the JLBL-1 loop under non-isothermal conditions, and CFD analyses were performed to study the associated hydromechanics and mass transfer. It was found that corrosion of the 316L SS is very severe. Corrosion can reach a maximum depth of approximately 10% of the tube wall thickness in the straight part of the specimen pipe, where the flow is fully developed. However, this ratio increases by as much as 50% at the inlet and outlet orifices, where abrupt changes in the diameter of the specimen lead to changes in the flow direction and/or recirculation. The corrosion profile is correlated with the near-wall turbulence level, but does not seem to correlate with the local shear stress and pressure profiles. In addition, morphological changes during the corrosion process influence the near-wall turbulence level. Moreover, it is shown that the near-wall turbulence level is extremely sensitive to the orifice angle, and that right-angled orifices should be avoided to suppress FAC.

Mass transfer analysis shows that the local turbulence level determines the mass concentration distribution in the near-wall region, thus significantly affecting the mass transfer coefficient. As a consequence, the mass transfer is an effective means of predicting the depth of FAC, which can be calculated with the assistance of CFD analyses with a corrosion model. The corrosion depth was calculated on the basis of the mass transfer coefficient obtained in the numerical simulation and was compared with that obtained in the loop. For the inlet orifice part, the predicted corrosion depth is comparable with the measured corrosion depth; for the outlet orifice part, the predicted corrosion depth is less than the measured corrosion depth, however, this gap was filled well after taking the wall roughness effects into consideration; for the straight tube part, the predicted corrosion depth is larger than the average experimental corrosion depth, which perhaps can be attributed to the fact that the iron concentration on the wall did not reach its saturation value, as assumed. Considering that many other factors beyond the range of the present study can lead to much more severe FAC, formation of an oxide layer is an extremely important means of preventing the rapid development of FAC. In the future, research on corrosion behavior under precisely controlled oxygen, flow speed, and temperature conditions will be carried out.

Author Contributions: T.W. carried out the numerical simulations; S.S. performed the experiments; T.W. analyzed the data; T.W. and S.S. were responsible for writing the paper.

Funding: This research received no external funding.

Acknowledgments: The first author sincerely appreciates Masatoshi Futakawa of J-PARC Center, JAEA, for providing valuable comments and discussions. The authors also extend their thanks to Hironari Obayashi and Toshinobu Sasa of J-PARC Center, JAEA, for useful and interesting discussions.

Conflicts of Interest: The authors declare no conflict of interest.

Nomenclature

a, b, c	constants in empirical equations linking Sh to Sc and Re
C_a	cavitation number
C_b, C_w, C_0	species concentration in the bulk fluid, at the wall, and in the first node, kmol/m^3
C_{O_2}	oxygen concentration in a liquid, kmol/m^3
D_h	hydraulic diameter, m
D_m	molecular diffusion coefficient, m^2/s
D_t	eddy or turbulent diffusivity, $D_t = \mu_t/\rho Sc_t$, m^2/s
D_{eff}	effective diffusion coefficient, $D_{eff} = D_m + D_t$, m^2/s
f_u, f_2	damping factor in the $k-\varepsilon$ LRN turbulence model
I_t	turbulence level in the near-wall region
J_{Fe}	mass flux of iron, kmol/(m^2·s)
k	turbulent kinetic energy, m^2/s^2
K_c	mass transfer coefficient, m/s
M_c	chemical component of a material
M_{Fe}	molar mass of iron, kg/kmol
P_k	generation of turbulence, kg/(m·s^3)
$P_k{}'$	additional term of turbulence generation in the $k-\varepsilon$ LRN turbulence model, kg/(m·s^3)
P_l	local pressure in a liquid, Pa
P_v	vapor pressure of a liquid, Pa
R_w	wall roughness, m
Re	Reynolds number, $Re = \rho V_l D_h/\mu$
Re_y, Re_t	turbulence Reynolds number, $Re_y = \rho y\sqrt{k}/\mu$, $Re_t = \rho k^2/\mu\varepsilon$
Sc	Schmidt number, $Sc = \mu/\rho D_m$
Sc_t	turbulent Schmidt number, $Sc_t = \mu_t/\rho D_t$
Sh	Sherwood number, $Sh = aRe^b Sc^c$
T	temperature, K
ΔT	temperature difference, K
$u+$	dimensionless velocity
V_l	characteristic flow speed of a liquid, m/s
y	distance from the wall, m
$y+$	dimensionless distance from the wall
y_0	distance of the first node from the wall, m
Greek symbols	
δ_d	thickness of diffusion boundary layer, m
δ_v	thickness of viscous sublayer, m
ε	dissipation of turbulent kinetic energy, m^2/s^3
μ	molecular dynamic viscosity, kg/(m·s)
μ_t	eddy or turbulent viscosity, kg/(m·s)
μ_{eff}	effective viscosity, $\mu_{eff} = \mu + \mu_t$, kg/(m·s)
ρ	density, kg/m^3
τ_w	wall shear stress, Pa

References

1. Bergman, T.L.; Incropera, F.P.; Devitt, D.P.; Lavine, A.S. *Fundamentals of Heat and Mass Transfer*, 7th ed.; John Wiley & Sons: Hoboken, NJ, USA, 2011; pp. 378–418.

2. Mahato, B.K.; Voora, S.K.; Schemilt, L.W. Steel pipe corrosion under flow conditions—I. an isothermal correlation for a mass transfer model. *Corros. Sci.* **1968**, *8*, 173–193. [CrossRef]
3. Corpson, H.R. Effects of velocity on corrosion. *Corrosion* **1960**, *16*, 130–136.
4. Silverman, D.C. Rotating cylinder electrode for velocity sensitivity testing. *Corrosion* **1984**, *40*, 220–226. [CrossRef]
5. Silverman, D.C. Rotating cylinder electrode—Geometry relationships for prediction of velocity-sensitive corrosion. *Corrosion* **1987**, *44*, 42–49. [CrossRef]
6. Nesic, S.; Postlethwaite, J. Relationship between the structure of disturbed flow and erosion-corrosion. *Corrosion* **1990**, *46*, 874–880. [CrossRef]
7. Nesic, S.; Postlethwaite, J. Hydrodynamics of disturbed flow and erosion-corrosion. Part I—Single-phase flow study. *Can. J. Chem. Eng.* **1991**, *69*, 698–703. [CrossRef]
8. Mahato, B.K.; Cha, C.Y.; Shemilt, L.W. Unsteady state mass transfer coefficients controlling steel pipe corrosion under isothermal flow conditions. *Corros. Sci.* **1980**, *20*, 421–441. [CrossRef]
9. Chang, Y.S.; Kim, S.H.; Chang, H.S.; Lee, S.M.; Choi, J.B.; Kim, Y.J.; Choi, Y.H. Fluid effects on structural integrity of pipes with an orifice and elbows with a wall-thinned part. *J. Loss Prev. Proc.* **2009**, *22*, 854–859. [CrossRef]
10. Utanohara, Y.; Nagaya, Y.; Nakamura, A.; Murase, M. Influence of local flow field on flow accelerated corrosion downstream from an orifice. *J. Power Energy Syst.* **2012**, *6*, 18–33. [CrossRef]
11. Utanohara, Y.; Nagaya, Y.; Nakamura, A.; Murase, M.; Kamahori, K. Correlation between flow accelerated corrosion and wall shear stress downstream from an orifice. *J. Power Energy Syst.* **2013**, *7*, 138–146. [CrossRef]
12. Crawford, N.M.; Cunningham, G.; Spence, S.W.T. An experimental investigation into the pressure drop for turbulent flow in 90° elbow bends. *Proc. Inst. Mech. Eng. Part E J. Process Mech. Eng.* **2007**, *221*, 77–88. [CrossRef]
13. Nesic, S. Using computational fluid dynamics in combating erosion-corrosion. *Chem. Eng. Sci.* **2006**, *61*, 4086–4097. [CrossRef]
14. Poulson, B. Complexities in predicting erosion corrosion. *Wear* **1999**, *233–235*, 497–504. [CrossRef]
15. El-Gammal, M.; Mazhar, H.; Cotton, J.S.; Shefski, C.; Pietralik, J.; Ching, C.Y. The hydrodynamic effects of single-phase flow on flow accelerated corrosion in a 90-degree elbow. *Nucl. Eng. Des.* **2010**, *240*, 1589–1598. [CrossRef]
16. El-Gammal, M.; Ahmed, W.H.; Ching, C.Y. Investigation of wall mass transfer characteristics downstream of an orifice. *Nucl. Eng. Des.* **2012**, *242*, 353–360. [CrossRef]
17. Takano, T.; Ikarashi, Y.; Uchiyama, K.; Yamagata, T.; Fujisawa, N. Influence of swirling flow on mass and momentum transfer downstream of a pipe with elbow and orifice. *Int. J. Heat Mass Transf.* **2016**, *92*, 394–402. [CrossRef]
18. Ikarashi, Y.; Taguchi, S.; Yamagata, T.; Fujisawa, N. Mass and momentum transfer characteristics in and downstream of 90° elbow. *Int. J. Heat Mass Transf.* **2017**, *107*, 1085–1093. [CrossRef]
19. Gromov, B.F.; Belomitcev, Y.S.; Yefimov, E.I.; Leonchuk, M.P.; Martinov, P.N.; Orlov, Y.I.; Pankratov, D.V.; Pashkin, Y.G.; Toshinsky, G.I.; Chekunov, V.V.; et al. Use of lead-bismuth coolant in nuclear reactors and accelerator-driven systems. *Nucl. Eng. Des.* **1997**, *173*, 207–217. [CrossRef]
20. Zhang, J. Lead-Bismuth Eutectic (LBE): A coolant candidate for Gen. IV advanced nuclear reactor concepts. *Adv. Eng. Mater.* **2014**, *16*, 349–356. [CrossRef]
21. Zhang, J. A review of steel corrosion by liquid lead and lead-bismuth. *Corros. Sci.* **2009**, *51*, 1207–1227. [CrossRef]
22. Zhang, J.; Hoseman, P.; Maloy, S. Models of liquid metal corrosion. *J. Nucl. Mater.* **2010**, *404*, 82–96. [CrossRef]
23. Zhang, J.; Li, N. Review of the studies on fundamental issues in LBE corrosion. *J. Nucl. Mater.* **2008**, *373*, 351–377. [CrossRef]
24. Schroer, C.; Wedemeyer, O.; Novotny, J.; Skrypnik, A.; Konys, J. Long-term service of austenitic steel 1.4571 as a container material for flowing lead–bismuth eutectic. *J. Nucl. Mater.* **2011**, *418*, 8–15. [CrossRef]
25. *STAR-CD methodology, Ver. 4.26*; Siemens Product Lifecycle Management Inc.: Plano, TX, USA, 2016.
26. Launder, B.E.; Spalding, D.B. The numerical computation of turbulent flows. *Comp. Meth. Appl. Mech. Eng.* **1974**, *3*, 269–289. [CrossRef]
27. Nesic, S.; Postlethwaite, J. Calculation of wall-mass transfer rates in separated aqueous flow using a low Reynolds number $k-\varepsilon$ model. *Int. J. Heat Mass Transf.* **1992**, *35*, 1977–1985. [CrossRef]

28. Lien, F.S.; Chen, W.L.; Leschziner, M.A. Low-Reynolds-Number Eddy-Viscosity Modelling Based on Non-Linear Stress-Strain/Vorticity Relations. In Proceedings of the 3rd Symposium on Engineering Turbulence Modelling and Measurements, Crete, Greece, 27–29 May 1996.
29. He, X.; Li, N.; Mineev, M. A kinetic model for corrosion and precipitation in non-isothermal LBE flow loop. *J. Nucl. Mater.* **2001**, *297*, 214–219. [CrossRef]
30. Simon, N.; Terlain, A.; Flament, T. The compatibility of austenitic materials with liquid Pb-17Li. *Corros. Sci.* **2001**, *43*, 1041–1052. [CrossRef]
31. Yamaki, E.; Ginestar, K.; Martinelli, L. Dissolution mechanism of 316L in lead-bismuth eutectic at 500 °C. *Corros. Sci.* **2011**, *53*, 3075–3085. [CrossRef]
32. Schad, M. To the corrosion of austenitic steel in sodium loops. *Nucl. Tech.* **1980**, *50*, 267–288. [CrossRef]
33. Distefano, J.R.; Hoffman, E.E. Corrosion Mechanisms in Refractory Metal-Alkali Metal Systems. *At. Energy Rev.* **1964**, *2*, 3–33.
34. Keating, A. A Model for the Investigation of Two-Phase Erosion-Corrosion in Complex Geometries. Maters's Thesis, University of Queensland, Brisbane, QLD, Australia, 1999.
35. Davis, C.; Frawley, P. Modelling of erosion-corrosion in practical geometries. *Corros. Sci.* **2009**, *51*, 769–775. [CrossRef]
36. Kikuchi, K.; Karata, Y.; Saito, S.; Fukazawa, M.; Salsa, T.; Okinawa, H.; Wake, E.; Miura, K. Corrosion–erosion test of SS316 in flowing Pb–Bi. *J. Nucl. Mater.* **2003**, *318*, 348–354. [CrossRef]
37. Kikuchi, K.; Saito, S.; Kurata, Y.; Futakawa, M.; Sasa, T.; Oigawa, H.; Wakai, E.; Umeno, M.; Mizubayashi, H.; Miura, K. Lead-bismuth eutectic compatibility with materials in the concept of spallation target for ADS. *JSME Int. J. B-Fluid Therm.* **2004**, *47*, 332–339. [CrossRef]
38. OECD NEA. *Handbook on Lead-Bismuth Eutectic Alloy and Lead Properties, Materials Compatibility, Thermal-Hydraulics and Technologies*; 2015 Edition; OECD NEA: Boulogne-Billancourt, France, 2015.
39. Li, N. Active control of oxygen in molten lead–bismuth eutectic systems to prevent steel corrosion and coolant contamination. *J. Nucl. Mater.* **2002**, *300*, 73–81. [CrossRef]
40. Balbaud-Celerier, F.; Barbier, F. Influence of the Pb-Bi hydrodynamics on the corrosion of T91 martensitic steel and pure iron. *J. Nucl. Mater.* **2004**, *335*, 204–209. [CrossRef]
41. Abella, J.; Verdaguer, A.; Colominas, S.; Ginestar, K.; Martinelli, L. Fundamental data: Solubility of nickel and oxygen and diffusivity of iron and oxygen in molten LBE. *J. Nucl. Mater.* **2011**, *415*, 329–337. [CrossRef]
42. Syrett, B.C. Erosion-corrosion of copper-nickel alloys in sea water and other aqueous environments—A literature review. *Corrosion* **1976**, *32*, 242–252. [CrossRef]
43. Heitz, E. Chemo-mechanical effects of flow on corrosion. *Corrosion* **1991**, *47*, 135–145. [CrossRef]
44. Brennen, C.E. *Cavitation and Bubble Dynamics*; Oxford University Press: Oxford, UK, 1995.
45. Yada, H.; Kanagawa, A.; Hattori, S. Cavitation inception and erosion in flowing system of water and liquid metal. *Trans. Jpn. Soc. Mech. Eng. Ser. B* **2011**, *78*, 811–820. (In Japanese) [CrossRef]
46. Fujisawa, N.; Uchiyama, K.; Yamagata, T. Mass transfer measurements on periodic roughness in a circular pipe and downstream of orifice. *Int. J. Heat Mass Transf.* **2017**, *105*, 316–325. [CrossRef]
47. Shan, F.; Liu, Z.; Liu, W.; Tsuji, Y. Effects of the orifice to pipe diameter ratio on orifice flows. *Chem. Eng. Sci.* **2016**, *152*, 497–506. [CrossRef]
48. Postlethwaite, J.; Lotz, U. Mass transfer at erosion-corrosion roughened surfaces. *Can. J. Chem. Eng.* **1988**, *66*, 75–78. [CrossRef]
49. Fazio, C.; Ricapito, I.; Scaddozzo, G.; Benamati, G. Corrosion behaviour of steels and refractory metals and tensile features of steels exposed to flowing PbBi in the LECOR loop. *J. Nucl. Mater.* **2003**, *318*, 325–332. [CrossRef]
50. Poulson, B. Mass transfer from rough surfaces. *Corros. Sci.* **1990**, *30*, 743–746. [CrossRef]
51. Balbaud-Celerier, F.; Barbier, F. Investigation of models to predict the corrosion of steels in flowing liquid lead alloys. *J. Nucl. Mater.* **2001**, *289*, 227–242. [CrossRef]
52. Malang, S.; Smith, D.L. *Modeling of Liquid Metal Corrosion/Deposition in a Fusion Reactor Blanket (No. ANL/FPP/TM-192)*; Argonne National Lab.: Lemont, IL, USA, 1984.
53. Sugawara, T.; Nishihara, K.; Obayashi, H.; Kurata, Y.; Oigawa, H. Conceptual design study of beam window for accelerator-driven system. *J. Nucl. Sci. Technol.* **2010**, *47*, 953–962. [CrossRef]

54. Wan, T.; Naoe, T.; Wakui, T.; Futakawa, M.; Obayashi, H.; Sasa, T. Study on the evaluation of erosion damage by using laser ultrasonic integrated with a wavelet analysis technique. *J. Phys. Conf. Ser.* **2017**, *842*, 012010. [CrossRef]
55. Yoon, M.; Hwang, J.; Lee, J.; Sung, H.J.; Kim, J. Large-scale motions on a turbulent channel flow with the slip boundary condition. *Int. J. Heat Mass Transf.* **2016**, *61*, 96–107. [CrossRef]
56. Rivai, A.K.; Saito, S.; Tezuka, M.; Kato, C.; Kikuchi, K. Effect of cold working on the corrosion resistance of JPCA stainless steel in flowing Pb–Bi at 450 °C. *J. Nucl. Mater.* **2012**, *431*, 97–104. [CrossRef]
57. Lambrinou, K.; Charalampopoulou, E.; Donck, T.V.; Delville, R.; Schryvers, D. Dissolution corrosion of 316L austenitic stainless steels in contact with static liquid lead-bismuth eutectic (LBE) at 500 °C. *J. Nucl. Mater.* **2017**, *490*, 9–27. [CrossRef]
58. Kurata, Y.; Masatoshi, F.; Saito, S. Corrosion behavior of steels in liquid lead–bismuth with low oxygen concentrations. *J. Nucl. Mater.* **2008**, *373*, 164–178. [CrossRef]
59. Kurata, Y. Corrosion experiments and materials developed for the Japanese HLM systems. *J. Nucl. Mater.* **2011**, *415*, 254–259. [CrossRef]

Article

Effect of Cold Deformation on Microstructures and Mechanical Properties of Austenitic Stainless Steel

Deming Xu [1,2], Xiangliang Wan [1,2], Jianxin Yu [3], Guang Xu [1,2] and Guangqiang Li [1,2,*]

1 The State Key Laboratory of Refractories and Metallurgy, Wuhan University of Science and Technology, Wuhan 430081, China; xudeming.wust@foxmail.com (D.X.); wanxiangliang@wust.edu.cn (X.W.); xuguang@wust.edu.cn (G.X.)
2 Key Laboratory for Ferrous Metallurgy and Resources Utilization of Ministry of Education, Wuhan University of Science and Technology, Wuhan 430081, China
3 Center of Analysis and Measurement, Harbin Institute of Technology, Harbin 150001, China; yujianxin03242@163.com
* Correspondence: liguangqiang@wust.edu.cn; Tel.: +86-027-6886-2665

Received: 14 June 2018; Accepted: 2 July 2018; Published: 6 July 2018

Abstract: In this paper, the effect of cold deformation on the microstructures and mechanical properties of 316LN austenitic stainless steel (ASS) was investigated. The results indicated that the content of martensite increased as the cold rolling reduction also increased. Meanwhile, the density of the grain boundary in the untransformed austenite structure of CR samples increased as the cold reduction increased from 10% to 40%, leading to a decreased size of the untransformed austenite structure. These two factors contribute to the improvement of strength and the decrease of ductility. High yield strengths (780–968 MPa) with reasonable elongations (30.8–27.4%) were achieved through 20–30% cold rolling. The 10–30% cold-rolled (CR) samples with good ductility had a good strain hardening ability, exhibiting a three-stage strain hardening behavior.

Keywords: austenitic stainless steel; cold deformation; microstructures; mechanical properties

1. Introduction

Austenitic stainless steels (ASSs) that generally have excellent corrosion resistance, good plasticity, and a high strain hardening coefficient are widely used in food, petrochemicals, and the nuclear industry. However, their low yield strength limits their application in structural engineering and automotive industry [1,2]. The development of high strength-high ductility ASSs is considered an attractive approach to expanding their use and has been extensively studied.

There are various methods to improve the strength of ASSs, including cold deformation strengthening, bake hardening strengthening, grain size strengthening, et cetera. [3]. Due to the high strain hardening coefficient of ASSs, cold deformation strengthening with a simpler and easier process is considered an effective method and more suitable for industrial application [4,5]. The high yield strengths of 570–926 MPa and 554–735 MPa with reasonable elongations of ~43–22% and ~35–19% were achieved in 304 [6] and 316 L [7] ASSs through 10–30% cold rolling, respectively. However, the plastic straining induced ductility loss was more pronounced than the strength increase for the ASSs as the cold rolling reduction further increased, exhibiting a typical "banana-shaped" strength-ductility trade-off [8].

The mechanical properties of cold-rolled (CR) ASSs are determined by their deformation microstructures. Normally, the austenite phase in ASSs is not a stable phase and will transform into strain-induced martensite during the plastic deformation process [9]. Martensite phase generally exhibits a higher strength and lower plasticity. According to the mix law described by Huang et al [10] for two phase materials, the strengths of two phase materials are determined by the volume fractions and the corresponding strengths of the two phases. The transformation from austenite phase into martensite phase will improve the strength of CR ASSs [6,11]. The phase transformation in ASSs is controlled by the stability of austenite that is determined by M_{d30} temperature (where 50% ά-martensite is present after 30% tensile deformation) [12]. Due to the higher content of alloy elements, some ASSs, like 316 ASSs, have a lower M_{d30} temperature and higher stability of austenite that inhibit the phase transformation during plastic deformation. Previous studies indicated that only 46–67% of austenite phase in 316 ASSs can be transformed into strain-induced martensite, even after 80–90% cold rolling [13–15]. Interestingly, these steels also exhibited the tendency of significant loss in plastic straining induced ductility [13–15], which cannot be explained by the mix law. Therefore, it is not enough to only consider the effect of volume fraction of the strain-induced martensite on the mechanical properties of ASSs, and the untransformed austenite evolution and its effect on mechanical properties should be considered at the same time. However, the evolution of untransformed austenite phase with increasing cold rolling reduction is less studied. The influence of untransformed austenite in CR ASSs on mechanical properties is not clear and needs in-depth studies.

In this study, the electron backscattered diffraction (EBSD) was chosen to study the evolution of untransformed austenite phase in CR 316LN ASS as the cold reduction increased from 10% to 40%. Evaluating the evolution of untransformed austenite in CR ASSs by using EBSD is rarely reported and is the innovation of this paper. Furthermore, transmission electron microscopy (TEM) was used to study the evolution of the deformation microstructures as the cold rolling reduction increased. The effect of cold deformation on the mechanical properties of 316LN ASS was studied by analyzing the relationship between deformation microstructures and mechanical properties.

2. Experimental Procedures

The chemical composition, the stacking fault energy (SFE) based equation of Brofman and Ansell [12], and the M_{d30} temperature based on Nohara's equation [16] of 3mm-thick 316LN ASS used in this study, are shown in Table 1. The strips were cold rolled in a pilot plant with thickness reductions from 10% to 90% reduction at room temperature. The volume fraction of martensite in CR samples was determined by X-ray diffraction (XRD, Panalytical, Almelo, The Netherlands). The Vickers hardness of samples was determined by measuring more than 20 points in different regions of the samples using a microhardness tester with a 0.5 kg load (HV-1000B, Matsuzawa, Tokyo, Japan). The structures in original 316LN ASS and CR samples were evaluated using an OM (Axioplan2 Imaging Zeiss, Göttingen, Germany). For the CR samples, an etchant of two solutions at a 1:1 ratio was used. The first solution consists of 0.20 g sodium-metabisulfate in 100 mL distilled water and the second consists of 10 mL hydrochloric acid in 100 mL distilled water. The original samples were mechanically polished and then electrochemically etched with 60% nitric acid solution. Microstructures of CR samples were examined by TEM (JEM-2100, JEOL, Tokyo, Japan). Thin foils were prepared by twin-jet electropolishing of 3 mm disks, punched from the specimens using a solution of 10% perchloric acid in acetic acid as the electrolyte. EBSD was utilized to evaluate the microstructures and grain boundary of 0–40% CR samples. For that purpose, the samples were electrochemically etched with 20% perchloric acid alcohol solution operated at 25 °C with an applied potential of 15 V. The size of the austenite structure in EBSD micrographs was determined by analyzing grain boundary misorientation over 2°. The original and CR strips were machined to make tensile samples with a profile of 140 × 20 mm and a gage length of 65 mm. The uniaxial tensile tests were carried out at room temperature at an engineering strain rate of 5×10^{-4} s^{-1} and each specimen was tested three times to obtain an average value for each mechanical property.

Table 1. Chemical compositions, stacking fault energy (SFE), and M_{d30} of the investigated steel (wt, %).

C	Si	Mn	N	Cr	Ni	Mo	SFE, mJ/m^2	M_{d30}, °C
0.04	0.34	1.15	0.048	18.06	8.33	0.051	18.9	6.2

3. Results

3.1. The Microstructural Evolution with Increasing Cold Rolling Reduction

Optical micrographs of 316LN commercial ASS and 10%, 30%, 50%, 70%, and 90% CR samples are presented in Figure 1. As shown in Figure 1a, the coarse-grained austenite existed in this commercial steel. The content of strain-induced martensite increased as the cold rolling reduction increased, as shown in Figure 1b–f. Many shear bands were discovered in the austenite phase of the 10% CR sample (Figure 1b). When the cold reduction increased to 30%, some strain-induced martensite formed at shear bands. When the cold reduction increased to 50%, the untransformed austenite was elongated along the rolling direction (Figure 1d). The content of martensite further increased and the formed martensite segmented the untransformed austenite when the cold reduction was increased to 70%, as shown in Figure 1e. When the cold reduction increased to 90%, the large block of untransformed austenite still existed. Meanwhile, the formed martensite seemed to be divided into fine martensite and mixed with fine austenite structure (white circle in Figure 1f). The shear bands were basically replaced by martensite and became difficult to distinguish, as shown in Figure 1f.

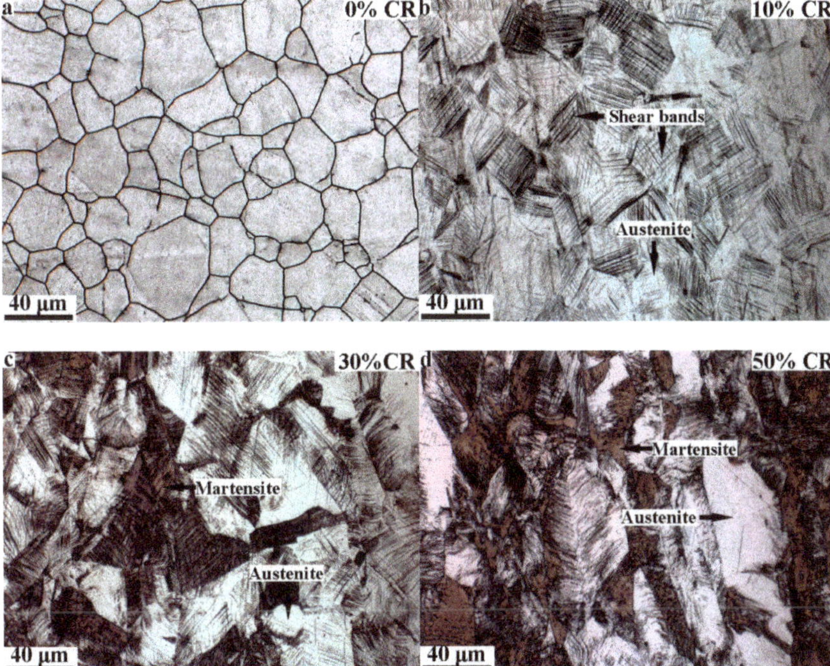

Figure 1. *Cont.*

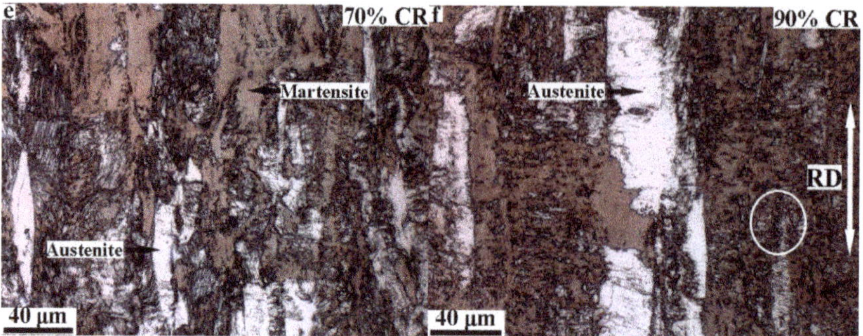

Figure 1. Optical micrographs of microstructures for (**a**) original; (**b**) 10% CR; (**c**) 30% CR; (**d**) 50% CR; (**e**) 70% CR; and (**f**) 90% CR 316LN ASSs. The cold rolling direction (RD) is shown in (**f**). The straw color phase is the strain-induced martensite and the white phase is the austenite matrix.

The XRD patterns of 316LN commercial ASS, 10%, 30%, 50%, 70%, and 90% CR samples are presented in Figure 2a. According to the XRD patterns, only ά-martensite formed during cold deformation and the increased cold reduction led to the development of ά-martensite peaks. The measurement of volume fractions of ά-martensite in CR ASSs based on XRD has been reported in many studies [17–19]. The measured results based on the equation of Dickson [17] are presented in Figure 2b. It can be observed that the volume fraction of ά-martensite increased with increasing cold reduction and about 46% ά-martensite was achieved in the 90% CR sample.

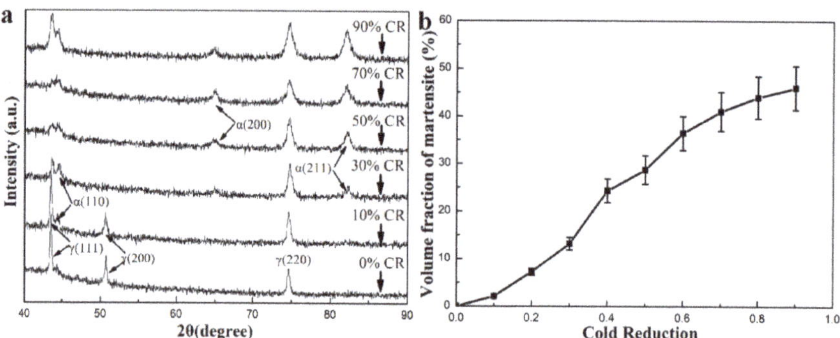

Figure 2. (**a**) X-ray diffraction patterns of 316LN ASS, 10%, 30%, 50%, 70%, and 90% CR samples and (**b**) the volume fractions of ά-martensite in CR samples as a function of cold reduction.

Figure 3 illustrates that the microstructures evolve with the increasing cold rolling reduction. As shown in Figure 3a,b, shear bands and mechanical twins existed in the 10% CR sample and the dislocations gathered at shear bands and twins' boundaries. When the cold reduction increased to 30%, the strain-induced martensite was observed and the dislocation movement was suppressed by the martensite boundary, which resulted in the gathering of dislocations around the martensite boundary (Figure 3c). Meanwhile, dislocation boundaries (DB) were formed to segment the untransformed austenite structure (Figure 3d). When the cold reduction increased to 50%, many martensite laths formed (Figure 3e). Furthermore, the dislocation-cell-type martensite was observed in samples when the cold reduction was increased to 90% (Figure 3f).

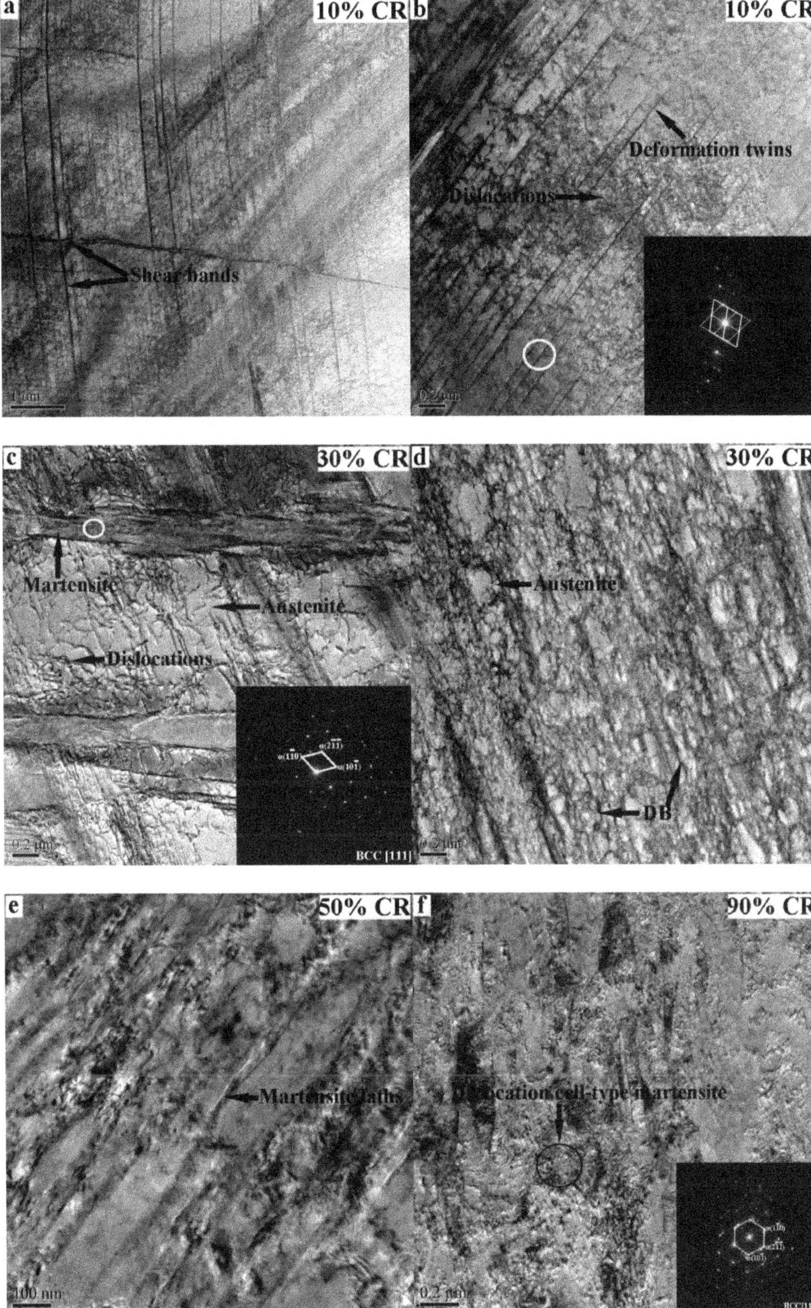

Figure 3. Representative bright field transmission electron micrographs of deformation microstructures in (**a**,**b**) 10% CR; (**c**,**d**) 30% CR; (**e**) 50% CR; and (**f**) 90% CR samples. The diffraction pattern confirmed (**b**) deformation twin, (**c**) martensite lath, and (**f**) dislocation-cell-type martensite. DB in (**b**) means dislocations boundary.

The original and 10–40% CR samples were characterized by EBSD and are presented in Figure 4. The main structure of the original sample was coarse austenite grains with many annealing twins (Figure 4a). A small amount of strain-induced martensite and grain boundaries formed in austenite grains after 10% cold rolling. The formed grain boundary was mainly a low angle grain boundary (0° < LAGB < 15°). The density of the grain boundary and volume fraction of martensite increased with the increasing cold rolling reduction. Figure 4f indicates that the densities of both LAGB and the high angle grain boundary (HAGB > 15°) increased with the increasing cold rolling reduction. The difference between LAGB and HAGB is that the density of LAGB first increased rapidly and then increased slowly when the cold reduction increased from 10% to 40%, while for HAGB, the increase rate change is the opposite. Sizes of austenite structure, percentage fractions, and densities of the austenite grain boundary with the different misorientation ranges based on Figure 4 are listed in Table 2. As shown in Table 2, the sizes of the austenite structure decreased with the increasing cold reduction. Figure 5a–d show the IPF map and Figure 5e–h show the corresponding misorientation variation along the black lines in untransformed austenite grains of 10% CR, 20% CR, 30% CR, and 40% CR samples, respectively. The results indicate that the main misorientation of point to point in untransformed austenite grains of the 10% CR sample was below 15°. When the cold reduction increased to 30% and 40%, the densities of high misorientation (>15°) increased obviously. The average numbers of misorientation in untransformed austenite grains of CR samples increased rapidly as the cold reduction increased from 10% to 40%, as shown in Figure 5e–h.

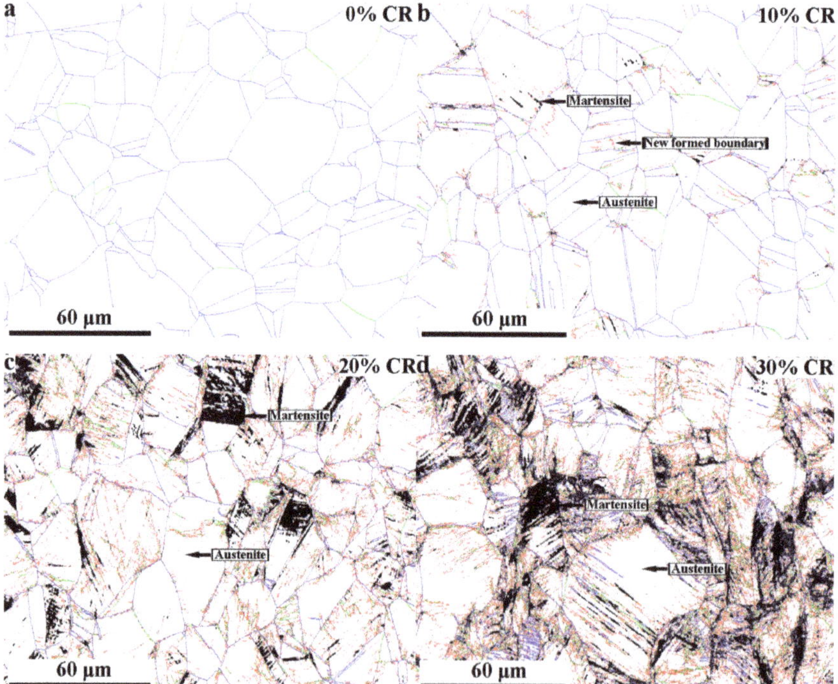

Figure 4. *Cont.*

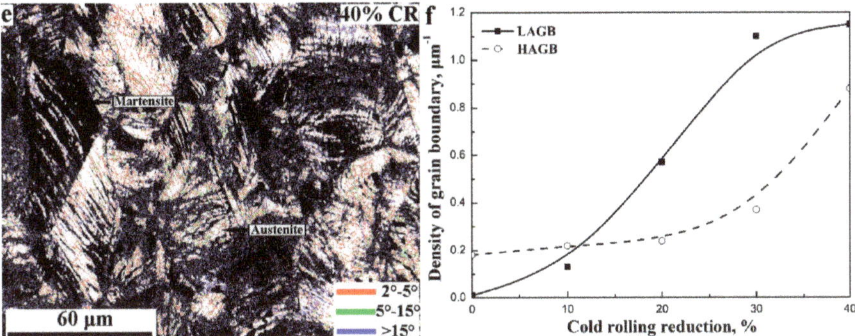

Figure 4. Electron backscattered diffraction (EBSD) micrographs grain boundary reconstruction maps for austenite phase of (**a**) original; (**b**) 10% CR; (**c**) 20% CR; (**d**) 30% CR; and (**e**) 40% CR samples, respectively; (**f**) The variation density of low angle grain boundary (LAGB) and high angle grain boundary (HAGB) as a function of cold rolling reduction. The white area is the austenite structure and the black area is martensite.

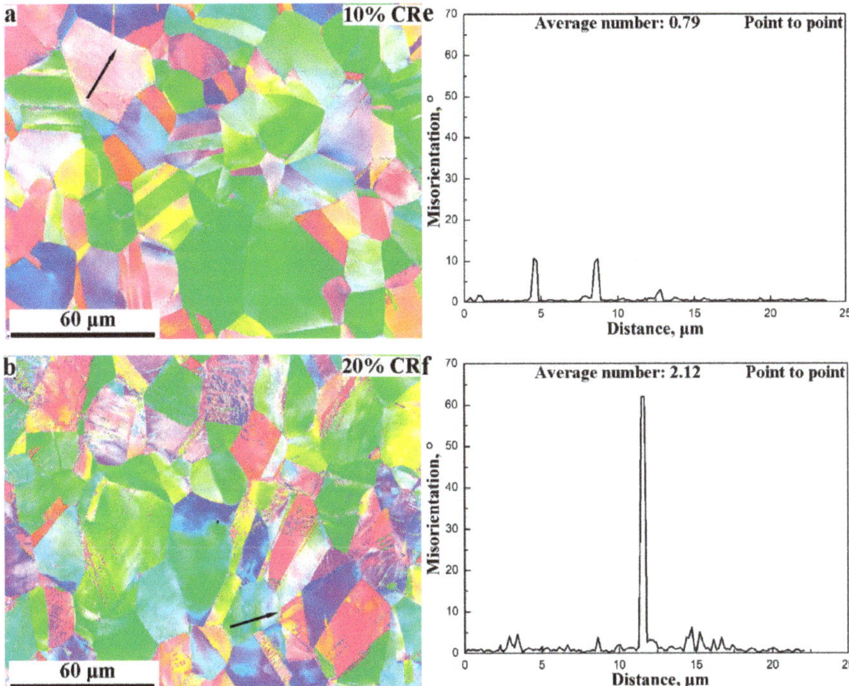

Figure 5. *Cont.*

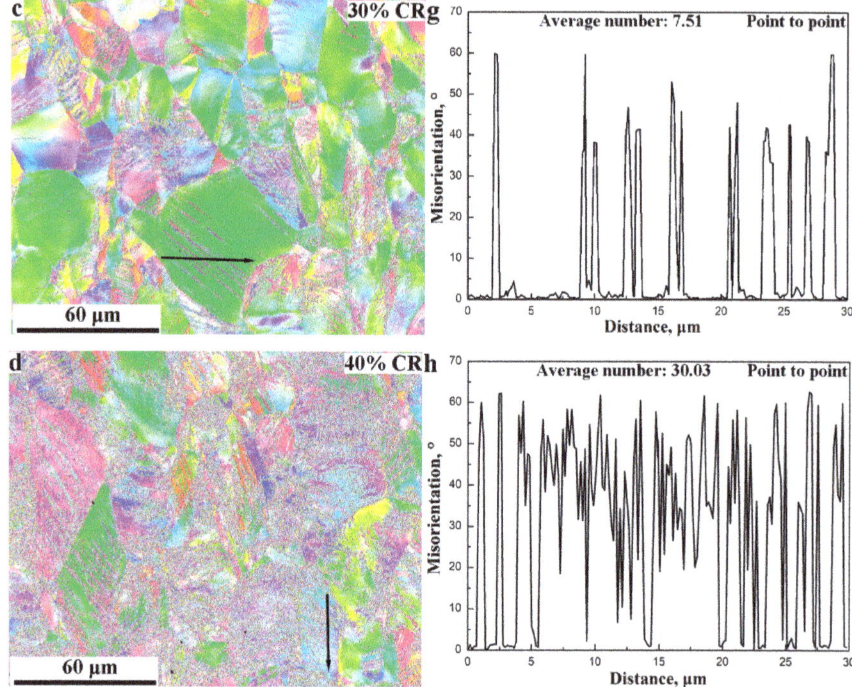

Figure 5. (**a–d**) IPF images for (**a**) 10% CR; (**b**) 20% CR; (**c**) 30% CR; and (**d**) 40% CR samples, respectively. (**e–h**) Misorientation variations along the black lines (black lines in figures) in untransformed austenite grain for (**e**) 10% CR; (**f**) 20% CR; (**g**) 30% CR; and (**h**) 40% CR samples, respectively.

Table 2. Austenite structure sizes, percentage fractions, and densities of the austenitic grain boundary with different misorientation ranges of samples with varying cold reductions.

Cold Rolling Reduction, %	Structure Size, µm	Percentage Fraction, %		Density, Length/Area, μm^{-1}	
		2–15°	15–65°	2–15°	15–65°
0	9.0	5.2	94.8	0.01	0.18
10	7.3	35.9	64.1	0.13	0.22
20	2.8	73.1	26.9	0.57	0.24
30	1.3	72.8	27.2	1.1	0.37
40	0.87	56.7	43.3	1.15	0.88

3.2. The Mechanical Properties of CR Samples

The engineering stress-engineering strain (σ_E-ε_E) curves of 10%, 30%, 50%, 70%, and 90% CR samples are shown together with those of the original sample in Figure 6a. The mechanical properties that changed due to the cold rolling reduction are listed in Table 3. It was found that the original sample exhibited a low average yield strength of 281 MPa and high average elongation of 52%. After 30% cold rolling, its average yield strength increased to 968 MPa with an average elongation of 27.4%. When the cold reduction increased to 90%, the yield strength rose to 1870 MPa, nearly seven times that of the original sample, but the elongation decreased to 1.3%. Figure 6b shows that the average tensile/yield strength increased, but the average elongation decreased, when the cold rolling reduction increased. The average Vickers hardness increased as the cold rolling reduction increased.

When the cold reduction increased to 90%, the hardness value increased to 549.4, which is about three times that of the original sample.

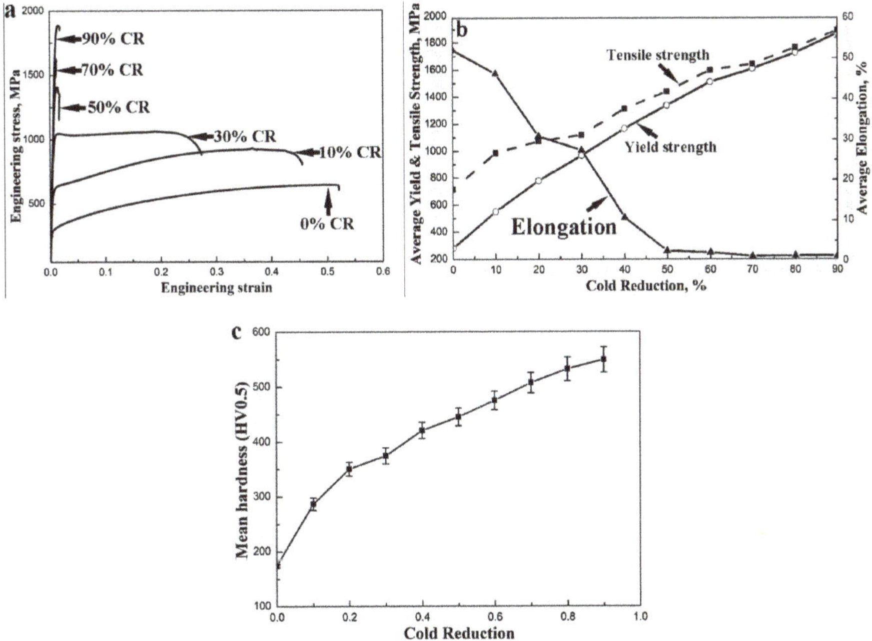

Figure 6. (**a**) Engineering stress-engineering strain behavior for original, 10%, 30%, 50%, 70%, and 90% CR samples; (**b**) variation average yield/tensile strengths and elongations; and (**c**) Vickers hardness as a function of cold rolling reduction.

Table 3. Variation of martensite volume fractions, Vickers harnesses, mean tensile and yield strengths, and mean elongations with % cold rolling reduction for 316LN ASS.

Cold Rolling Reduction, %	Martensite Volume Fraction, %	Vickers Hardness, HV0.5	Mean Tensile Strength, MPa	Mean Yield Strength, MPa	Mean Elongation, %
0	0	174.2	644	281	52.0
10	2.1	287.0	931	551	46.2
20	7.2	360.2	1021	780	30.8
30	13.2	374.2	1071	968	27.4
40	24.3	420.8	1274	1167	10.7
50	28.7	445.4	1407	1337	2.5
60	36.4	475.3	1576	1510	2.1
70	41.1	507.2	1623	1608	1.2
80	44.0	532.3	1750	1725	1.3
90	46.0	549.4	1887	1870	1.3

3.3. Strain Hardening Behaviors of CR Samples

The 10% CR and 30% CR samples with good ductility were selected to study the strain hardening behaviors and their strain hardening rates (SHRs; $d\sigma_T/d\varepsilon_T$) were plotted as a function of true strain, presented in Figure 7. The SHR plots of selected samples can be divided into three stages, as shown in Table 4. The SHRs first decreased (Stage A), then increased (Stage B), before finally decreasing again (Stage C) with increasing true strain.

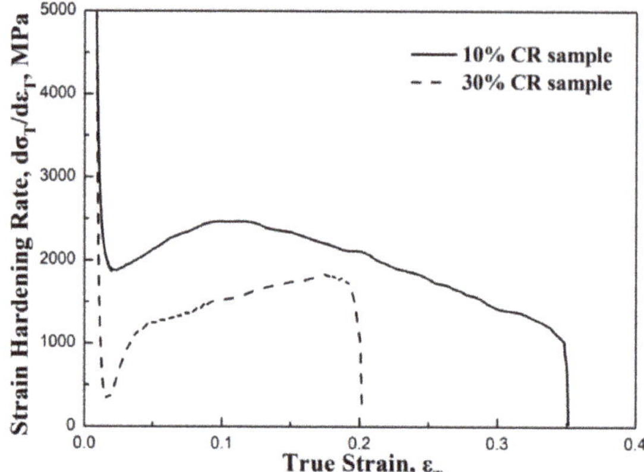

Figure 7. Strain hardening rate for 10% CR (A1, B1, C1) and 30% CR samples (A2, B2, C2).

Table 4. Values of plastic strain of the samples at each stage. The data are taken from Figure 7.

Sample	Stage A	Stage B	Stage C
10% CR	$\varepsilon_T < 0.02$	$0.023 < \varepsilon_T < 0.09$	$0.09 < \varepsilon_T < 0.35$
30% CR	$\varepsilon_T < 0.017$	$0.017 < \varepsilon_T < 0.18$	$0.18 < \varepsilon_T < 0.2$

4. Discussion

In this paper, the effect of cold deformation on the mechanical properties of 316LN ASS was studied. It is well-known that the microstructural characteristics are a major factor determining the mechanical properties of steels. Therefore, the effect of cold deformation on microstructural evolution and the relationship between the microstructures and mechanical properties of CR 316LN ASS were discussed. The ductility of the ASSs is related to their strain hardening behavior. To obtain high strength 316LN ASS with a reasonable ductility, the strain hardening behaviors of CR samples were discussed.

4.1. The Effect of Cold Deformation on the Microstructures

The deformation microstructures formed in ASSs during the plastic deformation process depend on their deformation mechanisms, which are determined by their SFEs. The SFE of the 316LN ASS at room temperature is ~18.9 mJ/m^2 (Table 1) and both the strain induced martensite and the mechanical twin could form during plastic deformation [13,20]. When the cold rolling reduction increased, the shear bands comprised of slip bands and mechanical twins formed in austenite grains and then the strain-induced martensite nucleated at the intersection of shear bands [21–23]. The formed ά-martensite replaced the shear bands and the shear bands gradually disappeared as the cold rolling reduction increased. Furthermore, the martensite lath could be broken and mixed with untransformed austenite to form dislocation-cell-type martensite that has a higher density of dislocation (Figure 3f) [24].

Previous studies [25,26] have indicated that heavily cold worked metals could be subdivided by grain boundaries and dislocation boundaries. The cold work inducing the formation of the grain boundary was also found in CR AZ91 alloy [27] and 308L stainless steel [28]. In this 316LN ASS, in addition to dislocation grain boundaries, the boundaries of mechanical twins and strain-induced martensite formed during the CR process can also subdivide untransformed austenite, leading to the decrease of the untransformed austenite structure size. The EBSD results in Figure 4 revealed a similar tendency, where the increased density of grain boundaries led to a decreasing untransformed austenite structure size. Hughes and Hansen's study [29] indicated that the continued subdivision of grains into crystallites surrounded by dislocation boundaries leads to a large orientation spread based on dislocation accumulation processes. The main grain boundaries formed in the 10% CR sample were LAGB that may originate from the dislocation boundary formed by dislocation accumulated at twins' boundaries or shear bands. The rapidly increasing density of LAGB as the cold reduction increased from 10% to 30% could be attributed to the increased dislocation boundaries caused by the increasing density of dislocation. The difficult dislocation boundaries formed could show an operation of different slip system combinations within the individual crystallites, leading to the possibility of different parts of a grain rotating towards different stable end orientations [29]. If such end orientations are not far apart, large misorientations within the original grain can build up. Considering that the density of the dislocation boundary in the untransformed austenite grains increases with the increasing cold rolling reduction, the probability of large misorientations within the original grains increases with the increasing cold rolling reduction, leading to the increasing density of HAGB. The results in Figures 4 and 5 illustrate this viewpoint very well as the variations of HAGB densities are in good agreement with the variations of the average numbers of misorientations in the original austenite grains.

4.2. Effect of Microstructures on Mechanical Properties

The previous section has shown that cold deformation could promote strain-induced martensite formation. Compared with the austenite with a face centered cubic (fcc) crystal structure, the ά-martensite with a body-centered cubic (bcc) crystal structure has fewer slip directions and a larger lattice resistance, which means that ά-martensite is more difficult to slip in the process of plastic deformation [30]. The strain-induced martensite formed during cold deformation could effectively improve the strength of the samples. Furthermore, although there are large blocks of untransformed austenite existing in CR samples, the results of TEM and EBSD have indicated that a high density grain boundary existed in these austenite structures. The grain boundary can hinder the movement of dislocations, which could reduce the stress concentration and improve the strength of the samples. Meanwhile, the increasing volume fraction of martensite and dislocation density as the cold rolling reduction increased resulted in the increasing Vickers hardness [31].

However, the high density dislocation in martensite and its uneasy sliding contribute to its poor plasticity. The high density of the grain boundary in the untransformed austenite structure will significantly hinder the movement of dislocations. Furthermore, austenite structure refinement could enhance the stability of austenite, which inhibits the martensite transformation during the plastic deformation, resulting in a decrease in the ductility [32].

4.3. Effect of CR Microstructures on Strain Hardening Behavior

The 10% CR and 30% CR samples presenting three-stage strain hardening curves and high SHRs reflect the high strain hardening ability. However, there are subtle differences between the samples in the SHRs, which are caused by their different microstructures. The strain range at stage A in 10% CR samples is larger than that in 30% CR samples. The variation of SHRs at an early stage is usually affected by the nucleation and slip of dislocation in ASSs [33–35]. Compared with the 10% CR sample, the 30% CR sample has more dislocations and a higher density grain boundary, which are not conducive to dislocation movement during the tensile process, resulting in its smaller strain range at stage A. The strain range at stages B and C was considered beneficial to plasticity in ASSs that was attributed to the transformation induced plasticity (TRIP) or twinning induced plasticity (TWIP) effect [36]. The tensile induced martensite or mechanical twin will nucleate and grow at stages B and C [37]. When the cold reduction increased from 10% to 30%, the austenite structure size decreased from 7.3 μm to 1.3 μm (Table 2). Grain refinement can increase the stability of austenite and inhibit the tensile induced martensite formation during plastic deformation [38], making the strain range at stages B and C in the 30% CR sample smaller than that of the 10% CR sample. With the increase of cold reduction, the finer austenite structure in samples is not conducive to the nucleation and movement of dislocation. Furthermore, the high content of martensite in the samples reduces the capacity to accommodate dislocations. These factors lead to the rapid decrease of the strain hardening ability of the samples when the cold rolling reduction increased.

5. Conclusions

The effects of cold rolling reduction on the microstructures and mechanical properties of 316LN ASS were studied. The conclusions can be summarized as follows:

(1) The yield strength of commercial 316LN ASS increased from 281 MPa to 780–968 MPa and it maintained a reasonable elongation value of 30.8–27.4% through thickness reduction of 20–30% cold rolling.
(2) The size of untransformed austenite in CR samples decreased when cold reduction increased from 10% to 40%. The decreased size of untransformed austenite was attributed to the increasing boundary density of dislocations, mechanical twins, and strain-induced martensite that formed during the CR process.
(3) The grain refinement of untransformed austenite phase and increasing content of strain-induced martensite resulted in an increased yield/tensile strength and decreased ductility of 316LN ASS under the influence of increasing the cold rolling reduction.
(4) The CR 316LN ASSs with high yield strengths and reasonable elongations had a good strain hardening ability and exhibited a three-stage strain hardening behavior.

In this paper, the increased yield/tensile strength and decreased elongation of CR 316LN ASS are attributed to grain refinement of the untransformed austenite phase and increasing content of strain-induced martensite. However, which of these two factors plays a major role is not clear. We plan to obtain a fully deformation austenite structure and fully strain-induced martensite through controlling rolling temperature and reduction in the next work. The effect of a fully deformation austenite structure or fully strain-induced martensite on the mechanical properties of ASSs will be studied in the next work.

Author Contributions: X.W. and G.L., supervisor, conceived and designed the experiments; D.X., doctoral student, conducted experiments, analyzed the data and wrote the paper; J.Y., conducted experiments; G.X., conducted experiments.

Funding: This research was funded by National Natural Science Foundation of China (No. 51501134).

Conflicts of Interest: The authors declare no conflict of interest.

References

1. European Steel Technology Platform (ESTEP). *Strategic Research Agenda. A Vision for the Future of the Steel Sector*; European Commission: Brussels, Belgium, 2005.
2. Andersson, R.; Schedin, E.; Magnusson, C.; Ocklund, J.; Persson, A. Stainless steel components in automotive vehicles. In Proceedings of the 4th European Stainless Steel Science and Market Congress, Paris, France, 10–13 June 2002; p. 57.
3. Karjalainen, L.P.; Taulavuori, T.; Sellman, M.; Kyröläinen, A. Some Strengthening Methods for Austenitic Stainless Steels. *Steel Res. Int.* **2008**, *79*, 404–412. [CrossRef]
4. Maki, T. Stainless steel: Progress in thermomechanical treatment. *Curr. Opin. Solid State Mater. Sci.* **1997**, *2*, 290–295. [CrossRef]
5. Padilha, A.F.; Plaut, R.L.; Rios, P.R. Annealing of cold-worked austenitic stainless steels. *Iron Steel Inst. Jpn.* **2003**, *43*, 135–143. [CrossRef]
6. Milad, M.; Zreiba, N.; Elhalouani, F.; Baradai, C. The effect of cold work on structure and properties of AISI 304 stainless steel. *J. Mater. Process. Technol.* **2008**, *203*, 80–85. [CrossRef]
7. Hou, X.; Zheng, W.; Song, Z.; Long, J. Effect of cold working on mechanical behavior and microstructure of 316L stainless steel. *J. Iron Steel Res.* **2013**, *25*, 53–57. (In Chinese)

8. Raabe, D.; Ponge, D.; Dmitrieva, O.; Sander, B. Nanoprecipitate-hardened 1.5 GPa steels with unexpected high ductility. *Scr. Mater.* **2009**, *60*, 1141–1144. [CrossRef]
9. Padilha, A.F.; Rios, P.R. Decomposition of austenite in austeniticstainless steels. *Iron Steel Inst. Jpn.* **2002**, *42*, 325–337. [CrossRef]
10. Huang, G.L.; Matlock, D.; Krauss, G. Martensite formation, strain rate sensitivity, and deformation behavior of type 304 stainless steel sheet. *Metall. Trans. A* **1989**, *20*, 1239–1246. [CrossRef]
11. Hedayati, A.; Najafizadeh, A.; Kermanpur, A.; Forouzan, F. The effect of cold rolling regime on microstructure and mechanical properties of AISI 304L stainless steel. *J. Mater. Process. Technol.* **2010**, *210*, 1017–1022. [CrossRef]
12. Brofman, P.J.; Ansell, G.S. On the effect of carbon on the stacking fault energy of austenitic stainless steels. *Metall. Trans. A* **1978**, *9*, 879–880. [CrossRef]
13. Xu, D.M.; Li, G.Q.; Wan, X.L.; Xiong, R.L.; Xu, G.; Wu, K.M.; Somani, M.C.; Misra, R.D.K. Deformation behavior of high yield strength-high ductility ultrafine-grained 316LN austenitic stainless steel. *Mater. Sci. Eng. A* **2017**, *688*, 407–415. [CrossRef]
14. Wu, H.; Niu, G.; Cao, J.; Yang, M. Annealing of strain-induced martensite to obtain micro/nanometre grains in austenitic stainless. *Mater. Sci. Technol.* **2017**, *33*, 480–486. [CrossRef]
15. Eskandari, M.; Najafizadeh, A.; Kermanpur, A. Effect of strain-induced martensite on the formation of nanocrystalline 316L stainless steel after cold rolling and annealing. *Mater. Sci. Eng. A* **2009**, *519*, 46–50. [CrossRef]
16. Nohara, K.; Ono, Y.; Ohashi, N. Composition and grain size dependencies of strain-induced martensitic transformation in metastable austenitic stainless steels. *Iron Steel Inst. Jpn.* **1977**, *63*, 772–782. [CrossRef]
17. Dickson, M.J. The significance of texture parameters in phase analysis by X-ray diffraction. *J. Appl. Crystallogr.* **1969**, *2*, 176–180. [CrossRef]
18. Johannsen, D.L.; Kyrolainen, A.; Ferreira, P.J. Influence of annealing treatment on the formation of nano/submicron grain size AISI 301 Austenitic stainless steels. *Metall. Mater. Trans. A* **2006**, *37*, 2325–2338. [CrossRef]
19. Egea, A.J.S.; Rojas, H.A.G.; Celentano, D.J.; Peiró, J.J. Mechanical and metallurgical changes on 308L wires drawn by electropulses. *Mater. Des.* **2016**, *90*, 1159–1169. [CrossRef]
20. Lee, T.H.; Shin, E.; Oh, C.S.; Ha, H.Y.; Kim, S.J. Correlation between stacking fault energy and deformation microstructure in high-interstitial-alloyed austenitic steels. *Acta Mater.* **2010**, *58*, 3173–3186. [CrossRef]
21. Murr, L.E.; Staudhammer, K.P.; Hecker, S.S. Effects of strain state and strain rate on deformation-induced transformation in 304 stainless steel: Part II. Microstructural study. *Metall. Trans. A* **1982**, *13*, 627–635. [CrossRef]
22. Choi, J.Y.; Jin, W. Strain induced martensite formation and its effect on strain hardening behavior in the cold drawn 304 austenitic stainless steels. *Scr. Mater.* **1997**, *36*, 99–104. [CrossRef]
23. Meyers, M.A.; Xu, Y.B.; Xue, Q.; Pérez-Prado, M.T.; McNelley, T.R. Microstructural evolution in adiabatic shear localization in stainless steel. *Acta Mater.* **2003**, *51*, 1307–1325. [CrossRef]
24. Misra, R.D.K.; Nayak, S.; Mali, S.A.; Shah, J.S.; Somani, M.C.; Karjalainen, L.P. On the Significance of Nature of Strain-Induced Martensite on Phase-Reversion-Induced Nanograined/Ultrafine-Grained Austenitic Stainless Steel. *Metall. Trans. A* **2010**, *41*, 3. [CrossRef]
25. Bellier, S.P.; Doherty, R.D. The structure of deformed aluminium and its recrystallization-investigations with transmission Kossel diffraction. *Acta Metall.* **1977**, *25*, 521–538. [CrossRef]
26. Hughes, D.A.; Hansen, N. High angle boundaries and orientation distributions at large strains. *Scr. Mater.* **1995**, *33*, 315–321. [CrossRef]
27. Sheng, Y.; Hua, Y.; Wang, X.; Zhao, X.; Chen, L.; Zhou, H.; Wang, J.; Berndt, C.C.; Li, W. Application of high-density electropulsing to improve the performance of metallic materials: Mechanisms, microstructure and properties. *Materials* **2018**, *11*, 185. [CrossRef] [PubMed]
28. Egea, A.J.S.; González-Rojas, H.A.; Celentano, D.J.; Perió, J.J.; Cao, J. Thermomechanical analysis of an electrically assisted wire drawing process. *J. Manuf. Sci. Eng.* **2017**, *139*, 111017. [CrossRef]
29. Hughes, D.A.; Hansen, N. High angle boundaries formed by grain subdivision mechanisms. *Acta Mater.* **1997**, *45*, 3871–3886. [CrossRef]
30. Cui, Z.Q.; Liu, B.X. *Principles of Metallography and Heat Treatment*; Harbin Institute of Technology Press: Harbin, China, 2004. (In Chinese)

31. Zhang, N.; Wang, Y. Dislocations and hardness of hard coatings. *Thin Solid Films* **1992**, *214*, 4–5. [CrossRef]
32. Yan, F.K.; Liu, G.Z.; Tao, N.R.; Lu, K. Strength and ductility of 316L austenitic stainless steel strengthened by nano-scale twin bundles. *Acta Mater.* **2012**, *60*, 1059–1071. [CrossRef]
33. Ding, H.; Ding, H.; Song, D.; Tang, Z.; Yang, P. Strain hardening behavior of a TRIP/TWIP steel with 18.8% Mn. *Mater. Sci. Eng. A* **2011**, *528*, 868–873. [CrossRef]
34. Challa, V.S.A.; Wan, X.L.; Somani, M.C.; Karjalainen, L.P.; Misra, R.D.K. Significance of interplay between austenite stability and deformation mechanisms in governing three-stage work hardening behavior of phase-reversion induced nanograined/ultrafine-grained (NG/UFG) stainless steels with high strength-high ductility combination. *Scr. Mater.* **2014**, *86*, 60–63.
35. Matsuoka, Y.; Iwasaki, T.; Nakada, N. Effect of grain size on thermal and mechanical stability of austenite in metastable austenitic stainless steel. *Iron Steel Inst. Jpn.* **2013**, *53*, 1224–1230. [CrossRef]
36. Takaki, S.; Fukunaga, K.; Syarif, J. Effect of grain refinement on thermal stability of metastable austenitic steel. *Mater. Trans.* **2004**, *45*, 2245–2251. [CrossRef]
37. Challa, V.S.A.; Wan, X.L.; Somani, M.C.; Karjalainen, L.P.; Misra, R.D.K. Strain hardening behavior of phase reversion-induced nanograined/ultrafine-grained (NG/UFG) austenitic stainless steel and relationship with grain size and deformation mechanism. *Mater. Sci. Eng. A* **2014**, *613*, 60–70. [CrossRef]
38. Barbier, D.; Gey, N.; Allain, S.; Bozzolo, N.; Humbert, M. Analysis of the tensile behavior of a TWIP steel based on the texture and microstructure evolutions. *Mater. Sci. Eng. A* **2009**, *500*, 196–206. [CrossRef]

 *metals*

Article

Corrosion Behavior Difference in Initial Period for Hot-Rolled and Cold-Rolled 2205 Duplex Stainless Steels

Tao Gao, Jian Wang *, Qi Sun and Peide Han *

College of Materials Science and Engineering, Taiyuan University of Technology, Taiyuan 030024, China; u.best.ok@163.com (T.G.); imsunqi@163.com (Q.S.)

* Correspondence: wangjian@tyut.edu.cn (J.W.); hanpeide@126.com (P.H.); Tel.: +86-351-601-8843 (J.W. & P.H.)

Received: 14 April 2018; Accepted: 25 May 2018; Published: 01 June 2018

Abstract: Precipitate phases often play an important role on the corrosion resistance of 2205 Duplex stainless steels (DSS). In the present paper, the microstructure and the corrosion resistance of the hot-rolled and cold-rolled 2205 steels aged for different times at 850 °C was investigated through XRD, SEM, and potentiodynamic polarization. It was discovered that the Chi(χ) phase and Sigm(σ) phase were precipitated in turn following the aging treatment of the hot-rolled and cold-rolled steels, but the precipitate amount in the cold-rolled samples was significantly higher when compared to the hot-rolled samples. The corrosion resistance of the solution-annealed cold-rolled samples was similar to the hot-rolled samples, but the corrosion resistance of the cold-rolled sample with precipitate was weaker when compared to the hot rolled sample following aging treatment. Pitting preferentially initiates in the Cr-depleted region from the σ phase in the aged hot-rolled 2205, becoming increasingly severe during aging for a long lime. Adversely, the initiation of pitting corrosion might occur at the phase boundary, defects, and martensite in the aged cold-rolled 2205. The σ phase was further selectively dissolved through the electrochemical method to investigate the difference in microstructure and corrosion behavior of the hot-rolled and cold-rolled 2205 duplex stainless steels.

Keywords: pitting; sigma phase; 2205; duplex stainless steel

---

1. Introduction

The 2205 duplex stainless steels (DSS), with excellent corrosion performance, as well as good mechanical properties, have been used as engineering alloys for many years and provide wide applications in many industrial fields, especially in aggressive environments [1–6]. The effects of microstructural modifications of 2205 DSS on their mechanical and corrosion resistance were intensively investigated in the past. Few works highlighted the changes in corrosion resistance performances that are caused by the phase proportions of austenite and ferrite [7,8]. Several researchers discussed the passive film properties of the 2205 DSS during electrochemical corrosion [9–12]. Certain studies also indicated that the pitting corrosion of 2205 DSS often initiated from the σ phase due to the depletion of both Cr and Mo [13,14].

By contrast, due to the high contents of alloying elements in 2205 DSS, the secondary phases, such as σ, χ, Cr_2N, α', and $M_{23}C_6$ were also easily precipitated in between approximately 300–1000 °C, leading to a detrimental effect on both mechanical properties and corrosion resistance behavior [15,16]. The χ phase, as a precursor of the σ phase, gradually disappeared with the σ phase precipitation. The $M_{23}C_6$ phase usually formed on the austenite grain boundaries during isothermal heating at 950 °C. The Cr_2N were often found in ferrite subsequently to rapid cooling from the higher annealing temperature of 1050–1250 °C. When compared to these phases, the σ phase was more easily precipitated and had increased effect on the material properties [17–19]. Certain researchers had focused on the σ

phase precipitation. Chen et al. found that the σ phase was often precipitated at the α/γ interphase boundaries following the sample aging at 900 °C for 5 min [20]. Elmer et al. in situ observed the dissolution of the sigma phase in the 2205 duplex stainless steel, while the σ phase could be detected subsequently to aging at 850 °C only for 81s, but it was completely dissolved as the temperature increased to 985 °C [21]. Sieurin et al. found that the sensitive temperature of the σ phase in the 2205 DSS steel was between 650 °C and 920 °C, whereas the "nose temperature" was approximately 850 °C in the TTP (temperature-time-precipitation) diagram [22].

In fact, it was inevitable that certain treatments, such as welding and other thermal treatments, could cause the precipitates to form in 2205 DSS when it would be exposed to the sensitive temperatures of the precipitates [23–25]. The cold rolling is an industrial technique for alloy hardening, also producing a high amount of deformation and increasing the grain energy, finally affecting the precipitation behavior and the microstructure of the materials [26]. Cho et al. observed that the cold deformation promoted the σ phase precipitation in 2205 DSS, as compared to the non-cold-rolled materials [27]. Breda et al. revealed the strain-induced martensite occurrence in the cold-rolled 2205 DSS [28]. In contrast, the researches regarding the effect of precipitates on the corrosion behavior of the hot-rolled and cold-rolled 2205 DSS steels have rarely been contrasted.

The purpose of the present research was to investigate the effects of hot-rolling or cold-rolling treatments on the microstructures and the corrosion resistance of 2205 DSS steels. The microstructure and the chemical composition of the 2205 DSS was investigated with an optical microscope and a scanning electron microscope, while potentiodynamic polarization and electrochemical impedance spectroscopy were employed to detect the corrosion resistance of the 2205 DSS steels for the corresponding corrosion resistance properties forecasting. Besides, the microstructures prior to and following corrosion resistance testing were contrasted to distinguish the corrosion mechanism of constituent phases. The final purpose of this paper was to provide a scientific basis for the hot working process optimization, the microstructure constituent prediction, and the corrosion resistance prediction.

2. Materials and Methods

The material under study was a 2205 DSS steel in the form of a 4-mm-thick hot-rolled bar that was supplied by the Taiyuan Iron & Steel Company Ltd. (TISCO, Taiyuan, China). It was solution that was annealed at 1050 °C and water quenched. The corresponding chemical composition is presented in Table 1. Following solution treatment, the specimens were cold rolled with reductions of 50%. Given the "nose temperature", aging treatments were carried out on the specimens at 850 °C for different times ranging from 10 min to 4 h. Subsequently, each specimen was mounted on epoxy resin, mechanically ground with SiC papers down to 3000 grit, polished to mirror finish, as well as cleaned with distilled water and ethanol.

Table 1. Chemical composition of 2205 duplex stainless steel (wt %).

Cr	Ni	Mo	Mn	Si	N	P	S	C	Fe
22.36	5.21	3.18	1.37	0.65	0.15	0.014	0.0008	0.020	Bal.

Prior to use, all of the samples were etched with a mixed solution of $K_2S_2O_5$ (0.3 g), HCl (20 mL), and H_2O (80 mL). The microstructures of the samples were observed with an optical microscope (OM, DMRM, LEICA, Shanghai Optical Instrument Factory, Shanghai, China). The different phases of the specimens were determined through X-ray diffraction analysis (XRD, X'Pert Powder, PANalytical, Almelo, the Netherlands) and the corresponding phase composition of the alloy was analyzed through electron dispersive X-ray spectroscopy (EDS, Octane SDD, EDAX Inc., Mahwah, NJ, USA) of the scanning electron microscopy (SEM, EVO18, Carl Zeiss Jena, Oberkochen, Germany). The phase proportion was analyzed with statistical methods with the Image-Pro Plus image manipulation software (Image ProPlus 6.0, Media Cybemetics, Rockville, MD, USA, 2006).

The electrochemical experiments were performed with a 3.5 wt % NaCl solution, at room temperature and atmospheric pressure. A three-electrode corrosion cell that was equipped with a saturated calomel electrode (SCE) reference electrode and a platinum foil counter electrode was utilized. The specimen with an exposure area of 1 cm^2 was used as a working electrode. Prior to experimentation, the samples were allowed to stabilize at an open circuit potential for 30 min, until the fluctuation potential reached 10 mV. The polarization curves were recorded potentiodynamically at 0.5 $mV{\cdot}s^{-1}$, while the potential scanning range was from below 200 mV of the open-circuit potentials to the potential when the current indicated that stable pitting had occurred.

The samples with the σ phase were treated through electrolysis in a 10 wt % KOH aqueous solution with a voltage of 2 V, until the σ phase was completely dissolved. Consequently, the sample surface was treated with alcohol and wascoated with a cyanoacrylate adhesive for the residual σ phase insulation. Following the cyanoacrylate glue complete solidification, the cotton swabs that were dipped in acetone were used to gently wipe the cyanoacrylate glue. Finally, an appropriate polishing processing was required to ensure the surface smoothness. The specific process is presented in Figure 1. It could be observed that the σ phases were dissolved through this method. Also, the second layer of σ phases was not exposed.

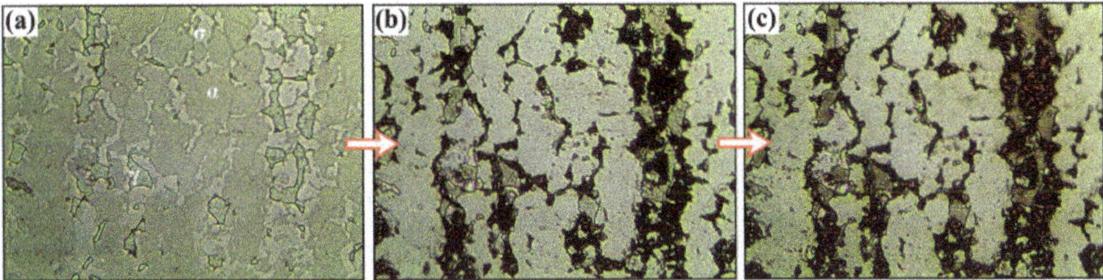

Figure 1. Process of preparing specimens without σ phase, (**a**) original specimen; (**b**) specimen electrolytically corroded by 10 wt % KOH; and, (**c**) specimen covered by cyanoacrylate glue.

3. Results

Figure 2a presents the microstructures of the solution-annealed hot-rolled 2205 steels. These consisted of elongated austenite islands in the ferrite matrix and no apparent intermetallic phases were observed. During aging at 850 °C for 10 min, the χ phase preferentially nucleated at the boundary of ferrite and grew through the adjacent ferrite, as presented in Figure 2b. This was discussed in the authors' previous work [29]. Furthermore, it could be observed form Figure 2c,d, that, as the aging time increased to 3 and 4 h, the σ phase originating from the transformation of $\alpha \rightarrow \gamma2 + \sigma$ would gradually appear and being coarse, where the $\gamma2$ was the secondary austenite. Therefore, it could be considered that high amounts of the σ phase existed in the matrix during aging for increased durations. Besides, a low amount of χ phase was also detected and distributed at the grains of the σ phase. Figure 2e,f presents the EDS line-scan profile of the hot-rolled 4 h aged samples, where the σ phase with a relatively low Cr and Mo contents could cause the generation of a chromium depleted region.

Figure 3 present the metallographic structure of 2205 duplex stainless steel with cold deformation subsequently to solid-solution and aging treatments. Figure 3a presents the microstructure of the solution-annealed cold-deformed sample under the optical microscope, and it can be seen that the austenitic phase with a lighter color was distributed within the ferrite phase, but the microstructures of the cold-rolled samples became more elongated along the cold rolling direction, as compared to the hot-rolled samples. Moreover, the austenite grains of the cold-deformed samples were fine and non-uniform, with a narrow strip shape, as a result of different local deformation. Following aging for 10 min at 850 °C, the bright white χ phase could also be observed in Figure 3b. As the aged time increased to 3 h and 4 h, as presented in Figure 3c,d, the σ phase precipitates began to appear at the boundary of the ferrite and austenite phases, but the corresponding grain size was lower when compared to the hot rolled samples. Figure 3e,f presents the EDS line-scan profiles of the cold-rolled sample aged for 4 h, suggesting that the precipitated χ and σ phases were enriched in Cr and Mo. This led to the uneven distribution of the alloying elements, such as Cr and Mo, within the matrix.μm

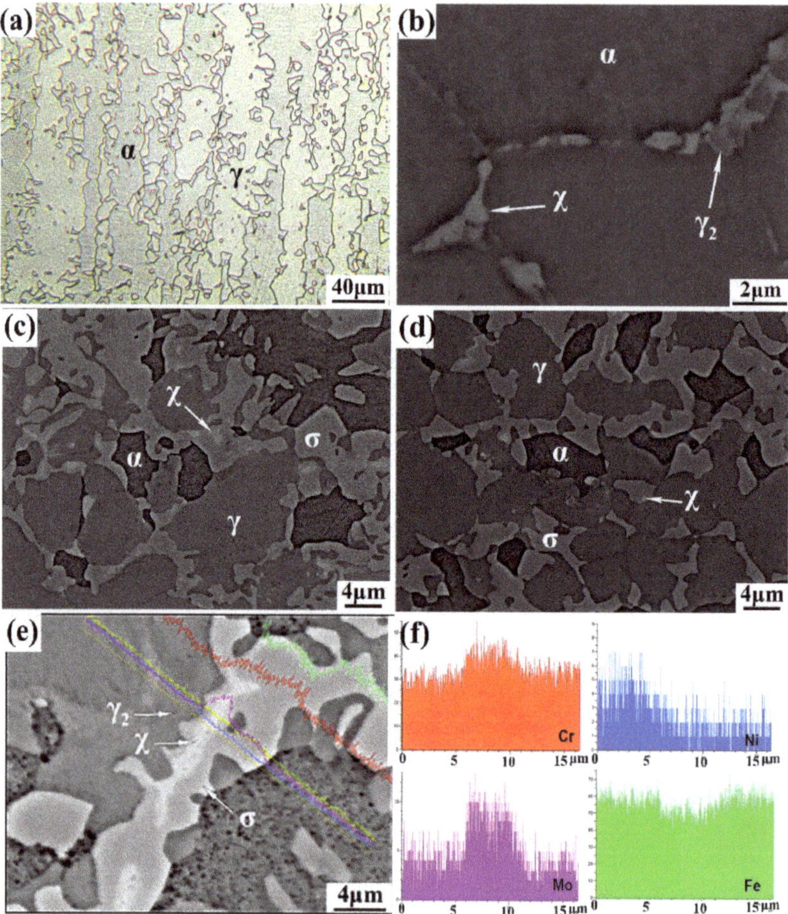

Figure 2. Microstructure of cold-rolled 2205 duplex stainless steel: (**a**) optical microscope (OM) morphology of hot-rolled specimen; (**b**) scanning electron microscopy (SEM)-scattered electron imaging (BSE) morphology of 10 min aged; (**c**) 3 h aged and (**d**) 4 h aged specimens at 850 °C; and, (**e**,**f**) electron dispersive X-ray spectroscopy (EDS) line-scan profile of hot-rolled 4 h aged specimen.

In Figure 4, the σ phase volume fractions of the cold-rolled and hot-rolled 2205 steels were plotted as a function of aging time. As the aging time increased, the amount of σ phase in the cold-rolled samples became gradually higher when compared to the hot-rolled samples at the same aging time. In particular, the precipitations of the cold-rolled samples reached approximately 38.2% for the specimen that was aged at 850 °C for 4 h, corresponding to 23.9% in the hot-rolled samples. The XRD diffraction spectra and local magnifications of the cold-rolled and hot-rolled samples are presented in Figure 5. It could be observed that the peak intensity of α(110), relative to the γ(111) decreased with aging time. Also, the sample that was aged for 4 h exhibited a low-sized ferrite peak, indicating that a major fraction of the ferrite was transformed into the σ phase. The diffraction peak intensity of the σ phase in the cold-rolled sample was significantly higher when compared to the hot rolled sample following aging for 4 h, which suggested that an additional amount of the σ phase was precipitated in the cold-rolled sample. It could also be observed from Figure 5 that the peak intensity of the α(200) and α(211) in the cold-rolled solution-annealed samples was significantly higher when compared to the hot-rolled samples.

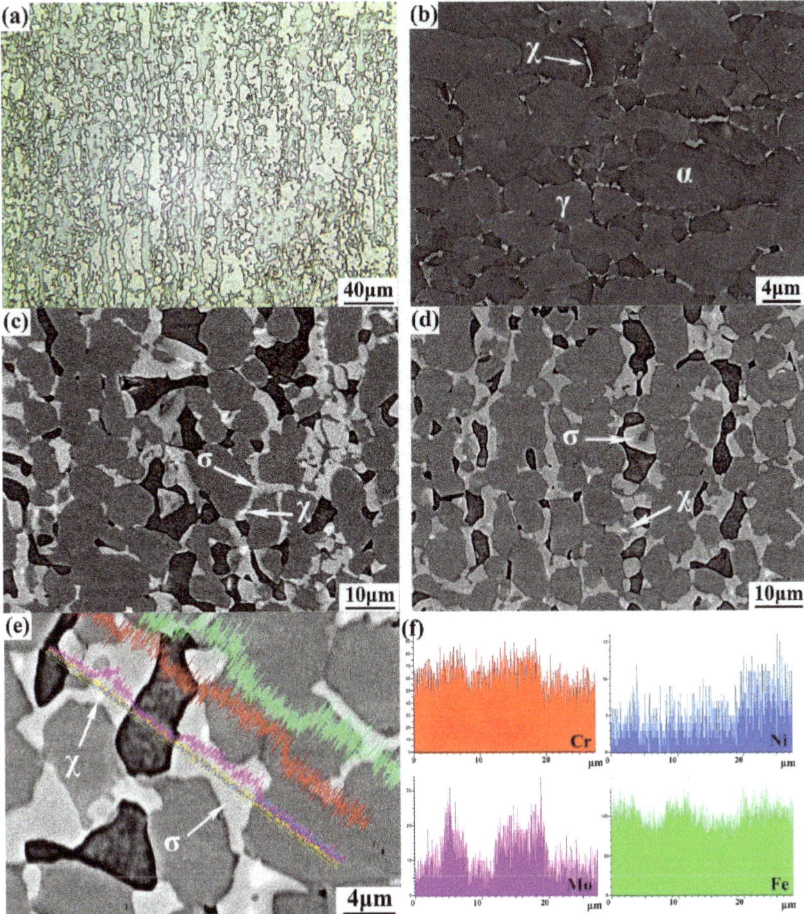

Figure 3. Microstructure of cold-rolled 2205 duplex stainless steel: (**a**) OM morphology of cold-rolled specimen; (**b**) SEM-BSE morphology of 10 min aged; (**c**) 3 h aged and (**d**) 4 h aged specimens at 850 °C; and, (**e**,**f**) EDS line-scan profile of cold-rolled 4 h aged specimen.

Figure 6 presents the potentiodynamic polarization curves comparison between the cold-rolled and hot-rolled 2205 duplex stainless steels subsequently to different aging treatments with a 3.5 wt % NaCl solution. Moreover, Tables 2 and 3 present the E_{pit} results in comparison from the potentiodynamic polarization curves for the hot-rolled and cold-rolled 2205 steels, respectively. It can be observed that the polarization curves of the hot-rolled and cold-rolled samples without aging were similar, while both of the pitting potentials (E_{pit}) exceeded 1000 mV, indicating that the pitting resistance was highly consistent for the solid-solution annealed cold-rolled and hot-rolled steels. Adversely, for the cold-rolled sample, the E_{pit} presented in Figure 6b apparently decreased following aging for 10 min only, while no current density at the end of the polarization curve existed, suggesting that the sample surface was not repassivated. As the aging time increased to 20 min, the E_{pit} was reduced to 643 mV, which was lower than the E_{pit} of the hot-rolled samples aged for 3 h. Through the aging time further increase to 40 min, the E_{pit} decreased to 548 mV, where the value was even lower than the hot-rolled samples aged for 4 h. When the aging time of the cold-rolled sample increased to 1 h or beyond, the E_{pit} further decreased, and remained it below 500 mV.

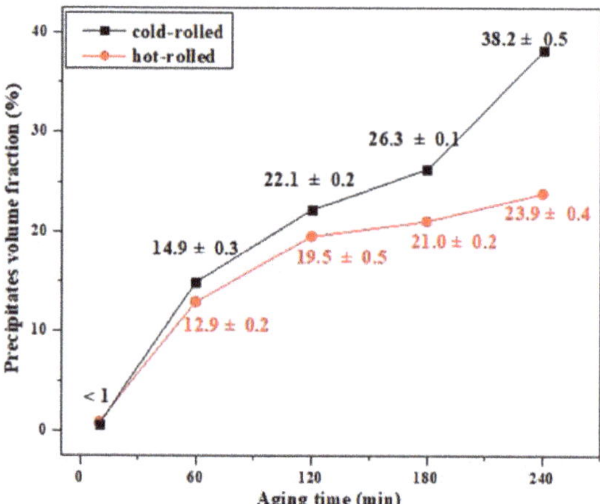

Figure 4. Changes in σ phase volume fraction of cold-rolled and hot-rolled of 2205 duplex stainless steel with prolonging aging time.

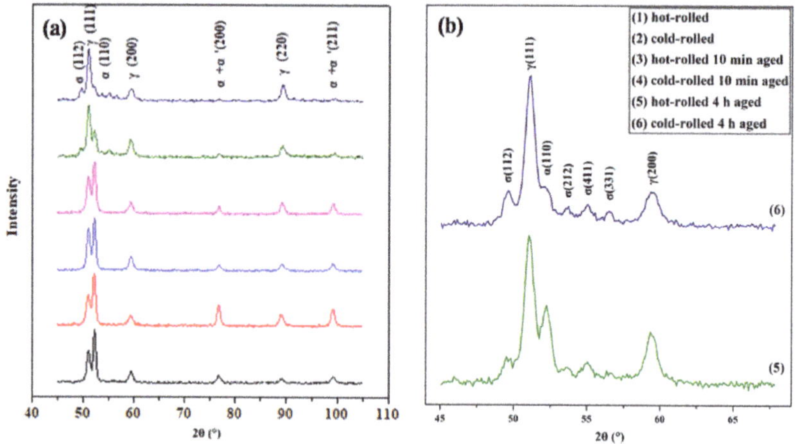

Figure 5. Comparison of X-ray diffraction patterns between cold-rolled and hot-rolled 2205 duplex stainless steel without aging treatment, and after aging for 10 min, 4 h at 850 °C, (**a**) overall patterns; and, (**b**) amplification patterns of cold-rolled and hot-rolled 4 h aged specimens.

Table 2. E_{pit} resulted from potentiodynamic polarization curves for hot-rolled 2205.

Hot-Rolled	Solution Treatment	Aging for 15 min	Aging for 1 h	Aging for 3 h	Aging for 4 h
E_{pit} (mV, vs. SCE)	1005 ± 3	989 ± 3	901 ± 2	771 ± 3	617 ± 5

Table 3. E_{pit} resulted from potentiodynamic polarization curves for cold-rolled 2205.

Cold-Rolled	Solution Treatment	Aging for 10 min	Aging for 20 min	Aging for 40 min	Aging for 1 h	Aging for 3 h	Aging for 4 h
E_{pit} (mV, vs. SCE)	1002 ± 4	828 ± 5	643 ± 3	548 ± 3	441 ± 5	508 ± 4	443 ± 2

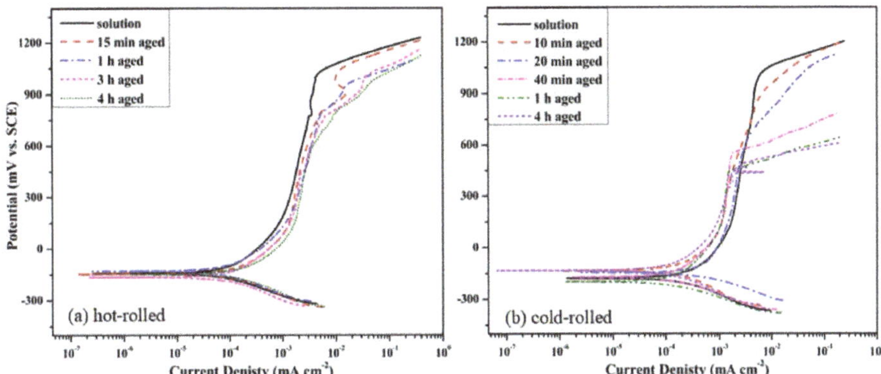

Figure 6. Comparison of potentiodynamic polarization curves between cold-rolled and hot-rolled 2205 duplex stainless steel after different aging treatment, (**a**) cold-rolled specimens, and (**b**) hot-rolled specimens.

Figure S1 presents the real impedance vs. the imaginary impedance plot at each frequency for the 2205 duplex stainless steels with different aging times with the 3.5% NaCl solution. It could be observed that the Nyquist diagrams of the cold-rolled and hot-rolled samples exhibited a depressed semicircle with a capacitive arc. Moreover, the diameter of the capacitive semicircle in the cold-rolled sample was lower as compared to the hot-rolled samples, indicating that the passive film stability of the cold-rolled samples was worse when compared to the hot-rolled samples. The equivalent circuit presented in Figure S2 was proposed for the EIS data fitting to quantify the electrochemical parameters. In this equivalent circuit, R_s is the solution resistance and R_t stands for the charge-transfer resistance. CPE1 represents the capacitance of the double electrical layers. CPE2 symbolizes the capacitance of the passive film on the metal surface. R_f is the passive film resistance. The electrochemical impedance parameters of the cold-rolled and hot-rolled samples that were obtained from the fitting of the EIS diagrams are presented in Tables S1 and S2, respectively. The passive film resistance (R_f) of the cold-rolled samples exceeded the surface charge transfer resistance (R_t), which occurred similarly for the hot-rolled samples. The passive film of the latter played a major role in the corrosion resistance. Furthermore, the R_f of the cold-rolled samples was significantly lower as compared to the hot-rolled samples, indicating that the passive films of the samples treated by cold rolling were significantly weaker compared to the hot-rolled samples. Moreover, the R_f change in the cold-rolled samples aged for a short time was not apparent, but the values of R_f for the samples following aging for 1 h or longer times significantly decreased, indicating that the passivation film became quite weaker subsequently to aging for a long time. The R_t also decreased as the aging time increased, which demonstrated that, the migration of the charged particles in the double layer between the electrode and the electrolyte solution gradually became easy.

It is well-known that the corrosion resistance has always been considered in regard to the microstructures of materials. For the hot-rolled 2205 duplex stainless steel, the precipitation of σ phase gradually increased as the aging time increased, while the corrosion resistance of the samples significantly decreased. In order to explain the correlation between the microstructure and corrosion resistance, the corrosion morphology of the hot-rolled 2205 duplex stainless steels aging for 10 min and 4 h, following potentiodynamic polarization, was characterized through SEM-BSE with an EDS system. In the results, each residual phase was confirmed. As presented in Figure 7a, in the hot-rolled 2205 aged for shorter aging times, pitting nucleation preferably occurred on the grain boundaries or on the ferrite/austenite interfaces. Furthermore, a severe pitting corrosion of the samples aged for 4 h was observed, as presented in Figure 7b. Tables 4 and 5 confirmed the chemical compositions of the phases presented in Figure 7.

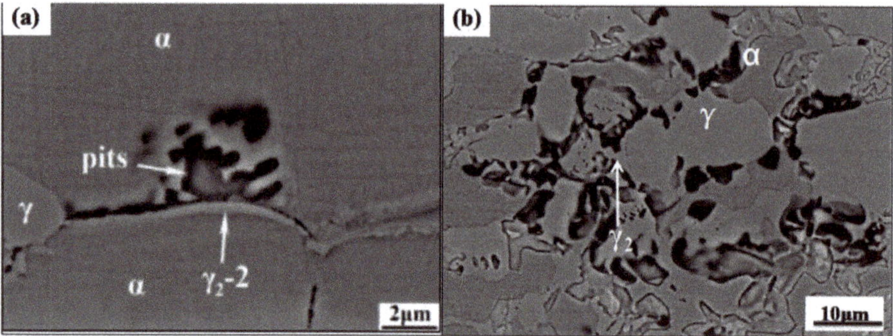

Figure 7. Corrosion morphology of hot-rolled 2205 duplex stainless steel after potentiodynamic polarization by SEM-BSE, (**a**) 10 min aged; and, (**b**) 4 h aged.

Table 4. Chemical composition of phases showed in Figure 7a (wt %).

Position	Cr	Ni	Mo	Si	Mn	Fe
α	24.73	3.53	4.57	0.75	1.50	Bal.
γ	21.55	7.21	3.05	0.63	1.53	Bal.
γ_2	21.28	6.30	2.94	0.61	1.55	Bal.

Table 5. Chemical composition of phases showed in Figure 7b (wt %).

Position	Cr	Ni	Mo	Si	Mn	Fe
α	24.14	2.25	2.49	0.62	1.36	Bal.
γ	21.81	6.47	2.98	0.53	1.51	Bal.
γ_2	20.89	6.91	1.92	0.52	1.28	Bal.

The corrosion morphology comparisons demonstrated that the corrosion of cold-rolled samples had apparent selectivity, in which the order of corrosion of each phase apparently differed from the hot rolled samples. Figure 8 present the corrosion morphology of the cold-rolled 2205 duplex stainless steel aged for 10 min, 1 h and 4 h; subsequently, to potentiodynamic polarization, through SEM-BSE. The pitting corrosion of the cold rolled samples was more evenly distributed on the sample surfaces when compared to the hot rolled samples, and it was easy to be concentrated in certain areas. Figure 8a presents that the phase boundary of cold-rolled 2205 preferred to be corroded in the 3.5 wt % NaCl solution, whereas it was also worth being noted that stripe patterns appeared in the corrosion morphology of the sample that was aged for 10 min. The corresponding EDS analysis is presented in Table 6. The chemical composition of the stripe was the same as the surrounding austenite, which might be α'-martensite. Furthermore, as presented in Figure 8b–d, the surfaces of the cold-rolled 2205 samples still remained a high amount of shallow white σ phases following corrosion. Therefore, differently from the hot-rolled 2205 sample, the cold-rolled 2205 DSS, subsequent to aging treatment, might be preferentially corroded from the phase boundary and the α'-martensite in the 3.5 wt % NaCl solution. Also, the precipitates were basically not subjected to corrosion in the initial process.

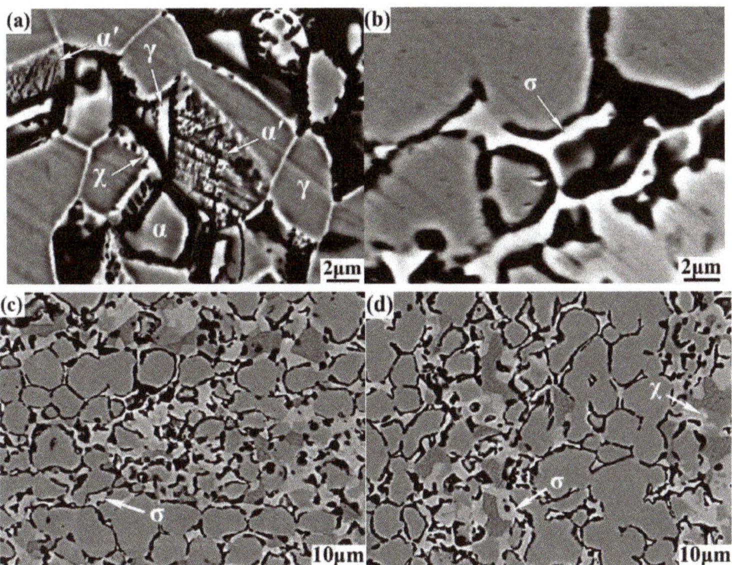

Figure 8. Corrosion morphology of cold-rolled 2205 duplex stainless steel after potentiodynamic polarization by SEM-BSE, (**a**) 10 min aged; (**b**,**c**) 1 h aged; and, (**d**) 4 h aged.

Table 6. Chemical composition of strain-induced martensite and γ phase showed in Figure 8a (wt %).

position	Cr	Ni	Mo	Si	Mn	Fe
α′	21.27	6.37	3.31	0.75	1.57	Bal.
γ	21.43	6.22	2.92	0.64	1.79	Bal.

To investigate in detail the corrosion behavior of the hot-rolled and cold-rolled 2205 DSS, the σ phase was selectively dissolved for the sample fabrication without the σ phase through the electrochemical method. The potentiodynamic polarization curves among the original specimens and the specimens without an σ phase are presented in Figure 9. For the hot-rolled materials, the E_{pit} of the sample without σ phase would increase to the solution-annealed values. For the cold-rolled 2205 DSS, the precipitation content of the σ phase increased, whereas the precipitation speed of the precipitates also increased. This appeared to undermine the corrosion resistance of the cold-rolled 2205 DSS, as compared to the hot-rolled materials. However, even if the σ phase was eliminated from the cold-rolled samples, the E_{pit} increased from 443 mV to 630 mV, being quite lower when compared to the non-aging sample (1002 mV). This suggested a more complicated influence factor on the corrosion resistance of the cold-rolled samples. The passive current of the samples without σ phases was higher when compared to the solid solution samples, which might be caused by the formation of a compact structure between the steel and the epoxy resin.

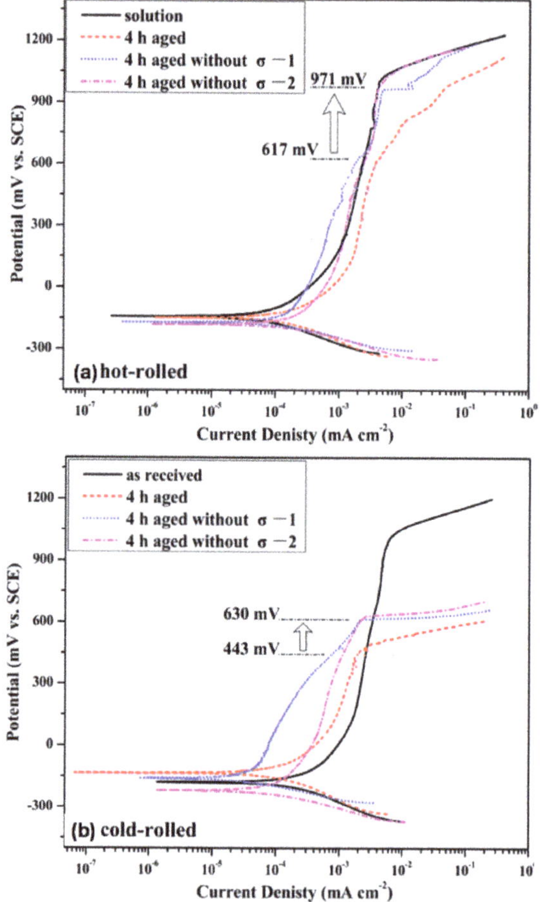

Figure 9. Comparison of potentiodynamic polarization curves between original specimens and specimens without σ phase, (**a**) hot-rolled specimens; and, (**b**) cold-rolled specimens.

4. Discussion

The 2205 DSS, χ phase would preferentially precipitate at the ferrite-ferrite boundaries, following aging at 850 °C. As the aging time increased, the σ phase would precipitate at both ferrite boundaries and ferrite-austenite boundaries. As the χ phase was a meta-stable phase, it would dissolve and transform into the σ phase along with the aging time increase. The σ phase could be transformed by a eutectoid reaction from ferrite and the dissolution of the χ phase. The σ phase precipitation could be accelerated by the cold deformation in the subsequent aging, which might be due to the increased defect and distortion energy during cold-rolling, promoting the ferrite transformation into the σ phase. Besides, it was proved that the austenite phase in duplex stainless steels would be transformed into ε-martensite or α′-martensite during cold deformation, also being directly transformed into the α′-martensite when the stacking fault energy was high [28]. Moreover, it is well known that ferrite has a body-centered cubic structure and the martensite has a body-centered tetragonal structure. Since the diffraction peak of α′-martensite was consistent with the ferrite peak, the peak intensity of ferrite increased when containing α′-martensite. Therefore, the XRD diffraction pattern that is presented in Figure 5 indicated the formation of α′-martensite following cold rolling. By contrast, due to the non-diffusive phase transition and unchanged chemical composition of the α'-martensitic transformation, the direct observation of α'-martensitic phase appeared difficult.

It has been proved that pitting occurring in chromium-depleted areas are associated with the precipitated phase [30,31]. When combined with Figures 2f and 7a, the ferrite was almost occupied by the σ phase and the secondary austenite, which often has relatively low Cr content than the σ or γ phases. Therefore, it could be concluded that the original grain-boundaries were the more susceptible to pitting, thus inducing the growth of the σ precipitates. During the initial stage, the pitting nucleation was located at the boundary of α/γ and around the σ phase, due to low Cr content and stability of the passive film in this position. With the transformation of α into σ and $\gamma 2$, the $\gamma 2$ was also corroded, allowing for the pitting to further extend within the α. Therefore, the pitting corrosion of the hot-rolled 2205 steels preference to occur around the σ phase was inevitable, due to the relatively high Cr and Mo contents of the σ phase. Moreover, pitting tended to further increase with the precipitation of an additional amount of σ phase in the hot-rolled 2205.

Besides the effect of σ phase on the corrosion resistance of the cold-rolled 2205 Dss, the influence factors of its corrosion were complicated. On the one hand, the cold-deformation of 2205 DSS easily generated dislocations, deformation twinning, and dislocation-twin interactions, where the nucleation of the pitting attack was quite likely to occur. Moreover, a more defective passive film on the cold-rolled 2205 DSS was likely to be formed at the defects of the grains. On the other hand, the strain-induced martensite in the cold-rolled 2205 might be active than the other precipitate and matrix phases. Consequently, it was also easily corroded in the early stage, which was similar to the reported role of strain-induced martensite in cold-worked 304 steels [32]. Therefore, the precipitated σ phase was the main factor affecting the corrosion resistance of the hot-rolled samples, which might be ascribed to the formation of a Cr-depleted zone around the σ phase, also preferring to be corroded. When the σ phase was removed, the corrosion resistance of the samples would be apparently restored, which suggested the chromium redistribution in the matrix of the hot-rolled 2205. Adversely, the σ phase elimination in the cold-rolled 2205 could not restore the E_{pit} to the level of non-aging. Further research is required to be conducted to clarify the influence factor of the corrosion resistance for the cold-rolled 2205 DSS.

5. Conclusions

The differences in the precipitate-induced selective corrosion in the hot-rolled and the cold-rolled 2205 duplex stainless steel were evaluated through microstructural analyses of the samples aged at 850 °C for different times, when combined with electrochemical methods. The Chi and Sigma phases precipitated in turn following aging treatment for the hot-rolled and cold-rolled materials, but the precipitation rate in the cold rolled sample was faster, as a result of the stored internal energy and increased defect density. The precipitation amount of sigma phase in the cold-rolled samples was higher as compared to the hot-rolled samples under the same aging conditions. Also, strain-induced martensite structure was also produced by the transformation of austenite under cold deformation.

The corrosion resistance of the solution-annealed cold-rolled samples was similar to the hot-rolled samples, but the corrosion resistance of the cold-rolled samples was weaker when compared to the hot-rolled material after aging treatment. The pitting preferentially nucleated in the Cr-depleted region around the σ phase for the hot-rolled 2205 aged samples, while the corrosion resistance of the hot-rolled 2205 without σ phase increased closely to the level of the solution-annealed 2205. By contrast, not only the σ phase, but also the strain-induced martensite, as well as the defects, which were induced by cold deformation, had an important effect on the corrosion resistance of the cold-rolled 2205 DSS. Correspondingly, the initiation of pitting corrosion occurred at the phase boundary, defects, and martensite for the cold-rolled 2205 aged sample, leading to a lower corrosion resistance, when compared to the hot-rolled 2205.

Supplementary Materials: The following are available online at http://www.mdpi.com/2075-4701/8/6/407/s1, Figure S1: Comparison of electrochemical impedance spectroscopy between cold-rolled and hot-rolled 2205 duplex stainless steel in 3.5% NaCl, Figure S2: Equivalent circuit of electrochemical impedance spectroscopy for

2205 duplex stainless steel aging in 3.5% NaCl solution, Table S1: Fitting results of equivalent circuit of cold-rolled 2205 duplex stainless steel after long-term aging in 3.5% NaCl solution, Table S2: Fitting results of equivalent circuit of hot-rolled 2205 duplex stainless steel after long-term aging in 3.5% NaCl solution.

Author Contributions: T.G. performed research, analyzed the data and wrote the paper; Q.S. helped in the experimental part; J.W. and P.H. assisted in the data analysis and revised manuscript.

Acknowledgments: This research was supported by the National Natural Science Foundation of China (Grant No. 51371123) and Shanxi Province Science Foundation for Youths (201601D202033).

Conflicts of Interest: The authors declare no conflict of interest.

References

1. Pohl, M.; Storz, O.; Glogowski, T. Effect of intermetallic precipitations on the properties of duplex stainless steel. *Mater. Charact.* **2007**, *58*, 65–71. [CrossRef]
2. Sato, Y.S.; Nelson, T.W.; Sterling, C.J.; Steel, R.J.; Pettersson, C.-O. Microstructure and mechanical properties of friction stir welded SAF 2507 super duplex stainless steel. *Mater. Sci. Eng. A* **2005**, *397*, 376–384. [CrossRef]
3. Wang, H. Investigation of a Duplex Stainless Steel as Polymer Electrolyte Membrane Fuel Cell Bipolar Plate Material. *J. Electrochem. Soc.* **2005**, *152*, B99–B104. [CrossRef]
4. Charles, J. Duplex stainless steels, a review after DSS'07 in Grado. *Rev. Metall.* **2008**, *105*, 455–465. [CrossRef]
5. Ghosh, S.K.; Mondal, S. Effect of heat treatment on microstructure and mechanical properties of duplex stainless steel. *Trans. Indian Inst. Met.* **2008**, *61*, 33–37. [CrossRef]
6. Liu, J.M.; Liu, J.; Fan, G.W.; Du, D.F.; Li, G.P.; Chai, C.J. Effect of Solution Treatment on Microstructure and Properties of the SAF2507 Super Duplex Stainless Steel. *Mater. Sci. Forum* **2012**, *724*, 3–6. [CrossRef]
7. Naghizadeh, M.; Moayed, M.H. Investigation of the effect of solution annealing temperature on critical pitting temperature of 2205 duplex stainless steel by measuring pit solution chemistry. *Corros. Sci.* **2015**, *94*, 179–189. [CrossRef]
8. Lacerda, J.C.D.; Cândido, L.C.; Godefroid, L.B. Corrosion behavior of UNS S31803 steel with changes in the volume fraction of ferrite and the presence of chromium nitride. *Mater. Sci. Eng. A* **2015**, *648*, 428–435. [CrossRef]
9. Cheng, X.Q.; Li, X.G.; Dong, C.F. Study on the passive film formed on 2205 stainless steel in acetic acid by AAS and XPS. *Int. J. Miner. Met. Mater.* **2009**, *16*, 170–176. [CrossRef]
10. Luo, H.; Dong, C.F.; Xiao, K.; Li, X.G. Characterization of passive film on 2205 duplex stainless steel in sodium thiosulphate solution. *Appl. Sur. Sci.* **2011**, *258*, 631–639. [CrossRef]
11. Lv, J.; Liang, T.; Wang, C.; Dong, L. Comparison of corrosion properties of passive films formed on coarse grained and ultrafine grained AISI 2205 duplex stainless steels. *J. Electrochem. Soc.* **2015**, *757*, 263–269. [CrossRef]
12. Lv, J.; Liang, T.; Wang, C.; Guo, T. Influence of sensitization on passive films in AISI 2205 duplex stainless steel. *J. Alloy. Compd.* **2016**, *658*, 657–662. [CrossRef]
13. Deng, B.; Wang, Z.; Jiang, Y.; Wang, H.; Gao, J.; Li, J. Evaluation of localized corrosion in duplex stainless steel aged at 850 °C with critical pitting temperature measurement. *Electrochim. Acta* **2009**, *54*, 2790–2794. [CrossRef]
14. Hoseinpoor, M.; Momeni, M.; Moayed, M.H.; Davoodi, A. EIS assessment of critical pitting temperature of 2205 duplex stainless steel in acidified ferric chloride solution. *Corros. Sci.* **2014**, *80*, 197–204. [CrossRef]
15. Chan, K.; Tjong, S. Effect of Secondary Phase Precipitation on the Corrosion Behavior of Duplex Stainless Steels. *Materials* **2014**, *7*, 5268–5304. [CrossRef] [PubMed]
16. Huang, C.S.; Shih, C.C. Effects of nitrogen and high temperature aging on σ phase precipitation of duplex stainless steel. *Mater. Sci. Eng. A* **2005**, *402*, 66–75. [CrossRef]
17. Santos, D.C.D.; Magnabosco, R.; Moura-Neto, C.D. Influence of sigma phase formation on pitting corrosion of an aged uns s31803 duplex stainless steel. *Corrosion* **2013**, *69*, 900–911. [CrossRef]
18. Park, C.J.; Rao, V.S.; Kwon, H.S. Effects of sigma phase on the initiation and propagation of pitting corrosion of duplex stainless steel. *Corrosion* **2005**, *61*, 76–83. [CrossRef]
19. Pohl, M.; Storz, O.; Glogowski, T. σ-phase morphologies and their effect on mechanical properties of duplex stainless steels. *Int. J. Mater. Res.* **2008**, *99*, 1163–1170. [CrossRef]

20. Chen, T.H.; Weng, K.L.; Yang, J.R. The effect of high-temperature exposure on the microstructural stability and toughness property in a 2205 duplex stainless steel. *Mater. Sci. Eng. A* **2002**, *338*, 259–270. [CrossRef]
21. Elmer, J.W.; Palmer, T.A.; Specht, E.D. In situ observations of sigma phase dissolution in 2205 duplex stainless steel using synchrotron X-ray diffraction. *Mater. Sci. Eng. A* **2007**, *459*, 151–155. [CrossRef]
22. Sieurin, H.; Sandström, R. Sigma phase precipitation in duplex stainless steel 2205. *Mater. Sci. Eng. A* **2007**, *444*, 271–276. [CrossRef]
23. Lai, J.K.L.; Wong, K.W.; Li, D.J. Effect of solution treatment on the transformation behaviour of cold-rolled duplex stainless steels. *Mater. Sci. Eng. A* **1995**, *203*, 356–364. [CrossRef]
24. Neissi, R.; Shamanian, M.; Hajihashemi, M. The Effect of constant and pulsed current gas tungsten arc welding on joint properties of 2205 duplex stainless steel to 316L Austenitic Stainless Steel. *J. Mater. Eng. Perform.* **2016**, *25*, 2017–2028. [CrossRef]
25. Yurtisik, K.; Tirkes, S. Fatigue Cracking of hybrid plasma gas metal arc welded 2205 duplex stainless steel. *Materialprufung* **2014**, *56*, 800–805. [CrossRef]
26. Örnek, C.; Engelberg, D.L. Towards understanding the effect of deformation mode on stress corrosion cracking susceptibility of grade 2205 duplex stainless steel. *Mater. Sci. Eng. A* **2016**, *666*, 269–279. [CrossRef]
27. Cho, H.S.; Lee, K. Effect of cold working and isothermal aging on the precipitation of sigma phase in 2205 duplex stainless steel. *Mater. Charact.* **2013**, *75*, 29–34. [CrossRef]
28. Breda, M.; Brunelli, K.; Grazzi, F. Effects of Cold Rolling and Strain-Induced Martensite Formation in a SAF 2205 Duplex Stainless Steel. *Metall. Mater. Tran. A* **2015**, *46*, 577–586. [CrossRef]
29. Sun, Q.; Wang, J.; Li, H.B.; Hu, Y.D.; Bai, J.G.; Han, P.D. Chi Phase after Short-term Aging and Corrosion Behavior in 2205 Duplex Stainless Steel. *J. Iron Steel Res. Int.* **2016**, *23*, 1071–1079. [CrossRef]
30. Magnabosco, R. Alonso-Falleiros, N. Sigma Phase Formation and Polarization Response of UNS S31803 in Sulfuric Acid. *Corrosion* **2005**, *61*, 807–814. [CrossRef]
31. Warren, A.D.; Harniman, R.L.; Guo, Z.; Younes, C.M.; Flewitt, P.E.J.; Scott, T.B. Quantification of sigma-phase evolution in thermally aged 2205 duplex stainless steel. *J. Mater. Sci.* **2016**, *51*, 694–707. [CrossRef]
32. Xu, C.; Gang, H. Effect of deformation-induced martensite on the pit propagation behavior of 304 stainless steel. *Anti-Corros. Method Mater.* **2004**, *51*, 381–388. [CrossRef]

MDPI
St. Alban-Anlage 66
4052 Basel
Switzerland
Tel. +41 61 683 77 34
Fax +41 61 302 89 18
www.mdpi.com

Metals Editorial Office
E-mail: metals@mdpi.com
www.mdpi.com/journal/metals

www.ingramcontent.com/pod-product-compliance
Lightning Source LLC
LaVergne TN
LVHW071241170726
843515LV00006BA/1295

* 9 7 8 3 0 3 9 2 8 8 6 9 4 *